River Algae

Orlando Necchi Jr.
Editor

River Algae

Springer

Editor
Orlando Necchi Jr.
Department of Zoology and Botany
São Paulo State University
São José do Rio Preto, SP, Brazil

ISBN 978-3-319-31983-4   ISBN 978-3-319-31984-1 (eBook)
DOI 10.1007/978-3-319-31984-1

Library of Congress Control Number: 2016939587

Printed on acid-free paper

This Springer imprint is published by Springer Nature
The registered company is Springer International Publishing AG Switzerland

I dedicated this book to my wife Vânia,
my daughter Lisandra and my son Cauê,
for the support, understanding and love
along our journey that made
my personal and professional life plenty
of joy and accomplishments.

Foreword

Comments about algae in flowing waters have appeared in the scientific literature for at least a hundred and fifty years, with true scientific papers appearing towards the end of the nineteenth century and a rapid increase over the past 60 years. No journal has, however, dealt only with river algae, and the literature is scattered in many different types of publications, not just journals.

River Algae provides overviews of all the main groups of algae to be found in flowing waters, including the taxonomy of characteristic species, physiological and descriptive ecology and the contribution of algae to river ecosystem processes. The book provides readers not only with an easy way of finding out about the many studies, but also how this information relates to topics such as pollution and environmental monitoring. A broad knowledge of the literature is especially important when dealing with practical matters. For instance, in the case of river algae causing nuisance problems, their control has sometimes been hindered by researchers failing to read relevant information published in the past.

As discussed for diatom indices in the final chapter, the use of river algae for monitoring purposes has greatly increased over the past quarter of a century. However, monitoring with algae has often relied on correlations between occurrence and particular environmental variables. A broad knowledge of the literature should give researchers a much better understanding of the real drivers influencing the occurrence and abundance of the species involved.

Although having a problem to solve is often essential for getting funds for research, those with a wider understanding of river algae can also get great enjoyment from visiting a stream or river simply to look at the algae. Features can often be seen which stimulate interest and questions, especially in shallow fast-flowing streams. Answers to the questions may occur many years later as research progresses.

Durham, UK

Brian A. Whitton
School of Biological and Biomedical Sciences
Durham University

Preface

Algae inhabiting river habitats have never been approached in a single book but only as parts of more general publications dealing with ecology, taxonomy and bio-assessment of freshwater algae. As a result, river algae have been treated among other organisms and ecological topics of river ecosystems or as part of publications dealing with freshwater algae in general. The basic aim of this book is to provide a comprehensive approach of some relevant topics on algae inhabiting rivers. *River Algae* is focused on benthic communities aiming to show the importance of their role in the structure and function of communities in river ecosystems. Although the content is not formally divided into sections, the book covers two basic aspects: (a) the taxonomy of river-inhabiting algal groups, including phylogeny, distribution, collection, preservation and description of the most representative genera of algae in river benthic algal communities, and (b) the ecology of river algae, including ecological factors influencing abundance, distribution and diversity of river benthic algal communities and their use as bio-indicators. The book comprises 11 chapters written by renowned experts from Europe, North and South America providing updated information on taxonomy, ecology, methodology and uses of benthic river algae. The book is of potential interest to a wider audience, including researchers, students and professionals in diverse fields such as freshwater algae, limnology, aquatic biology, water quality assessment and biodiversity in river ecosystems.

São José do Rio Preto, SP, Brazil Orlando Necchi Jr.

Contents

Contributors

Joan Artigas Laboratoire Microorganismes, Génome et Environnement, Université Blaise-Pascal, Aubière Cedex, France

Ana Luiza Burliga Rhithron Associates, Inc., Missoula, MT, USA

Dale A. Casamatta Department of Biology, University of North Florida, Jacksonville, FL, USA

Natàlia Corcoll Catalan Institute for Water Research (ICRA), Scientific and Technologic Park of the UdG, Girona, Spain

Luc Ector Department Environmental Research and Innovation (ERIN), Luxembourg Institute of Science and Technology (LIST), Belvaux, Luxembourg

Petr Hašler Department of Botany, Faculty of Sciences, Palacký University Olomouc, Olomouc, Czech Republic

Carla Giselda Heinrich Laboratory of Limnology, University of Santa Cruz do Sul, Santa Cruz do Sul, RS, Brazil

J. Patrick Kociolek Museum of Natural History and Department of Ecology and Evolutionary Biology, University of Colorado, Boulder, CO, USA

Eduardo A. Lobo Laboratory of Limnology, University of Santa Cruz do Sul, Santa Cruz do Sul, RS, Brazil

Orlando Necchi Jr. Department of Zoology and Botany, São Paulo State University, São José do Rio Preto, SP, Brazil

Lorenzo Proia Catalan Institute for Water Research (ICRA), Scientific and Technologic Park of the UdG, Girona, Spain

Eugen Rott Biology Faculty, Institute of Botany, University of Innsbruck, Innsbruck, Austria

Sergi Sabater Catalan Institute for Water Research (ICRA), Scientific and Technologic Park of the UdG, Girona, Spain

Institute of Aquatic Ecology, Faculty of Sciences, University of Girona, Girona, Spain

Marilia Schuch Laboratory of Limnology, University of Santa Cruz do Sul, Santa Cruz do Sul, RS, Brazil

Alison R. Sherwood Department of Botany, University of Hawaii, Honolulu, HI, USA

Xisca Timoner Catalan Institute for Water Research (ICRA), Scientific and Technologic Park of the UdG, Girona, Spain

Institute of Aquatic Ecology, Faculty of Sciences, University of Girona, Girona, Spain

Elisabet Tornés Catalan Institute for Water Research (ICRA), Scientific and Technologic Park of the UdG, Girona, Spain

Institute of Aquatic Ecology, Faculty of Sciences, University of Girona, Girona, Spain

Morgan L. Vis Department of Environmental and Plant Biology, Ohio University, Athens, OH, USA

John D. Wehr Louis Calder Center—Biological Field Station, Fordham University, Armonk, NY, USA

Carlos Eduardo Wetzel Department Environmental Research and Innovation (ERIN), Luxembourg Institute of Science and Technology (LIST), Belvaux, Luxembourg

Chapter 1
An Overview of River Algae

Orlando Necchi Jr.

Abstract River-inhabiting algae are briefly presented in terms of predominant algal groups, morphological types, patterns of abundance and distribution, responses to environmental variables, ecological roles in river ecosystems, and biogeographic trends.

Keywords Algal groups • Benthic algae • Drainage basin • Distribution • Environmental variables • Morphology • River

The distinction between lentic (standing waters) and lotic (running waters) freshwater systems is not absolute because the former, such as lakes, have a regular flow-through of water and large rivers often have a relatively low flow for at least part of the year (Bellinger and Sigee 2010). However, there is an important distinction between them in relation to the algae present, with lentic systems typically dominated by planktonic algae, and lotic systems by benthic algae.

Benthic algae are directly associated with various types of substrata, including rocks, mud, organic and inorganic particles, macrophytes, and artificial objects. Two different kinds of communities are usually recognized: periphyton and benthic algae. Periphyton includes several microscopic organisms (algae, bacteria, and fungi) growing as biofilms or thicker matrices with flow, nutrient diffusion, and exposure to other organisms very different to those experienced by macroscopic algae (Stevenson 1996). Microalgae in the periphyton can be filamentous or colonial and prostrate or upright. In contrast, macroalgae typically exhibit a variety of morphologies in rivers (Sheath and Cole 1992). The focus of this book will be the benthic algae in rivers.

The macroalgae most commonly found in rivers are cyanobacteria (blue-green algae), green algae, red algae, diatoms, and brown algae. Yellow-green algae and yellow-golden algae and some flagellated algae are minor components of the benthic algal flora of rivers. Green algae tend to predominate (35–37 % of species) in floras of well-studied regions of the world (Sheath and Cole 1992; Necchi et al. 2000),

O. Necchi Jr. (✉)
Department of Zoology and Botany, São Paulo State University,
Rua Cristóvão Colombo, 2265, São José do Rio Preto, SP 15054-000, Brazil
e-mail: orlando@ibilce.unesp.br

O. Necchi Jr. (ed.), *River Algae*, DOI 10.1007/978-3-319-31984-1_1

with contributions of other algal groups varying considerably: cyanobacteria (24–35 %), red algae (14–20 %), diatoms, and yellow-green algae (14–21 %).

The following morphological types of river benthic macroalgae were defined by Sheath and Cole (1992): mats (flat body composed of tightly interwoven filaments), gelatinous colonies (flat or semierect thallus with numerous cells or filaments in a common matrix), gelatinous filaments (semierect individual filament within a matrix), free filaments (semierect individual filament without a matrix), tissue-like forms (semierect parenchymatous or pseudoparenchymatous thallus), tufts (short, radiating filaments), and crusts (flat thallus compacted tiers of cells). The proportion of morphological types reported in major surveys varied considerably (Sheath and Cole 1992; Necchi et al. 2000): mats (42–52 %), gelatinous colonies (18–23 %), gelatinous filaments (12–13 %), free filaments (8–9 %), tissue-like (3–7 %), tufts (4–6 %), and crusts (1 %). The majority of these forms can be considered as relatively well adapted to cope with the mechanical effect by the flow because filaments or cells are densely intermingled (mats), encased in a gelatinous matrix (gelatinous colonies or filaments), or grown within or near the boundary layer close to the substrata (tufts or crusts). These macroalgal forms are considered to be partial avoiders of flow-related breaking stress (Sheath and Hamrook 1988; Sheath and Cole 1992).

Sheath and Vis (2013) proposed three possible origins of these species for species of river-inhabiting algae: generalists, specialists, and marine invaders. Generalist lotic algae include members of the stream communities with broad ranges of occurrence for various environmental conditions; these taxa may also occur in lakes or ponds and become part of the outflow drift. Two macroalgal generalists are the filamentous green alga *Cladophora glomerata* and the mat-forming cyanobacteria *Phormidium retzii*. The former has been found as widely distributed in surveys in Europe (Whitton 1984, John et al. 2011) and North America (Sheath and Cole 1992), whereas the latter was present in most biomes in North America (Sheath and Cole 1992) and southeastern Brazil (Necchi et al. 2000). Specialist species with a rather narrow set of ecological requirements are limited in their distribution and might be restricted to certain geographic regions. Major surveys (e.g., Sheath and Cole 1992, Necchi et al. 2000) have listed a number of exclusive species of different algal groups found in only one or few sampling sites and can be classified within this category. Specialist taxa may be restricted to a particular habitat type, rather than a geographic area. Isolated spring-fed systems have the potential to contain certain algal species, as the red alga *Sheathia boryana* in Italy (Abdelahad et al. 2015). A relatively small fraction of the river algal flora is composed of marine invaders, i.e., taxa that have migrated or been transported from marine and brackish habitats. These species may move into freshwater watersheds by upstream migration from estuaries. The red algae *Bangia* is considered a marine invader of freshwater habitats (Müller et al. 2003).

No general combinations of environmental factors have been determined as favorable for macroalgae abundance as a whole. There are variations for particular regions or algal groups in terms of water variables (temperature, oxygen, specific conductance), substrata, and irradiance levels (Necchi et al. 2000). In contrast, some general ecological patterns have been reported for benthic macroalgal communities in rivers. Macroalgae in rivers are often patchy in distribution and it has been found that species richness is positively correlated with abundance, indicating that the most abundant communities are also the most diverse (Sheath et al. 1986, 1989; Necchi et al. 2000). Dominance of few species is widely reported in many studies on river macroalgae. Only a small number of species are widespread in regions or biomes (Sheath et al. 1986, 1989; Necchi et al. 2000). The rapidly fluctuating physical and chemical conditions of river systems have been considered as a major force determining the occurrence and abundance of macroalgal species (Sheath and Burkholder 1985). Few species are able to successfully colonize and grow in such conditions.

Benthic algae are the most successful primary producers exploiting stream habitats and they are the main source of energy for higher trophic levels in many streams (Biggs 1996). In addition, because of their rapid response to environmental change they are also useful as indicators of river water quality. Benthic algae have high turnover and opportunistic life-history strategies that have enabled them to exploit stream habitats, many of which are harsh environments. However, some patterns in biomass and community structure are discernable and can be explained by gradients in accrual (related to nutrient and light resources) and loss (related to disturbance and grazing) variables that operate over small-to-large spatial scales (from habitats to biomes) and for periods from weeks to years.

According to the River Continuum Concept (Vannote et al. 1980) pristine river ecosystems are considered as a network of lotic water bodies with a continuum of longitudinally interconnected resource gradients in which the biological communities respond to these spatial changes in many ways. Algal primary production is predicted to be relatively low in upper parts of the drainage basin (first to third order streams) due to heavy shading by the canopy cover of surrounding vegetation. Productivity reaches its peak in mid-sized (fourth to sixth order streams) in response to higher underwater irradiance due to channel enlargement and a reduction in allochthonous organic matter. In larger rivers (> sixth order) the continuum concept predicts that, while the river basin would be open to full sun, increased depth, greater levels of fine particulate matter from upstream, and resuspended sediments limit algal production. In nonforested basins or those with frequent disturbances, benthic algal communities may display quite different patterns (Biggs 1996).

The following chapters focus on various aspects of the taxonomy, distribution, and abundance of benthic river algae, as well as their potential use as bioindicators.

References

Abdelahad N, Bolpagni R, Lasinio GJ et al (2015) Distribution, morphology and ecological niche of *Batrachospermum* and *Sheathia* species (Batrachospermales, Rhodophyta) in the fontanili of the Po plain (northern Italy). Eur J Phycol 50:318–329

Bellinger EG, Sigee DC (2010) Freshwater algae: identification and use as bioindicators. Wiley-Blackwell, Chichester

Biggs BJF (1996) Patterns in benthic algae of streams. In: Stevenson RJ, Bothwell ML, Lowe RL (eds) Algal Ecology: freshwater benthic ecosystems. Academic, San Diego, pp 31–56

John DM, Whitton BA, Brook AJ (2011) Freshwater algal flora of the British Isles. Cambridge University Press, Cambridge

Müller KM, Cole KM, Sheath RG (2003) Systematics of *Bangia* (Bangiales, Rhodophyta) in North America II. Biogeographic trends in karyology: chromosome numbers and linkage with gene sequence phylogenetic trees. Phycologia 42:209–219

Necchi O Jr., Branco CCZ, Branco LHZ (2000) Distribution of stream macroalgae in São Paulo State, southeastern Brazil. Algol Stud 97:43–5

Sheath RG, Burkholder JM (1985) Characteristics of softwater streams in Rhode Island. II. Composition and seasonal dynamics of macroalgal communities. Hydrobiologia 128:109–118

Sheath RG, Cole KM (1992) Biogeography of stream macroalgae in North America. J Phycol 28:448–460

Sheath RG, Hamrook J (1988) Mechanical adaptations to flow in freshwater red algae. J Phycol 24:107–111

Sheath RG, Hamilton PH, Hambrook JA et al (1989) Stream macroalgae of the eastern boreal forest region of North America. Can J Bot 67:3353–3362

Sheath RG, Vis ML (2013) Biogeography of freshwater algae. In: eLS. John Wiley and Sons, Ltd: Chichester. DOI: 10.1002/9780470015902.a0003279.pub3

Sheath RG, Vis ML, Hambrook JA et al (1986) Tundra stream macroalgae of North America: composition, distribution and physiological adaptations. Hydrobiologia 336:67–82

Stevenson RJ (1996) An introduction to algal ecology in freshwater benthic habitats. In: Stevenson RJ, Bothwell ML, Lowe RL (eds) Algal ecology: freshwater benthic ecosystems. Academic, San Diego, pp 3–30

Vannote RL, Minshall GW, Cummins KW et al (1980) The river continuum concept. Can J Fish Aquat Sci 37:130–137

Whitton BA (1984) Ecology of European Rivers. Blackwell Scientific, Oxford

Chapter 2
Blue-Green Algae (Cyanobacteria) in Rivers

Dale A. Casamatta and Petr Hašler

Abstract This chapter presents some of the more commonly encountered lotic cyanobacterial taxa. The cyanobacteria are a group of oxygenic prokaryotes present in nearly all aquatic ecosystems. While the ecological importance of this lineage is well known, much confusion exists pertaining to their systematic and taxonomic status. In order to facilitate generic-level identification, we separate the cyanobacteria into four major groupings: the Chroococcales (coccoid cells often in a mucilaginous envelop), the Oscillatoriales (filamentous forms lacking specialized cells), the Nostocales (filamentous with inducible specialized cells), and the Stigonematales (filamentous, obligatory specialized cells coupled with cell division in multiple planes). We discuss the major genera found in each lineage, the current state of the systematics, and the broad ecological roles and niches of these taxa. Dichotomous keys and images are presented to facilitate generic identifications.

Keywords Cyanobacteria • Genus • Identification • Lotic • Morphology • River • Taxonomy

Introduction

The cyanobacteria (also known as blue-green algae, cyanophyta, cyanoprokaryotes) are a group of photo-oxygenic bacteria found in aquatic, aerophytic and terrestrial habitats, and from pole to pole. They are among the most ancient lineages of prokaryotes and are some of the most ubiquitous organisms on Earth (Falcon et al. 2010). The cyanobacteria are incredible ecosystem engineers, accounting for ca. 20–30 % of global oxygen production (Pisciotta et al. 2010) and have been credited with elevating the atmospheric oxygen levels ca. 2.5–2.2 bya (Schopf 2000). Cyanobacteria are also known to fix atmospheric nitrogen, contributing greatly to

D.A. Casamatta (✉)
Department of Biology, University of North Florida, 1 UNF Drive, Jacksonville, FL, USA
e-mail: dcasamat@unf.edu

P. Hašler
Department of Botany, Faculty of Sciences, Palacký University Olomouc, Šlechtitelů 11, 771 46 Olomouc, Czech Republic

O. Necchi Jr. (ed.), *River Algae*, DOI 10.1007/978-3-319-31984-1_2

the global nitrogen budget (Karl et al. 2002), serve as the centerpiece of aquatic foodwebs (Scott and Marcarelli 2012), help to stabilize substrates, and are common photobionts.

Cyanobacteria are usually seen as harbingers of ecosystem degradation in lotic systems. Often the most visible components of freshwater harmful algal blooms (HABs), excessive cyanobacteria may lead to negative consequences such as light attenuation, biofouling, the accumulation of excess biomass and subsequent anoxia, and toxin production. Cyanobacterial blooms, typically in lentic ecosystems but also in lotic ones, are often triggered by anthropogenic factors, such as increased nitrogen and phosphorus loads related to land use (e.g., Beaver et al. 2014). Within lotic habitats, the presence of cyanobacteria are useful biomonitoring units (e.g., oligotrophic vs. eutrophic, Loza et al. 2013). Like their planktic counterparts, lotic cyanobacteria may be a source of cyanotoxins, but much less attention is paid to monitoring such benthic cyanobacteria (Seifert et al. 2007).

Given their vast preponderance, diversity, and ecosystem importance, cyanobacteria make excellent taxa for monitoring the health of aquatic ecosystems. However, two major impediments exist that preclude their usage to the wider aquatic community. First, the current state of cyanobacterial systematics is rather confusing. Second, many of the cyanobacteria are difficult to identify due to a limited amount of morphological characters, phenotypic plasticity, and small size.

Systematics and Taxonomy

Cyanobacteria have long been observed by phycologists in classic monographs (e.g., Agardh 1824; Kützing 1849; Nägeli 1849), but were first extensively documented by Bornet and Flahault (1886–1888) and later by Gomont (1892–1893). Their starting taxonomy was expanded by later researchers, most notably Geitler (1932), who helped revise many of the currently recognized genera familiar to most scientists. Recognizing the prokaryotic nature of the cyanobacteria, Stanier et al. (1978) advocated the transfer of the cyanobacteria from the *International Code of Botanical Nomenclature* (ICBN) to the *International Code of Nomenclature of Bacteria* (ICNB). While this appeal had much merit (after all, they are bacteria), it also meant that both codes could be utilized to name novel taxa, leading to an explosion of new names. The inclusion into the ICNB code allowed the cyanobacteria to be catalogued in the Bergey's Manual of Systemic Bacteriology, where Castenhoz (2001) broke them into five major lineages based, in part, on type of cell division and the presence of differentiated cells. While this was a major revision, many felt that this approach missed a tremendous amount of the actual biodiversity of this lineage. All of this began to change in the 1980s–1990s when Komárek and Anagnostidis began their revisionary work on the cyanobacteria as a whole (e.g., Anagnostidis and Komárek 1985). Eschewing the simplified version of systematics proposed by Castenhoz (2001), Komárek and Anagnostidis set about erecting smaller, monophyletic genera employing a polyphasic approach using a number of different characters including morphology, ecology, and genetic characters (most notably the 16S rDNA sequence as proposed by the ICNB). Their work, along with that of colleagues, has greatly increased our knowledge of cyanobacterial diversity and evolutionary relationships (see references below).

Difficulties with Identifications

One of the main difficulties in identifying the cyanobacteria is the fact that they may exhibit a tremendous amount of environmentally or culturally induced phenotypic plasticity (Casamatta and Vis 2004). Conversely, other lineages are obligatorily very simple, and thus exhibit a limited range of morphologies that may mask a tremendous amount of genetic or cryptic diversity (Casamatta et al. 2003).

The other difficulty in identifying the cyanobacteria lies in the small size (filaments typically range from 1 to 10 μm) and lack of easily discernable morphological characters, as most cyanobacteria exist as simple coccoid or bacillus cells. Some lineages have evolved more complex structures, perhaps even being considered “multicellular” (Schirrmeister et al. 2011), including the “overwintering” akinete and the heterocyte (a common type of differentiated cell dedicated to nitrogen fixation), both of which are commonly employed in phylogenetic assessments. However, given the ancient history and ecological permissively of cyanobacteria, it is clear that the described diversity is not reflected in the current phylogenetic character sets.

Habitats

Lotic habitats are replete with cyanobacteria. Many taxa are strictly or predominantly benthic (e.g., *Phormidium*, *Geitlerinema*) and thus easily collected as mats or solitary filaments. Other taxa begin as benthic colonies before migrating to a more planktic habitat (e.g., *Merismopedia*, *Pseudanabaena*). Still others are may be epiphytic (*Chamaesiphon*, *Leibleinia*), endogloeic (*Pseudanabaena*, *Synechococcus*), or epilithic (*Homoeothrix*, *Pleurocapsa*, *Chamaesiphon*, *Calothrix*). In fast flowing waters, taxa with the ability to adhere to surfaces are common (*Lyngbya*), while stagnant waters, or even water margins, may select for forms with elaborate sheaths (*Microcoleus*). Myriad other microhabitats are excellent places for additional genera, but it should be noted that many cyanobacteria are quite capable of motility (*Geitlerinema*) or posses gas vacuoles for rapid movement throughout the water column (*Microcystis*). Thus, nearly every lotic ecosystem will have representative cyanobacteria if one merely looks.

We would be remiss if we did not point out that there exists tremendous variability among lotic ecosystems, with concurrent differences present in the cyanobacterial community. For example, first-order streams are often greatly influenced by allochothonous carbon inputs, and often contain more rocky substrates, favoring cyanobacteria capable of surviving periodic episodes of desiccation and benthic growth habits (e.g., *Phormidium*). Further, differences in aspects such as substrates (e.g., limestone vs. muddy vs. sandy) may greatly impact the cyanobacterial community present. Additional factors, such as light levels, flow rates, the presence grazers, degree of disturbance, and anthropogenic impacts, just to name a few potential cofounding factors, all conspire to affect community composition. Conversely, the cyanobacterial community in higher order streams may appear more like their lentic counterparts, especially in deeper rivers with very slow flow, where planktic taxa frequently occur (e.g., *Microcystis*, *Woronichinia*, *Snowella*, *Anabaena/Dolichospermum*, *Planktothrix*; Hašler et al. 2008) and benthic species usually inhabit fine sediments such as mud or fine sand.

Another potential pitfall in crafting a guide to common cyanobacteria of lotic habitats concerns itself with the sometimes vast differences presented by geographic regions. The issue of endemism in cyanobacteria is one of great debate. Some taxa appear to be cosmopolitan (e.g., *Phormidium*, *Leptolyngbya*) within permissible habitats and appear in European and North American phycological keys. However, even wide-spread taxa may not actually represent the same organism, as many cyanobacteria are difficult to identify to species and may be masked by cryptic variation (e.g., *Phormidium retzii*, *sensu* Casamatta et al. 2003). In a recent survey, Sherwood et al. (2015) noted a large number of endemic cyanobacteria from Hawaii, indicating the need for more extensive sampling and descriptions to fully characterize the cyanobacteria. In addition, even within national boundaries there may exist vast habitat and concurrent cyanobacterial variability and heterogeneity (e.g., in the United States alone there would be alpine, desert, temperate, tropic, ephemeral, and other habitats). Thus, we caution that no guide will ever be able to encompass the tremendous amount of potential diversity, and thus we seek to present only some of the most common, ubiquitous members that one may encounter in a practical, brief, concise manner. For a more complete survey, we provide additional links to resources elsewhere.

Identification of Cyanobacteria

The current state of cyanobacterial systematics is in a state of some debate, but in order to facilitate identifications we have chosen to employ the scheme proposed by Anagnostidis and Komárek (1988, 1990), Komárek (2013), and Komárek and Anagnostidis (1986, 1989, 1998, 2005) (see Table 2.1).

Additional Keys and Resources

Alas, an exhaustive description of all common lotic cyanobacteria is not feasible, but fortunately many of the more "common" taxa have rather global distributions. There has been a flurry of new genera proposed over the last decade, representing an increase in types of habitats sampled and also more intensive scrutiny of less

Table 2.1 General orders of the cyanobacteria and main features (*sensu* references cited)

Order	Feature	Example genera
Chroococcales (Key 1)	Unicellular or colonial, never forming filaments (but rarely pseudofilamentous)	*Aphanocapsa*, *Chroococcus*, *Merismopedia*
Oscillatoriales (Key 2)	Filamentous, sometimes with false branching, no heterocytes, and akinetes	*Leptolyngbya*, *Lyngbya*, *Oscillatoria*, *Phormidium*
Heterocytous genera I (Key 3)	Filamentous, heterocytes and akinetes present (usually), sometimes with false branching	*Anabaena*, *Calothrix*, *Dichothrix*, *Nostoc*
Heterocytous genera II (Key 4)	Filamentous, division in multiple planes (true branching), heterocytes present	*Hapalosiphon*, *Nostochopsis*

esoteric habitats but with a more practiced eye for finely discernable genera. Many traditional "large" genera were found polyphyletic. Using a polyphasic approach many new monophyletic genera were established. For a more detailed listing of taxa, we recommend monographs from North America (e.g., Tilden 1910; Smith 1950; Prescott 1962; Whitford and Schumaker 1969; Komárek and Johansen 2015a, b) or Europe (Whitton 2005; Komárek and Anagnostidis 1998, 2005). While much of the taxonomy in these texts has changed, the works by Komárek and Anagnostidis (1998, 2005) and Komárek (2013) provide excellent starting points to the revised taxonomy and species identifications. The CyanoDB site (www.Cyanodb.cz, Komárek and Hauer 2015) is an excellent resource for approved cyanobacteria genera. For additional photomicrographs, we recommend the Hindák 2001 book.

Additional Lotic Cyanobacterial References

If describing cyanobacterial diversity in lotic ecosystems is challenging, understanding their ecology is even more difficult. Small, ephemeral, first-order streams through massive, ofttimes slow moving bodies of water all fall under the umbrella of lotic systems, with everything in between. The ecology of the constitutive cyanobacteria varies by surrounding land use, altitude, climate, flow rates, anthropogenic inputs, etc., just to name a few parameters. Thus, a comprehensive review of the ecological roles of cyanobacteria is beyond the purview of this chapter. In order to even partially ameliorate this shortcoming, we provide a list of references to further explore some of the vast ecological roles that cyanobacteria play in flowing waters: Sheath and Cole (1992), Steinman et al. (1992), Stevenson et al. (1996), Perona et al. (1998), Whitton and Potts (2000), Rott et al. (2006), Whitton (2012a, b), Loza et al. (2013), Manoylov (2014), and Stevenson (2014).

Key 1: Form-genera of the Chroococcales

1a	Cells not heteropolar	2
1b	Heteropolar cells or pseudofilaments divided into basal and apical end	9
2a	Flat, tabular-like colonies, cells spherical, or ovoid	*Merismopedia*
2b	Other than tabular-like colonies	3
3a	Spherical or hemispherical cells, irregularly arranged in diffluent mucilage	*Aphanocapsa*
3b	Cells and colonies otherwise	4
4a	Elliptical, oval to rod-like cells, irregularly in diffluent mucilage	*Aphanothece*
4b	Cells otherwise arranged (not in common mucilage)	5
5a	Cells solitary or in irregular clusters, but not mucilaginous colonies	*Synechococcus*
5b	Cells, mucilage, and colonies of different shape	6
6a	Cell spherical to hemispherical with distinct mucilaginous envelopes	7
6b	Cells of various shape, often irregular	8
7a	Spherical mature cells, daughter cells reach mother cell shape after division, layered mucilaginous envelopes, often colored	*Gloeocapsa*
7b	Spherical, hemispherical cells, do not reach cell shape after division	*Chroococcus*
8a	Mucilaginous colonies, spherical to irregular cells, often individually enveloped by distinct mucilage	*Chlorogloea*
8b	Usually pseudoparenchymatous colonies, pseudofilaments, irregular cells, produces baeocytes	*Pleurocapsa*
9a	Species does not form pseudofilaments	10
9b	Species forms pseudofilaments, cells spherical, cylindrical, or barrel shaped, exospores of different shape liberated separately	*Stichosiphon*
10a	Oval, elongated to rod-like cells, sheath (pseudovagina) not present, exospores produced at the upper part of mother cell	*Geitleribactron*
10b	Species enveloped by distinct sheath (pseudovagina)	11
11a	Cells oval, elongated, cylindrical, apically produced exospores	12
11b	Cells spherical, oval, club/pear like or irregular, produce baeocytes, mainly epiphytic	13
12a	One or more spherical or hemispherical exospores, epiphytic, or epilithic	*Chamaesiphon*
12b	Cells form gelatinous hairs at the apical part, one or more spherical to elongated exospores, epiphytic, or epilithic	*Clastidium*
13a	Cells usually live solitary or gathered in groups	*Cyanocystis*
13b	Cells usually form pseudoparenchymatous colonies	14
14a	Few cells enveloped by a common sheath	*Xenococcus*
14b	A common sheath envelopes the whole colony	*Xenotholos*

Aphanocapsa Nägeli (Fig. 2.1a)

Individuals are microscopic to macroscopic colonies, usually mucilaginous, spherical to irregular. Cells without individual gelatinous envelopes are spherical, temporarily hemispherical after division, irregularly arranged in homogenous colorless mucilage. Cells are usually green, blue-green, or olive-green. Cell content is homogenous or finely granulated, always without aerotopes. Binary fission in one

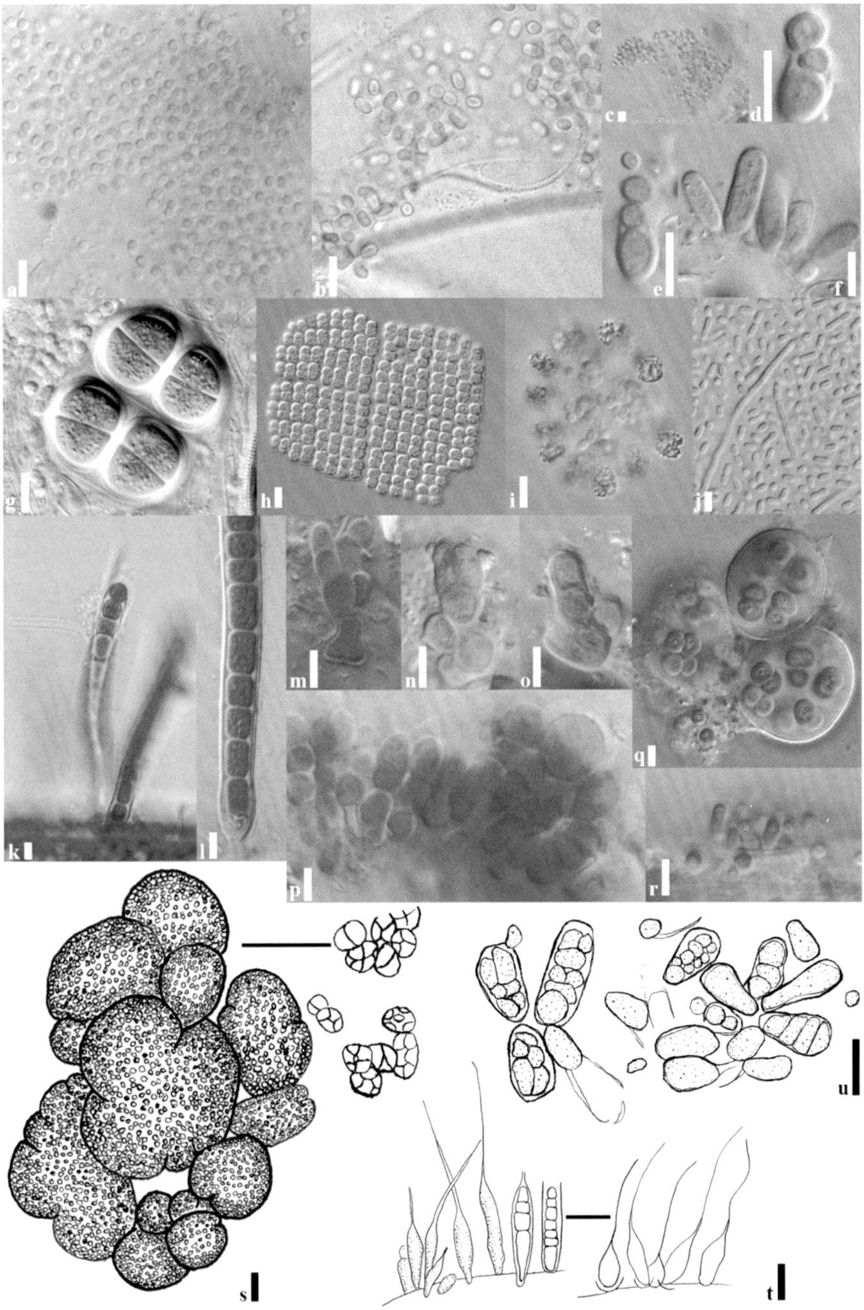

Fig. 2.1 (**a**–**u**) Examples of coccoid genera. (**a**) *Aphanocapsa*. (**b** and **c**) *Aphanothece*. (**d**–**f**) *Chamaesiphon*. (**g**) *Chroococcus*. (**h**) *Merismopedia*. (**i** and **j**) *Synechcoccus*: note the growth within the mucilage of other algae [endogloeic]. (**k** and **l**) *Stichosiphon*: note the pseudofilaments and pseudovaginal arrangement. (**m**–**p**) *Pleurocapsa*. (**q**) *Gloeocapsa*. (**r**) *Geitleribactron*. (**s**) *Chlorogloea*. (**t**) *Clastidium*. (**u**) *Cyanocystis*. Scale bars: (**a**–**c**, **g** and **h**, **k**–**q**) = 10 μm; (**d**–**f**, **i** and **j**, **r**–**u**) = 5 μm

or two perpendicular planes is typical for this genus. Cells after division can be gathered in small groups or subcolonies. Colonies of various species can differ in density of cells.

Occurrence: many species are epilithic on submerged stones or subaerophytic in spray and zones (*sensu* Rott et al. 2006) but may also be common as planktic members. Found extensively in lentic and lotic systems.

Expected taxa: *A. fonticola* Hansgirg, *A. grevillei* (Berkeley) Rabenhorst, *A. muscicola* (Meneghini) Wille, *A. rivularis* (Carmichael) Rabenhorst

Aphanothece Nägeli (Fig. 2.1b, c)

Individuals are microscopic (to macroscopic), usually consisting gelatinous colonies which can consist of small subcolonies. Cells are arranged within homogenous mucilage (usually colorless, sometimes yellowish, or brownish), occasionally slightly layered mucilage around cells, or small groups of cells can occur (e.g., *A. bullosa*). Cells divide into daughter cells by binary fission in one plane and after division grow into original cell shape (elongated, cylindrical to rod like) and size. Cell content is homogenous or slightly granulated, green, blue-green, violet, brownish, often with visible peripheral chromatoplasma.

Remarks: two major subgroups of *Aphanothece* differ in cell dimensions. The first is usually planktic and ≤1 μm wide, while the majority of benthic, metaphytic, or aerophytic species are in the second group (≥1 μm wide).

Occurrence: the majority of species is epilithic on submerged stones or subaerophytic in spray zones (*sensu* Rott et al. 2006), epipelic, epiphytic, metaphytic, in springs and streams.

Expected taxa: *A. stagnina* (Sprengel) A. Braun, *A. minutissima* (W. West) Komarková-Legnerová et Cronberg, *A. smithii* Komarková-Legnerová et Cronberg

Chamaesiphon A. Braun et Grunow in Rabenhorst (Fig. 2.1e, f)

Thallus formed by layered/shrub-like micro to macroscopic gelatinous colonies or can live solitary. Vegetative cells are variable in shape (spherical, elliptic, oval, pear like, or rod like) and can change during life cycle. Cell content is homogenous or finely granulated blue-green, olive-green, yellowish, reddish, pinkish, or violet, sometimes with visible peripheral chromatoplasma. Mother cells divide asymmetrically and form apically placed exocytes. Both mother cells and exocytes are surrounded by colorless or yellowish to brownish envelopes, which burst after exocytes liberation.

Remarks: species are differentiated by the shape and dimensions of cells.

Occurrence: mainly epilithic on submerged or wetted stones, epiphytic on filamentous algae or mosses, especially in unpolluted freshwaters, streams, and waterfalls.

Expected taxa: nearly all taxa (*sensu* Komárek and Anagnostidis 1998) can occur in streams and rivers, e.g., *C. minutus* (Rostafinski) Lemmermann, *C. incrustans* Grunow, *C. polymorphus* Geitler, *C. starmachii* Kann, *C. subglobosus* (Rostanfinski) Lemmermann.

Chroococcus Nägeli (Fig. 2.1g)

Solitary living cells in small groups (usually two to eight cells) or agglomerations of microscopic colonies. Spherical to hemispherical cells (1–50 μm) are surrounded by colorless to yellowish mucilaginous envelopes. The envelopes can be markedly distinct at margin, layered and wide or tightly attached to the cell surface. The genus was recently divided into two major lineages, with planktic species moved into *Limnococcus* and the remaining attached or metaphytic species remaining in *Chroococcus* (Komárková et al. 2010). Cell content is greenish, olive-green, reddish to violet, often granulated or divided into granulated centroplasma and homogenous or finely granulated peripheral chromatoplasma. Cell division depends on colony age; cells in young colonies divide by binary fission in three planes, in old colonies divide almost irregularly with respect to density of cells. Colonies reproduce by fragmentation into small parts or subcolonies.

Occurrence: mainly periphytic on stones and waterfalls or metaphytic, in cold to hot/thermal springs and streams.

Expected taxa: *C. tenax* (Kirchner) Hieronymous

Chlorogloea Wille (Fig. 2.1s)

Colonies are usually mucilaginous with markedly limited surface and variable in shape (spherical, hemispherical to irregular). Cells are variable in shape as well (spherical to almost irregular, 1–6 μm) usually irregularly arranged in colony, seldom at margin radial rows can occur. They are placed in individual colorless or greenish to reddish envelopes. Cell division in three perpendicular planes and reproduction by disintegration of colonies occur in this genus.

Occurrence: typically attached species on various substrates (stones, plants, and algae).

Expected taxa: *C. rivularis* (Hansgirg) Komárek et Anagnostidis, *C. microcystoides* Geitler

Clastidium Kirchner (Fig. 2.1t)

Cells live solitary or gathered in small groups. Cells are different in shape and size from spherical to oval/cylindrical or pear like (approximately 6–40 × 2–4 μm). Basal part of cell is oval while apical part is narrowed and can be extended into thin hair-like gelatinous projection (up to 75 μm long). Cell content is yellowish, blue-green, olive-green, or gray-green. Mucilaginous envelope (pseudovagina) is usually thin and firm, occasionally can be slightly lamellate. Spherical exocytes are produced in apical part of cells, then released and attached on substrate.

Occurrence: epiphytic (on algae) or epilithic in clear streams.

Expected taxa: *C. rivulare* (Hansgirg) Hansgirg, *C. setigerum* Kirchner, *C. sicyoideum* Li.

Cyanocystis Borzi (Fig. 2.1u)

Cells live solitary or gathered in hemispherical to flatted groups attached to substrate. Cell shape varies from spherical, hemispherical, oval, broadly oval to club/pear like, or elongated, often obviously heteropolar, usually 10–30 μm long and 5–20 μm wide, seldom more or less. Cell content is homogenous or finely granulated, blue-green, olive-green, yellowish, and reddish to violet. Mucilaginous envelope (pseudovagina) is usually thin, colorless, firm, and disintegrates after baeocyte liberation, which grow to the mother cell size and shape before next reproduction.

Occurrence: common in mountain streams, cosmopolitan, mainly epiphytic species on filamentous cyanobacteria and algae, seldom epilithic.

Expected taxa: *C. aquae-dulcis* (Reinsch) Kann, *C. mexicana* Montejano et al., *C. versicolor* Borzi

Geitleribactron Komárek (Fig. 2.1r)

Solitary living heteropolar cells or groups of cells attached to the substrate. Cells are ovoid, cylindrical to rod like or club like, straight or slightly bent, usually with rounded apical part and narrowed basis, blue-green, pale blue-green, olive-green to grayish. Protoplast is homogenous or finely granulated, often with clearly visible peripheral chromatoplasma. Mucilaginous envelope is not present. Cells divide in the middle or asymmetrically in upper part. Daughter cells attach to the substrate by their ends after liberation and grow into original size.

Occurrence: epiphytic genus, usually on filamentous cyanobacteria and algae.

Expected taxa: *G. crissa* Gold-Morgan et al., *G. periphyticum* Komárek

Gloeocapsa Kützing (Fig. 2.1q)

Microscopic to macroscopic colonies, which can gather in gelatinous mats. Cells are irregularly arranged in colonies, spherical, hemispherical, oval to slightly elongated, with homogenous to finely granulated content, blue surrounded by individual mucilaginous envelopes, colorless or colored yellow, yellow-brown, reddish, and blue to violet. Structure of envelopes is dependent on life cycle and can be distinct and layered or gelatinous and diffluent. Cells divide by binary fission in three perpendicular planes and then are irregularly arranged in the colony. Cells after division grow into original size and shape. In some, species was observed formation of resting spores (akinetes) surrounded by individual thick, firm, structured, and intensely colored envelopes. Nanocyte production was occasionally observed in several species.

Occurrence: mainly aerophytic and subaerophytic species, members of the genus can occur on walls of waterfalls, on wetted stones along banks.

Expected taxa: *G. aeruginosa* Kützing, *G. gelatinosa* (Meneghini) Kützing, *G. granosa* (Berkeley) Kützing, *G. punctata* Nägeli, *G. thermalis* Lemmermann.

Merismopedia Meyen (Fig. 2.1h)

Microscopic (occasionally large macroscopic), flat, tabular-like, large colonies often consist of small subcolonies. Cells are precisely oriented along two perpendicular planes in the colony. Cell density is variable from tightly to sparsely arranged cells. Cells are hemispherical, oval, usually with homogenous content, blue-green, olive-green to reddish. Mucilaginous envelopes vary from diffluent gelatinous to thin distinct layers. Cells divide by binary fission in two perpendicular planes and grow into original size before next division.

Occurrence: very common in metaphytic and benthic habitats, often becoming planktic.

Expected taxa: *M. convoluta* Brébisson, *M. elegans* A. Braun, *M. glauca* (Ehrenberg) Kützing

Pleurocapsa Thuret in Hauck (Fig. 2.1m–p)

Colonies are irregular, attached to stony substrates, pseudofilamentous, sometimes sarcinoid. Yellowish to brownish mucilaginous sheaths are thin and firm and cover cells or rows of cells or pseudofilaments. Cells are variable, ovoid, elongate, polygonal, or apically narrowed, 0.8–20 μm in diameter, homogenous or slightly granular, blue-green, gray-green, brownish, or violet. Cells divide irregularly in various planes. Reproduction by baeocytes was observed in several species.

Occurrence: typically epilithic, parts of colonies can be endolithic.

Expected taxa: *P. aurantiaca* Geitler, *P. concharum* Hansgirg, *P. minor* Hansgirg.

Stichosiphon Geitler (Fig. 2.1k, l)

Solitary or gathered in groups, single celled to heteropolar pseudofilamentous attached by narrowed basis (mucilaginous stalk) to the substrate. Mucilaginous sheath (pseudovagina) is colorless, firm, and distinct, opened at the apex. Cells are spherical, hemispherical, cylindrical, barrel shaped, or almost rectangular, homogenous to finely granulated, blue-green, olive-green to grayish. After repeated binary

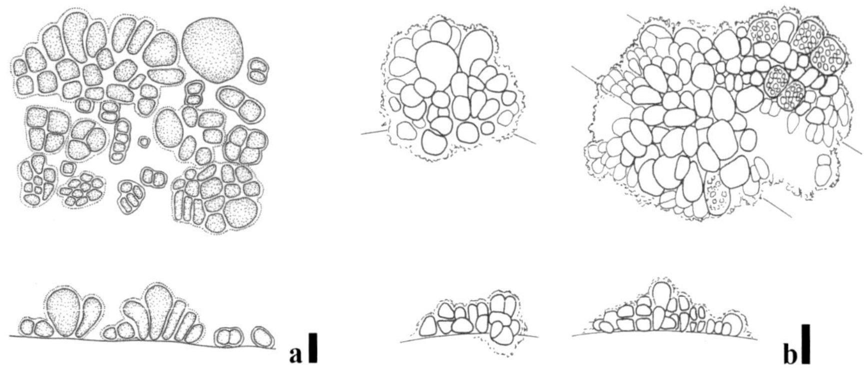

Fig. 2.2 (**a** and **b**) Examples of coccoid genera. (**a**) *Xenococcus*. (**b**) *Xenotholos*. Scale bars = 5 µm

fission exospores remain in the pseudovagina. Number of exospores is species dependent and ranges from 4 to more than 80, with shape a stable species level identifier (e.g., *S. filamentosus*, oblong exocytes; *S. gardneri*, narrowly oblong; *S. regularis*, widely ovate).

Occurrence: mainly epiphytic on submerged plants and cyanobacteria and algae (*Rhizoclonium*, *Cladophora*, *Oedogonium*, *Plectonema*, *Homoeothrix*), occasionally epilithic.

Expected taxa: *S. regularis* Geitler, *S. pseudopolymorphus* (Fritsch) Komárek, *S. filamentosus* (Ghose) Geitler

Synechcoccus Nägeli (Fig. 2.1i, j)

Cells cylindrical to oblong, elliptical, occasionally forming pseudofilaments (2–4, but up to 20 cells). Transverse cell division results in similar shaped or different daughter cells that may remain joined together. Cells typically small (0.5–6(11) µm wide), but may occasionally be much longer (1.5–20 µm long).

Occurrence: very common genus, may be planktic, occasionally associated with other algae, subaerial, or endogloeic.

Expected taxa: *S. elongates* (Nägeli) Nägeli, *S. nidulans* (Pringsheim) Komárek in Bourrelly

Xenococcus Thuret in Bornet et Thuret (Fig. 2.2a)

Sessile cells are solitary or gathered in monolayered groups of colonies, old colonies form irregular or pseudoparenchymatous mats. Colorless or yellowish, thin and firm sheath (envelope) covers small clusters including few cells (subcolonies). Cell shape varies from spherical or oval to irregular or pear shaped (1.5–5.5 µm in diameter, freshwater species), polarity of attached cells is possible to recognize. Cells divide irregularly in multiple planes, usually in a perpendicular plane to substrate. Baeocytes (reproducing cells) are formed by multiple fission and subsequently liberates from mother cells. The members of the genus can be misidentified with members of the genus *Xenotholos* (see description below).

Occurrence*:* freshwater species are usually epiphytic on filamentous algae such as *Cladophora* or *Tribonema*

Expected taxa*: X. bicudoi* Montejano et al., *X. lamellosus* Gold-Morgan et al., *X. minimus* Geitler, *X. willei* Gardner

Xenotholos Gold-Morgan et al. (Fig. 2.2b)

Cells usually aggregated into globular, often multilayered, colonies. Old colonies form pseudoparenchymatous thallus. Single cells occur at the beginning of life cycle and subsequently divide, grow, and form new colonies. Sheath (envelope) is usually firm, thin, colorless, and overlaps the whole colony. Cell shape varies from spherical, oval to irregular, or pyriform (1–20 μm in diameter), polarity of attached cells is possible to recognize. Cells divide by binary fission in more planes, often with layered aggregations, division tend to be synchronized with pseudofilamnetous formation. Baeocytes (reproducing cells) are formed by multiple fission and subsequently liberation from mother cells. The identification of *Xenotholos* and *Xenococcus* requires a detailed study of colonies, arrangement of cells and mucilage sheath.

Occurrence*:* mainly epiphytic, found on filamentous cyanobacteria and algae (e.g., *Blennothrix*, *Cladophora*, *Rhizoclonium*)

Expected taxa: *Xenotholos kerneri* Gold-Morgan et al., *Xenotholos caeruleus* Gold-Morgan et al., *Xenotholos amplus* Gold-Morgan et al.

Key 2: Form-genera of the Oscillatoriales

1a	Trichomes screw-like coiled	*Arthrospira*
1b	Trichomes straight	2
2a	Filaments attached as epiphytes (along the length)	3
2b	Filaments not attached	4
3a	Filaments attached along the entire length	*Leibleinia*
3b	Filaments attached at one end	*Heteroleibleinia*
4a	Multiple trichomes (2+) in a common sheath	*Microcoleus*
4b	Trichomes single (or rarely multiple)	5
5a	Trichomes attached at one end (epilithic, rarely endogloeic), heteropolar	*Homoeothrix*
5b	Trichomes otherwise	7
6a	Trichomes without sheaths (or, very fine), trichome straight (occasionally wavy), cells barrel shaped, typically narrow (1–3 μm)	8
6b	Trichomes otherwise	10
7a	Trichomes solitary (or in fine mats), short, cells cylindrical to barrel shaped, constricted (often markedly) at cross-walls	*Pseudanabaena*

7b	Trichomes otherwise	9
8a	Trichomes isopolar, in mats or clusters, rarely solitary, thin, fine, or firm sheath	*Leptolyngbya*
8b	Trichomes (very) motile, typically attenuated, typically distinctive apical cells	*Geitlerinema*
9a	Cells ± isodiametric, sheaths facultative, or obligatory	*Phormidium*
9b	Cells wider than long	11
10a	Trichomes in mats, sheaths absent (very rarely environmentally inducible)	*Oscillatoria*
10b	Sheathes thick, ± obligatory	*Lyngbya*

Arthrospira (Gomont) Stizenberger (Fig. 2.3a)

Individuals are typically encountered as solitary trichomes, often free floating in plankton, but occasionally as fine mats in the benthos. Trichomes are screw shaped or coiled, isopolar, and sheathes rare, but may be facultatively present (fine and thin). Cells are ± isodiametric with visible cross-walls. Morphologically similar to *Spirulina*, but with visible cross-walls (the separation between these genera has been conformed via molecular and morphological examination).

Remarks: many species have been recorded, but the intrageneric systematics is rather vague at this time. No known toxic forms and often employed in biotechnical applications (often incorrectly identified as "*Spirulina platensis*"). We also note that *Arthrospira* has been cited in the North American Water Quality Assessment and by researchers from California and Florida.

Occurrence: typically planktic, but often present in the benthos of a variety of aquatic habitats.

Expected taxa: *A. jenneri* (Gomont) Stizenbergeri, *A. platensis* Gomont

Geitlerinema (Anagnostidis et Komárek) Anagnostidis (Fig. 2.3m)

Trichomes thin (1–4 μm), straight (rarely bent or screw like), not (rarely) constricted, cells longer than wide, never posses aerotopes. End cells are often distinctive, being hooked, coiled, bent, typically acumate, or rounded. Highly motile with intensive gliding motility, but also rotation and waving. Thallus often vibrant blue-green, delicate, diffluent, thin mats, but sometimes individual trichomes. This genus is composed mostly of taxa formally members of *Oscillatoria*, but appears to be polyphyletic and in need of revision (Perkerson et al. 2010).

Occurrence: common in freshwaters, often forming thin, delicate, brightly colored mats in benthic habitats. Also an occasional epiphyte or present in subaerial habitats. Records concerning marine taxa probably correspond to a new, as of yet undescribed genus.

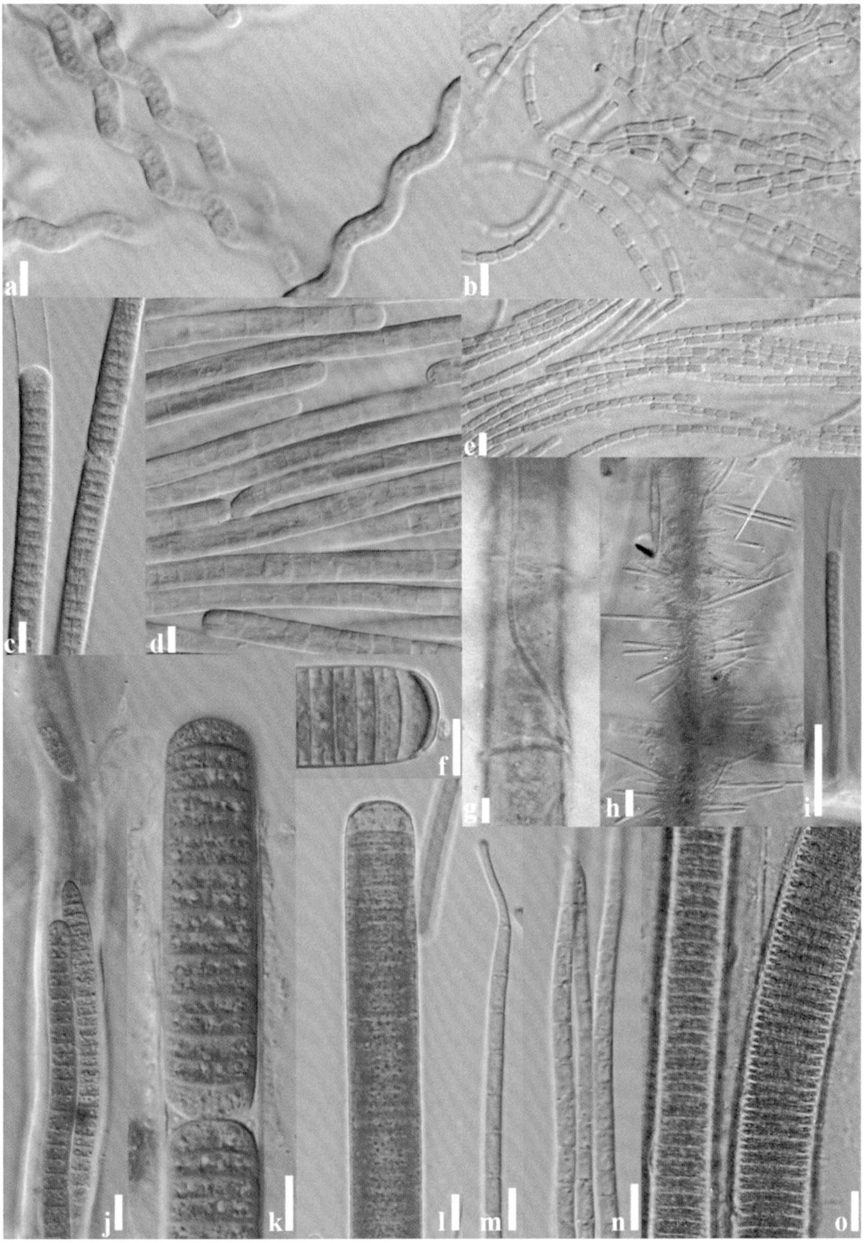

Fig. 2.3 (**a–o**) Examples of filamentous genera without heterocytes. (**a**) *Arthrospira*. (**b**) *Pseudanabaena*: note the isopolar nature. (**c**) *Phormidium*: note the difference between the filament and the trichome. (**d**) *Ammassolinea*. (**e**) *Pseudanabaena*. (**f**) *Oscillatoria*. (**g**) *Leibleinia*. (**h** and **i**) *Heteroleiblenia*: note the heteropolar tapering. (**j**) *Microcoleus*: note the copious mucilaginous sheath. (**k** and **l**) *Oscillatoria*: (**k**) note the fragmentation into hormogonia. (**m**) *Geitlerinema*. (**n**) *Leptolyngbya*. (**o**) *Lyngbya*: note the filament (cells with the sheath). Scale bars: (**a**, **f**, **j–l**, **o**) = 10 μm; (**b–e**, **g–i**, **m–n**) = 5 μm

Expected taxa: *G. amphibium* (Agardh ex Gomont) Anagnostidis, *G. splendidum* (Greville ex Gomont) Anagnostidis, *G. tenuis* (Stockmayer) Anagnostidis

Heteroleiblenia **(Geitler) Hoffmann (Fig. 2.3h, i)**

Heteropolar filaments attached at one end, at times forming tuft-like layers. Sheathes thin, colorless, firm. Cells ± isodiametric, with unornamented, obligatorily rounded end cells. Trichomes may be constricted or not at cross-walls. Resembles *Leibleinia*, but differs in mode of attachment (one end vs. the entire filament). Reproduction via trichome disintegration into hormocytes and hormogonia.

Occurrence: previously described as thin *Lyngbya* species, common, cosmopolitan components of both freshwater and marine habitats as epiphytes on other algae, plants, or inanimate substrates. Komárek and Anagnostidis (2005) caution that due to the paucity of molecular data the status of this genus remains open to debate.

Expected taxa: *H. kuetzingii* (Schmidle) Compere, *H. pusilla* (Hansgirg) Compere

Homoeothrix **(Thuret) Kirchner (Fig. 2.5e)**

Simple heteropolar filaments, rarely with false branching, appearing solitary, or occasionally in loose fascicles, but always attached basally to substrate. Sheathes typically thin, but may be firm, hyaline, or occasionally widened, often do not extend to the end of the filament. Trichomes 3–5 μm, tapering, typically to a thin, hair-like projection. Filaments constricted or not, cells ± isodiametric, blue-green to yellowish to grayish.

Occurrence: commonly attached to variable substrates in flowing and stagnant waters, may also be endogloeic (all taxa are attached, though). Colonies often have a collection of coccoid cells at the base (sometimes quite conspicuous). Some systematic confusion remains between *Tapinothrix* and *Homoeothrix*, so further investigation into the complex is warranted. Differs from *Heteroleibleinia* in polarity.

Expected taxa: *H. janthina* (Bornet et Flahault) Starmach, *H. varians* Geitler

Leibleinia **(Gomont) Hoffmann (Fig. 2.3g)**

Filaments solitary, attached along their length (or by a part in later stages, leaving free ends), always epiphytic. Obligatory sheathes thin, firm, colorless. Apical cells may be rounded, never with calyptra or thickened outer cell wall.

Remarks: it is a difficult genus to identify to species due to a paucity of morphological characters, many members of the genus were originally described as epiphytes of morphologically similar genera (e.g., typically *Lyngbya*, but also *Phormidium*). *Leibleinia* is a poorly understood genus, not yet studied in culture or

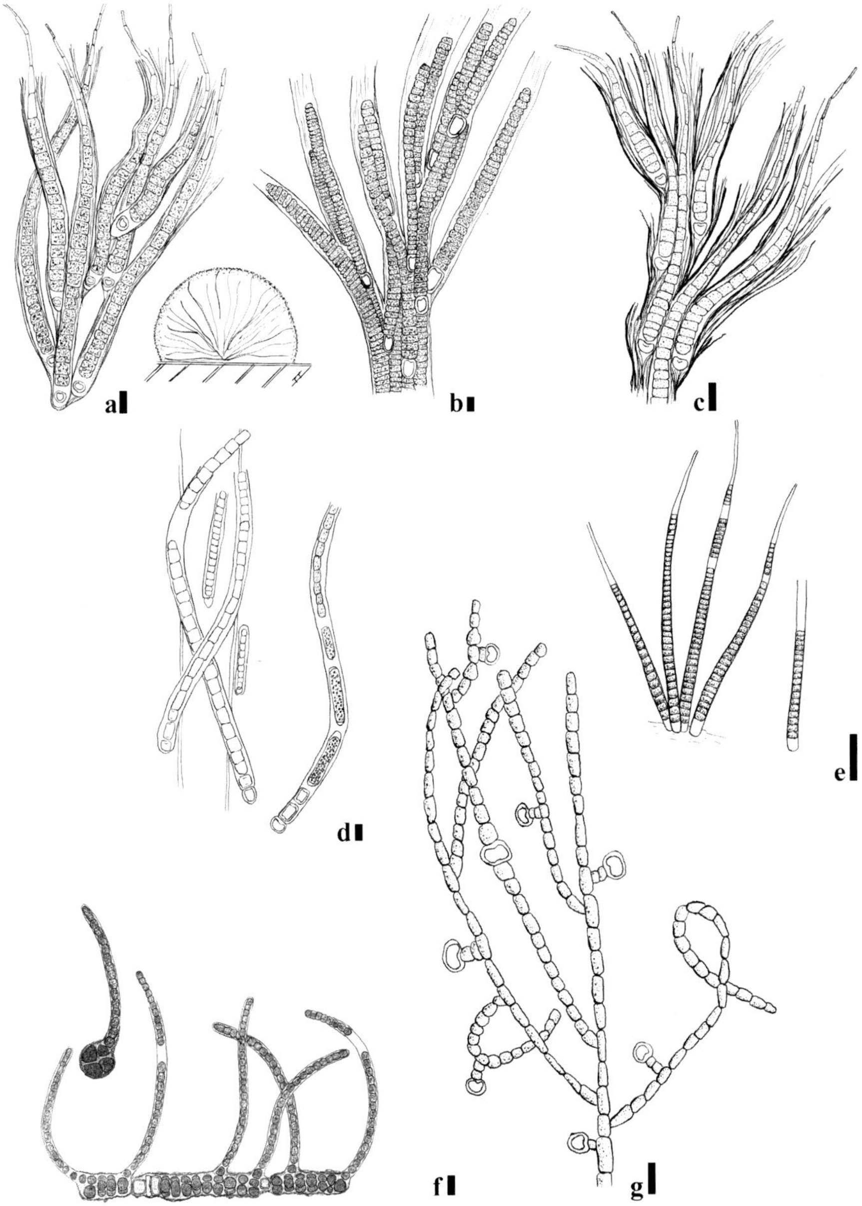

Fig. 2.5 (**a**–**g**) Examples of filamentous heterocytous genera. (**a**) *Rivularia*. (**b**) *Coleodesmium*. (**c**) *Dichothrix*. (**d**) *Microchaete*. (**e**) *Homoeothrix*. (**f**) *Fischerella*. (**g**) *Nostocopsis*. Scale bars = 10 μm

via molecular methods; Komárek and Anagnostidis (2005) advocated critical revisions in the future.

Occurrence: common in marine and freshwater habitats as epiphytes; it can be periodically common, but more common in less disturbed habitats.

Expected taxa: *L. epiphytica* (Hieronymous) Compere, *L. nordgaardii* (Wille) Anagnostidis et Komárek

Leptolyngbya Anagnostidis et Komárek (Fig. 2.3n)

Filaments typically forming mats or clusters, rarely solitary, occasionally solitary, free floating, or in fascicles. Filaments mostly straight, ± flexuous, wavy, curved, or rarely straight. Facultative thin, firm sheaths, with small trichomes (0.5–3.5 μm wide) and isodiametric, cylindrical cells, that may occasionally be shorter than wide. Peripheral thylakoids occasionally visible.

Occurrence: one of the most commonly encountered cyanobacterial genera, it is found in nearly all aquatic habitats, as well as subaerophytic and in soils, cosmopolitan taxa. One of the most specious genera, it is highly polyphyletic as currently described (Komárek and Anagnostidis 2005; Casamatta et al. 2005). Considered a highly "weedy" species, often associated with other cyanobacteria and algae. It is difficult to identify at species level due to lack of morphological characters and small size. Many species have been transferred from other genera (e.g., *Oscillatoria*, *Lyngbya*, *Phormidium*).

Expected taxa: *L. subtillissima* (Kützing ex Hansgirg) Komárek, *L. angustissima* (W. et G.S. West) Anagnostidis et Komárek, *L. valderiana* (Gomont) Anagnostidis et Komárek

Lyngbya (Gomont) C. Agardh (Fig. 2.3o)

Thallus often expansive, leathery, large, prostrate. Filaments straight or sometimes rarely wavy, rarely solitary. Obligatory sheathes that are firm, thin, or thick, may be lamellate, colorless to yellowish, or reddish, containing single motile trichome. Filaments rarely with false branching, typically ≥6.8 μm. Cells are discoid, always shorter than long (up to 1/15 as long as wide, but very rarely approaching isodiametric), only rarely with aerotopes. Apical cells usually with thickened outer cell or with a calyptra. Reproduction via trichome disintegration into short, motile hormogonia.

Occurrence: a specious genus with wide ecological tolerance (freshwater, marine, tropical to polar, etc.), many members are cosmopolitan marine taxa (e.g., *L. aestuarii*, *L. agardhii*). *Lyngbya* may be commonly encountered as benthic mats in lotic systems. Many species, especially planktic taxa, have been transferred into other genera (e.g., *Planktolyngbya* was erected to include small, planktic members).

Expected taxa: *L. martensiana* (Gomont) Meneghini, *L. maior* (Gomont) Meneghini

Microcoleus (Gomont) Desmazieres (Fig. 2.3j)

Thallus flat, prostrate on substrates. Sheathes usually colorless, firm, tapering, typically open at ends, containing multiple (often numerous) trichomes. Trichomes often densely aggregated, parallel in tight fascicles, may extend beyond the end of the sheathes. Without or slightly constricted, typically attenuated. Cells ± isodiametric, granular, with apical cells subconical to acutely conical.

Occurrence: a common genus, found in aquatic (freshwater and marine) and terrestrial habitats. Recently, some taxa have been transferred into new genera (e.g., *M. chthonoplastes* = *Coleofasciculus chthonoplastes sensu* Siegesmund et al. 2008).

Expected taxa: *M. lacustris* (Rabenhorst) Farlow, *M. subtorulosus* (Gomont) Gomont

Oscillatoria (Gomont) Vaucher (Fig. 2.3f, k, l)

Thallus flat, often conspicuous, smooth, often blackish blue-green, green to olive, thin. Trichomes often isopolar, straight, cylindrical, may be screw like or waved. Motile, with gliding, oscillation and rotation. Trichomes larger (typically ≥6.8 μm), constricted or not, cells discoid, always more than 2× shorter than wide (may be 3–11×), ± prominent granules, never with aerotopes. Sheathes absent, but rare under adverse conditions. Reproduction via trichome disintegration into short hormogonia, employing necridia.

Remarks: most members are benthic. *Oscillatoria* was previously perhaps the largest, most commonly encountered genus with many members have been transferred into new genera based on more narrowly defined requirements (e.g., presence of aerotopes, type of cell division, and cell dimensions) such as *Leptolyngbya*, *Limnothrix*, *Pseudanabaena*.

Occurrence: exceptionally common in lotic systems, often benthic, epipsammic, or epiphytic, but may become free floating after dislodging. Planctic isolates belong to other genera (e.g., *Planktothrix*). Widely reported in the literature, but many species have been transferred to other genera.

Expected taxa: *O. princeps* (Gomont) Vaucher, *O. simplicissima* Gomont, *O. tenuis* (Gomont) Agardh

Phormidium (Gomont) Kützing (Fig. 2.3c)

Thallus typically expanded, mucilaginous, often leathery, often conspicuous, often dark blue-green, but variable based on epiphtyes and habitat conditions. Sheathes often present, but highly environmentally plastic. Filaments straight or curved, but not with false branching (probably *Pseudophormidium*). Trichomes straight, flexed, curved, motile (often highly). Cells ± isodiametric, lacking aerotopes. End cells may be pointed, rounded, narrowed, with or without calyptra.

Occurrence: exceedingly common in flowing water; Sheath and Cole (1992) reported *P. retzii* as the most common macroalgal (forming macroscopic mats) taxon in North America. One of the largest genera of cyanobacteria in terms of species numbers, it is also perhaps the most problematic from a phylogenetic standpoint (Komárek and Anagnostidis 2005) and has been shown polyphyletic. Numerous species have been transferred to other genera based on molecular and morphological data (e.g., *P. autumnale* has been transferred to *Microcoleus autumnalis sensu* Strunecký et al. 2013, and some *Phormidium*-like taxa were described as new genus *Ammassolinea*, Fig. 2.2d, *sensu* Hašler et al. 2014). It resembles *Oscillatoria*, but is differentiated by types of trichome division and disintegration. It is found from pole to pole in all types of freshwaters, may also be subaerial.

Expected taxa: *P. retzii* (Gomont) Agardh, *P. nigrum* (Gomont) Anagnostidis et Komárek.

Pseudanabaena Lauterborn (Fig. 2.3b, e)

Trichomes typically solitary, occasionally in fine mats, sheathes rare, occasionally motile, rarely more than 30 cells. Cells typically cylindrical (rarely ± isodiametric), 1–3.5 μm wide, often barrel shaped, usually with conspicuous constrictions at cross-walls. End cells may be rounded or pointed, facultative aerotopes. Considered by some a weedy, junk genus composed of a polyphyletic assemblage of filaments with little morphological differentiation; recent molecular analyses have identified "true" *Pseudanabaena*, but a polyphyletic genus as a whole. Species rich yet poorly described from marine, freshwater, and subaerial habitats, many members were transferred into from other genera (most notably *Oscillatoria* and *Phormidium*). May possess aerotopes, but this may also be a similar genus (*Limnothrix*) and may necessitate taxonomic change. Superficially resembles *Komvophoron*, but differs in type of reproduction and cell dimensions.

Occurrence: very common in aquatic habitats in benthic and planktic habitats, also known as common endogloeic members.

Expected taxa: *P. catenata* Lauterborn, *P. galeata* Bocher, *P. mucicola* (Naumann et Huber-Pestalozzi) Schwabe

Key 3: Form-genera of the Nostocales

1a	Filaments heteropolar	2
1b	Filaments other than heteropolar	7
2a	Filaments form hairs at the apex	3
2b	Filaments do not form hairs at the apex	5
3a	Gelatinous spherical colonies, sometimes with internal cavity, filaments are radially arranged	*Rivularia*
3b	Species do not form gelatinous spherical colonies	4
4a	Single filaments, attached to the substrate by heterocyte, usually unbranched	*Calothrix*
4b	Single filaments, attached to the substrate by heterocyte, false branching	*Dichothrix*

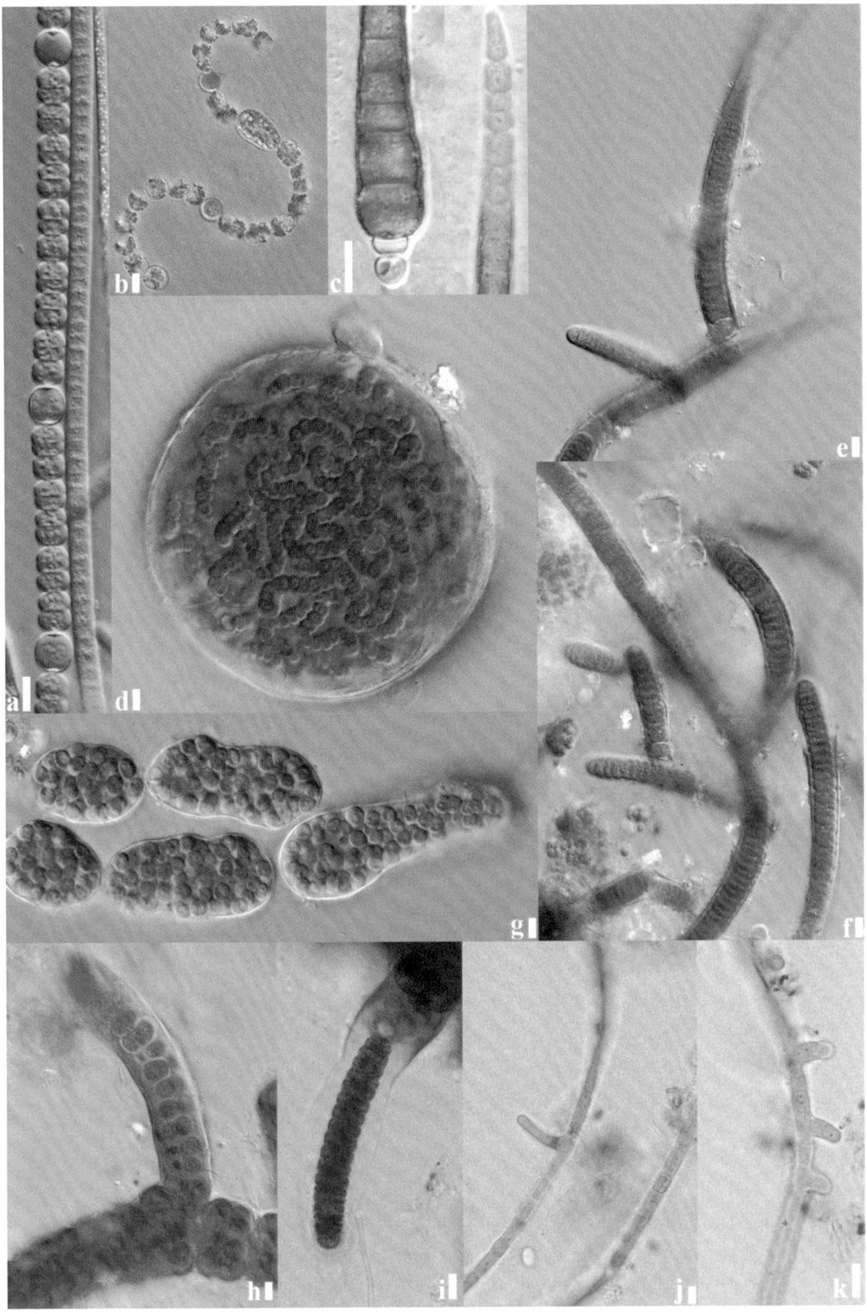

Fig. 2.4 (**a**–**k**) Examples of filamentous heterocytous genera. (**a** and **b**) *Anabaena*: (**a**) note the heterocytes and (**b**) overwintering akinetes. (**c**) *Calothrix* (**d** and **g**) *Nostoc*. (**e** and **f**) *Tolypothrix*: note the false branchings. (**h** and **i**) *Stigonema*. (**j** and **k**) *Hapalosiphon*. Scale bars: (**h** and **k**) = 10 μm; (**a**–**g**) = 5 μm

5a	Filaments usually without false branching, or occasionally false branched	*Microchaete*
5b	Filaments with frequently false branching	6
6a	More than one trichome within common sheath	*Coleodesmium*
6b	One trichome within sheath, common branching	*Tolypothrix*
7a	Paraheterocytic type of akinete development, akinete usually larger than vegetative cells, thin gelatinous mats, or single filaments	*Anabaena*
7b	Apoheterocytic type of akinete development, spherical to lobate microscopic, or macroscopic gelatinous colonies	*Nostoc*

Anabaena (Bornet et Flahault) Bory (Fig. 2.4a, b)

Single filaments, bundles, or micro- to macroscopic mats on submerged water plants. Trichomes are straight, bent or coiled, isopolar, paraheterocytic, constricted, or unconstricted at cross-walls. Sheath if present is usually diffluent and colorless. Cells are spherical, oval, barrel-like, cylindrical with homogenous to granular content, blue-green to greenish, planktic species contain gas vesicles. Heterocytes are intercalary, spherical, oval to cylindrical, homogenous, colorless to yellowish, with visibly thicken walls. Akinetes are intercalary and vary from almost spherical to, oval, reniform, cylindrical and grow solitary or in short chains (up to five). Cells divide perpendicularly to trichome axis. Trichomes reproduce by fragmentation and akinete germination in short parts (hormogonia).

Remarks: the genus was recently split into *Anabaena sensu stricto* (mainly periphytic species without gas vesicles), *Dolichospermum* (mainly planktic species with gas vesicles), and *Sphaerospermum* (planktic with gas vesicles). Depending on the order of the body of water, different genera may be present (e.g., higher order rivers or those with low flow rates may favor the occurrence of *Dolichospermum*, with low order streams containing *Anabaena*).

Occurrence: *Anabaena sensu stricto* occurs world wide, benthic or epiphytic, especially on submerged plants.

Expected taxa: *A. cylindrica* Lemmermann, *A. subcylindrica* Borge, *A. sphaerica* Bornet et Flahault, *A. oscillarioides* Bornet et Flahault.

Calothrix (Bornet et Flahault) Agardh (Fig. 2.4c)

Heteropolar filaments are attached to the substrate, living solitary, or occasionally false branched, in groups or mats. Basal parts of filaments are formed by heterocytes (seldom two), which can be spherical to elliptical or flattened. Occasional akinetes formed above the heterocyte. Vegetative cells above heterocytes are wider than at the apex where are narrowed continuing into hair-like gelatinous projection, unconstricted to slightly constricted at cross-walls. Cell content is homogenous to finely granulated, blue-green, pale blue-green, greenish, brownish, yellowish. Sheath is usually distinct, thin, firm mono to occasionally multilayered, colorless to yellowish or brownish. Cells divide perpendicularly to filament axis, meristematic zones occurs below hair projections. Apical parts of filaments liberate short motile hormogonia by help of necridic cells. Hormogonia possess heteropolar development and growth.

Occurrence: a cosmopolitan genus, mainly periphytic, epilithic to epiphytic.

Expected taxa: *C. fusca* Bornet et Flahault, *C. parietina* (Bornet et Flahault) Thuret, *C. braunii* Bornet et Flahault, *C. elenkinii* Kosinskaja, *C. stellaris* Bornet et Flahault, *C. simplex* Gardner.

Rivularia (Bornet et Flahault) Roth (Fig. 2.5a)

Microscopic to macroscopic (several centimeters), hemispherical to irregular colonies attached to substrate, occasionally internal cavity occurs, colonies usually gelatinous, or in limestone habitats can be encrusted with $CaCO_3$. Filaments are radially oriented, false branched, and sometimes grow in layer-like structure. Sheath is thin to thick with conspicuously widened apical end. Trichomes are heteropolar, widen at the base, have a basal heterocyte, and apices narrow to hair-like end, which usually grows out of the sheath. Trichomes are constricted or unconstricted at the cross-walls, falsely branched at the intercalar heterocytes. Akinetes do not occur (typical feature differing *Rivularia* from *Gloeotrichia*). Cells are cylindrical, barrel shaped to wider than longer without any aerotopes. Reproduction via hormogonia from the apical parts of trichomes.

Occurrence: benthic, often epiphytic, occasionally epilithic in stagnant, or running waters.

Expected taxa: *R. aquatica* De-Wildeman, *R. minuta* (Bornet et Flahault) Kützing, *R. biasolettiana* (Bornet et Flahault) Meneghini, *R. dura* (Bornet et Flahault) Roth, *R. borealis* Richer.

Coleodesmium Borzi (Fig. 2.5b)

Heteropolar filaments are attached to the substrate, usually branched forming fascicles or coherent mats. One or more trichomes (9–10 μm wide) are placed within firm, colorless or yellowish to brownish and often layered sheath. In contrast to morphologically similar genera such as *Calothrix*, trichomes do not form apical hairs. Cells are short, barrel shaped, constricted at the cross-walls, blue-green, with finely granulated or homogenous content. Basally formed heterocytes are elliptical, akinetes occur occasionally, 8 μm wide, one or few in chain. Trichomes reproduce by hormogonia formation (with help of necridic cells).

Occurrence: mainly epiphytic or epilithic in cold mountain streams, a small genus in terms of species.

Expected taxa: *C. floccosum* Borzi, *C. wrangelii* (Geitler) Borzi.

Dichothrix (Bornet et Flahault) Zanardini (Fig. 2.5c)

Members of the genus usually form fasciculate colonies with falsely branched filaments. Sheaths are thin, firm, lamellose, colorless, or yellowish. Trichomes in colonies are fasciculate, heteropolar, and branched, 5–30 μm wide. Spherical to elliptical heterocyte at the base of trichomes, which facilitates trichome branching. Apical portion is narrowed or protrudes into thin hairs. Cells are variable from short to barrel shaped to elongated or almost cylindrical with or without constrictions at the cross-walls. Akinetes usually do not occur (described in *D. gelatinosa*). At the subterminal part of the trichome is often a meristematic zone, with trichomes reproducing by formation of hormogonia using necridic cells.

Remarks: several *Dichothrix* stages can be similar to *Calothix* and false branching is an important diagnostic feature.

Occurrence: epilithic or epiphytic, often in cold clear streams

Expected taxa: *D. willei* Gardner, *D. olivacea* Bornet et Flahault, *D. sinensis* Jao.

Microchaete (Bornet et Flahault) Thuret (Fig. 2.5d)

Members of the genus live in single filaments or aggregated into small clusters growing erect on substrates. Filaments are heteropolar, occasionally false branched at heterocytes, constricted, or unconstricted at the cross-walls. Sheath is usually, colorless, thin, and narrow. Heteropolar trichomes with spherical, hemispherical, or oval heterocyte at the base. Basal part is not conspicuously widened, nor is the apical portion conspicuously narrowed (which separates it from *Calothrix*). Intercalar, isodiametric to cylindrical heterocytes occur often in pairs. Akinetes develop facultatively, usually in basal part of the trichome. Cells are of various shapes from wider than longer to cylindrical or barrel shaped. Apical parts of trichomes disintegrate into short hormogonia, which subsequently germinate by formation of heterocytes and possess a heteropolar mode of growth.

Occurrence: frequently epiphytic in freshwaters.

Expected taxa: *M. diplosiphon* Gomont ex (Bornet et Flahault), *M. tenera* (Bornet et Flahault) Thuret, *M. brunnescens* Komárek, *M. robinsonii* Komárek.

Nostoc (Bornet et Flahault) Vaucher (Fig. 2.4d, g)

Usually found as gelatinous microscopic to macroscopic colonies (spherical, lobate, irregular) of various colors (yellowish, green, olive-green, brownish). Filaments are isopolar, metameric, usually surrounded by distinct or diffluent sheath. Mucilage of old colonies sometime incorporates intensely colored pigments such scytomemin or gloeocapsin. Isopolar trichomes are usually constricted or slightly constricted at the cross-walls. Intercalar heterocytes are spherical, oval, rectangular, isodiametrical to slightly cylindrical. Akinetes are spherical, oval or slightly elongated, and develop apoheterocyticaly, often in short chains. Cells are spherical to barrel shaped, often distinctly granulated. Life cycle is complicated and includes various stages of different morphology from unifilamentous stages to large macroscopic multifilamentous colonies. Hormogonia occur after trichomes disintegration into short parts and mainly during akinete germination.

Remarks: the taxonomy of this genus currently under revision, e.g., a new genus *Desmonostoc* was recently erected based on *Nostoc muscorum* (Hrouzek et al. 2013)

Occurrence: mainly periphytic species, on submerged stones or plants, occasionally planktic or tychoplanktic, subaerophytic on wetted stones along banks and waterfalls.

Expected taxa: *N. linckia* (Bornet et Flahault) Bornet, *N. paludosum* (Bornet et Flahault) Kützing, *N. sphaericum* (Bornet et Flahault) Vaucher, *N. parmeliodes* (Bornet et Flahault) Kützing.

Tolypothrix (Bornet et Flahault) Kützing (Fig. 2.4e, f)

Single filaments or densely aggregated filamentous mats. Heteropolar filaments are often multilaterally branched, unconstricted to constricted at the cross-walls. Sheathes can be thin to wide (up to two times wider than trichome), firm, uni- to multilayerd, often colorless to yellowish. Trichomes can be slightly narrowed at the end, but more frequently they are cylindrical or sometime widened. Hairs are lacking. Colonies are attached by basal spherical, hemispherical, or oval heterocytes to the substrate or to the other trichome. Akinetes occur occasionally. Cells are barrel shaped, isodiametrical, or slightly cylindrical, green, blue-green, yellowish, homogenous to slightly granulated. Hormogonia are liberated from the apical parts of trichomes and can germinate from one or both ends.

Occurrence: benthic on submerged plants and stones, often subaerophytic on wetted stones and bark of trees.

Expected taxa: *T. tenuis* Kützing (Bornet et Flahault), *T. distorta* (Bornet et Flahault) Kützing, *T. penicillata* (Bornet et Flahault) Thuret, *T. limbata* (Bornet et Flahault) Thuret, *Tolypothrix crassa* W. et G.S. West.

Key 4: Form-genera of the Stigonematales

1a	Filaments uniseriate (rarely biseriate) heterocytes	2
1b	Filaments biseriate	3
2a	Heterocytes terminal on short lateral branches	*Nostochopsis*
2b	Heterocytes intercalary, thin, firm, close sheath	*Hapalosiphon*
3a	Narrow, erect branches with hormogonia at the apices	*Fischerella*
3b	Wide branches (nearly as wide as the main filament), arising from all sides, not producing hormogonia in the branch apices	*Stigonema*

Fischerella (Bornet et Flahault) Gomont (Fig. 2.5f)

Thallus felt-like, prostrate, creeping, sometimes erect. Filaments biseriate, highly branched (true branching), sheathes thick, sometimes lamellate, colorless, or yellowish to brownish. Erect, true branches typically unilateral. Cells barrel shaped to round, heterocytes globose, barrel shaped or quadrate, intercalary in basal trichome, cylindrical in branches.

Occurrence: typically subaerial or on moist soil, also encountered in shallow waters at margins epiphytic, epidendric, or entangled with other aquatic plants. It is not found in running waters *per se*, but often associated with moist habitats in the immediate vicinity.

Expected taxa: *F. ambigua* (Kützing *ex* Bornet and Flahault) Gomont, *F. muscicola* (Thuret) Gomont

Hapalosiphon (Bornet et Flahault) Nägeli (Fig. 2.4j, k)

Thallus attached (but later may be free floating), creeping, coiled filaments. Filaments are uniseriate, copious unilateral, true branches that are often as thick as the main branch. Heterocytes may be oblong to quadrate to cylindrical, intercalary. Sheathes thin, firm, usually colorless.

Occurrence: a common epiphyte in the littoral region of lakes and ponds, commonly attached to plants, also present in streams (in similar habitats).

Expected taxa: *Hapalosiphon tenuis* Gardner, *H. intricatus* W. et G.S. West

Nostochopsis (Bornet et Flahault) Wood (Fig. 2.5g)

Thallus gelatinous, attached, expansive, mucilaginous, and typically smooth; large (>4 cm), yellowish to yellow-green to olive-green. Uniseriate trichomes with ± isodiametric to barrel shaped cells, but may occasionally be elongated and ballooned. Many true branches of the main axis, terminating in either rounded apical cells or on short branches ending with an apical heterocyte. Copious heterocytes may be terminal, intercalary, or lateral.

Occurrence: a monotypic genus (*N. lobata*), often encountered in clean and small order streams throughout temperate to tropical regions attached to substrates.

Expected taxa: *N. lobata* (Bornet et Flahault) Wood

Stigonema (Bornet et Flahault) Agardh (Fig. 2.4h, i)

Thallus cushion like, turf like (but may become free floating or even encountered scattered among other algae). Trichomes multiseriate (seldom uniseriate) true branches that may become "erect." Filaments enclosed by a yellow to yellow-brown sheath, thick or thin, may be lamellated. Cells often ± round to cylindrical to barrel shaped may appear to not touch each other. Heterocytes mostly intercalary, but may be lateral. Hormogonia common under favorable conditions, often consisting of only a few (two) cells, morphologically dissimilar to the rest of the trichomes.

Occurrence: commonly reported from several biomes, most often in subaerial, humid, periodically wetted habitats, but also common from streams, seeps, waterfalls, and dripping walls. It has a worldwide distribution, being common in the tropics but also reported from alpine regions as well.

Expected taxa: *S. mamillosum* (Bornet et Flahault) Agardh, *S. minutum* (Bornet et Flahault) Hassall, *S. ocellatum* (Bornet et Flahault) Hassall

Acknowledgments The authors extend their thanks to Alyson R. Norwich for editorial assistance and for the line drawings in the plates. The authors extend thanks to two colleagues in particular (Jiří Komárek, University of South Bohemia, and Jeffrey Johanson, John Carroll University) for their pioneering work on cyanobacterial systematics. The authors acknowledge the works of Gardner (1927), Geitler (1932), Starmach (1966), Montejano et al. (1993), and Gold-Morgan et al. (1994), which served as inspiration and starting points for the redrawn some line drawings.

Cyanobacterial Terms

Akinete Thick-walled spore, often inducible, typically used to survive adverse environments. Frequently employed in species or generic identifications (e.g., *Anabaena*, Fig. 2.4a). May be apoheterocytic (develop from vegetative cells between heterocytes) or paraheterocytic (develop from vegetative cells outside of heterocytes).

Baeocytes Reproductive cells resulting from successive cell division within a mother cell without being liberated and enclosed by a sheath (e.g., *Pleurocapsa*, Fig. 2.1p).

Calyptra A thick covering at the tip of a trichome (e.g., *Phormidium*, Fig. 2.3c).

Endogloeic Growing within mucilage (typically of other algae, e.g., *Synechoccocus*, Fig. 2.1i).

Exocytes Reproductive cells released from apical portions of sessile, heteropolar cells (e.g., *Chaemosiphon*, Fig. 2.1e).

False branching A branch not formed as a result of cell division and does not result in multiple planes, leading to filaments which appear to pass each other (e.g., *Tolypothrix*, Fig. 2.4e).

Filament Linear arrangement of cells enveloped by a sheath (e.g., *Phormidium*, Fig. 2.3c).

Heterocyte A thick-walled cell, often inducible, used to fix atmospheric nitrogen. The size, shape, and placement are frequently employed in identifications (e.g., *Anabaena*, Fig. 2.4a).

Heteropolar Cyanobacterial "body plan" in which basal and apical regions (cells, filaments, trichomes) may be distinguished (e.g., *Heteroleibleinia*, Fig. 2.3h).

Hormogonia A desheathed, reproductive fragment of a trichome typically arising adjacent to necridic cells or heterocytes (e.g., *Oscillatoria*, Fig. 2.3k).

Isopolar Cyanobacterial "body plan" in which each end is identical (e.g., *Pseudanabaena*, Fig. 2.3b).

Pseudofilaments Row of cells incidentally arranged in a linear series, not a single physiological entity (e.g., *Stichosiphon*, Fig. 2.1k, l).

Pseudovagina Sheath for heteropolar cells or pseudofilaments open only at one apical end (e.g., *Stichosiphon*, Fig. 2.1k, l).

Sheath Mucilaginous layer that may surround trichomes or cells. Many varieties exist (e.g., thin, thick, and lamellate) and may be facultative based on environmental conditions (e.g., *Microcoleus*, Fig. 2.3j).

Trichome Filament excluding the sheath (e.g., *Arthrospira*, Fig. 2.3a).

References

Agardh CA (1824) Systema algarum. Literis Berlingianis, Lundae

Anagnostidis K, Komárek J (1985) Modern approach to the classification system of the Cyanophytes 1: Introduction. Algol Stud 38(39):291–302

Anagnostidis K, Komárek J (1988) Modern approach to the classification system of cyanophytes. 1. Introduction. Algol Stud 38:291–302

Anagnostidis K, Komárek J (1990) Modern approach to the classification system of cyanophytes. 3. Oscillatoriales. Algol Stud 50:327–472

Beaver JR, Manis EE, Loftin KA et al (2014) Land use patterns, ecoregion, and microcystin relationships in U.S. lakes and reservoirs: a preliminary evaluation. Harmful Algae 36:57–62

Bornet E, Flahault C, (1886–1888) Revision des Nostocacees heterocystees contenues dans les principaux herbiers de France. Ann Sci Nat Bot 3: 323–381; 4, 343–373; 5, 51–129; 7, 177–262

Casamatta DA, Vis ML (2004) Current velocity and nutrient level effects on the morphology of *Phormidium retzii* (Cyanobacteria) in artificial stream mesocosms. Algol Stud 113:87–99

Casamatta DA, Vis ML, Sheath RG (2003) Cryptic species in cyanobacterial systematics: a case study of *Phormidium retzii* (Oscillatoriales) using RAPD molecular markers and 16S rDNA sequence data. Aquat Bot 77:295–309

Casamatta DA, Johansen JR, Vis ML, Broadwater ST (2005) Molecular and ultrastructural characterization of ten polar and near-polar strains within the Oscillatoriales (Cyanobacteria). J Phycol 41:421–438

Castenhoz RW (2001) General characteristics of the cyanobacteria. In: Boone DR, Castenhoz RW (eds) Bergey's manual of systematic bacteriology, vol 1, 2nd edn. Springer, New York

Falcon LI, Magallon S, Castillo A (2010) Dating the cyanobacterial ancestor of the chloroplast. ISME J 4:777–783

Gardner LN (1927) New Myxophyceae from Porto Rico. Mem N Y Bot Gard 7:1–144

Geitler L (1932) Cyanophyceae. In: Rabenhorst L (ed) Kryptogamenflora van Deutschland, Osterreich und der Schweiz, vol XIV. Akademische, Leipzig, pp 1–1196

Gold-Morgan M, Montejano G, Komárek J (1994) Freshwater epiphytic cyanoprokaryotes from Central Mexico, II. Heterogeneity of the genus *Xenococcus*. Arch Protistenkd 144:383–405

Gomont M (1892) Monographie des Oscillariees (Nostocacees homocystees). Ann Sci Nat Bot Paris Ser 15:263–368, 16, 91–264

Hašler P, Štěpánková J, Špačková J, Neustupa J, Kitner M, Hekera P, Veselá J, Burian J, Poulíčková A (2008) Epipelic cyanobacteria and algae: a case study from Czech ponds. Fottea 8:133–146

Hašler P, Dvořák P, Poulíčková A et al (2014) A novel genus *Ammassolinea* gen. nov. (Cyanobacteria) isolated from sub-tropical epipelic habitats. Fottea 14:241–248

Hindák F (2008) Color atlas of cyanophytes. Veda, Bratislava, p 251

Hrouzel P, Lukesová A, Mares J, Ventura S (2013) Description of the cyanobacterial genus Desmonostoc gen. nov. including D. muscorum comb. nov. as a distinct, phylogenetically coherent taxon related to the genus. Nostoc Fottea 13:201–213
Karl D, Michaels A, Bergman B et al (2002) Dinitrogen fixation in the world's oceans. Biogeochemistry 57(58):47–98
Komárek J (2013) Cyanoprokaryota. 3. Heterocytous Genera. Süsswasserflora von Mitteleuropa, 19/3. Springer, Heidelberg
Komárek J, Anagnostidis K (1986) Modern approach to the classification system of cyanophytes. 2. Chroococcales. Algol Stud 43:157–226
Komárek J, Anagnostidis K (1989) Modern approach to the classification system of cyanophytes. 4. Nostocales. Algol Stud 56:247–345
Komárek J, Anagnostidis K (1998) Cyanoprokaryota I: Chroococcales, Süsswasserflora von Mitteleuropa, 19/1. G. Fischer, Stuttgart
Komárek J, Anagnostidis K (2005) Cyanoprokaryota II: Oscillatoriales, Süsswasserflora von Mitteleuropa, 19/2. G. Fischer, Stuttgart
Komárek J, Hauer T (2015) CyanoDB.cz—On-line database of cyanobacterial genera. Word-wide electronic publication, Univ. of South Bohemia and Inst. of Botany. http://www.cyanodb.cz. Accessed 4 Dec 2015
Komárek J, Johansen J (2015a) Coccoid cyanobacteria. In: Wehr JD, Sheath RG, Kociolek JP (eds) Freshwater algae of North America. Ecology and classification. Academic, San Diego, pp 75–133
Komárek J, Johansen J (2015b) Filamentous cyanobacteria. In: Wehr JD, Sheath RG, Kociolek JP (eds) Freshwater algae of North America. Ecology and classification. Academic, San Diego, pp 135–235
Komárková J, Jezberová J, Komárek O et al (2010) Variability of *Chroococcus* (Cyanobacteria) morphospecies with regard to phylogenetic relationships. Hydrobiologia 639:69–83
Kützing TF (1849) Species algarum. FA Brockhaus, Leipzig
Loza V, Perona E, Mateo P (2013) Molecular fingerprinting of cyanobacteria from river biofilms as a water quality monitoring tool. Appl Environ Microbiol 79:1459–1472
Manoylov KM (2014) Taxonomic identification of algae (morphological and molecular): species concepts, methodologies, and their implications for ecological bioassessment. J Phycol 50:409–424
Montejano G, Gold M, Komárek J (1993) Freshwater epiphytic cyanoprokaryotes from Central Mexico. I. Cyanocystis and xenococcus. Arch Protistenkd 143:237–247
Nägeli C (1849) Gattungen einzelliger Algaen. Friedrich Schulthess, Zürich
Perkerson RB, Perkerson E, Casamatta DA (2010) Phylogenetic examination of the cyanobacterial genera *Geitlerinema* and *Limnothrix* (Pseudanabaenaceae) using 16S rDNA gene sequence data. Algol Stud 134:1–16
Perona E, Bonilla I, Mateo P (1998) Epilithic cyanobacterial communities and water quality: an alternative tool for monitoring eutrophication in the Alberche River (Spain). J Appl Phycol 10:183–191
Pisciotta JM, Zou Y, Baskakov IV (2010) Light-dependent electrogenic activity of cyanobacteria. PLoS One 5:e10821
Prescott GW (1962) Algae of the western Great Lakes area, 2nd edn. Brown, Dubuque
Rott E, Cantonati M, Füreder L et al (2006) Benthic algae in high altitude streams of the Alps—a neglected component of the aquatic biota. Hydrobiologia 562:195–216
Schirrmeister BE, Antonelli A, Bagheri HC (2011) The origin of multicellularity in cyanobacteria. BMC Evol Biol 11:45–49
Schopf JW (2000) The fossil record: tracing the roots of the cyanobacterial lineage. In: Whitton BA, Potts M (eds) The ecology of the cyanobacteria. Kluwer, Dordrecth, pp 13–35
Scott JT, Marcarelli AM (2012) Cyanobacteria in freshwater benthic environments. In: Whitton BA, Potts M (eds) The ecology of the cyanobacteria II. Springer, Dordrecht, pp 271–289

Seifert M, McGregor G, Eaglesham G et al (2007) First evidence for the production of cylindrospermopsin and deoxy-cylindrospermopsin by the freshwater benthic cyanobacterium *Lyngbya wollei* (Farlow ex Gomont) Speziale and Dyck. Harmful Algae 6:73–80

Sheath RG, Cole KM (1992) Biogeography of stream macroalgae in North America. J Phycol 28:448–460

Sherwood AR, Carlile AL, Vaccarino MA et al (2015) Characterization of Hawaiian freshwater and terrestrial cyanobacteria reveals high diversity and numerous putative endemics. Phycol Res 63:85–92

Siegesmund MA, Johansen JR, Karsten U et al (2008) *Coleofasciculus* gen. nov. (Cyanobacteria): morphological and molecular criteria for revision of the genus *Microcoleus* Gomont. J Phycol 44:1572–1585

Smith GM (1950) Freshwater algae of the United States of America, 2nd edn. McGraw-Hill Book Co., New York

Stanier RY, Sistrom WR, Hansen TA et al (1978) Proposal to place the nomenclature of the cyanobacteria (blue-green algae) under the rules of the International Code of Nomenclature of Bacteria. Int J Syst Bacteriol 28:335–336

Starmach K (1966) Cyanophyta-Sinice, Glaukophyta-Glaukofity. In: Starmach K (ed) Flora Słodkowodna Polski. Państwowe Wydawnictwo Naukowe, Warszawa, pp 1–808

Steinman AD, Mulholland PJ, Hill WR (1992) Functional responses associated with growth form in stream algae. J N Am Benthol Soc 11:229–243

Stevenson J (2014) Ecological assessments with algae: a review and synthesis. J Phycol 50:437–461

Stevenson RJ, Bothwell MI, Lowe RL (1996) Algal ecology: freshwater benthic ecosystems. Elsevier, San Diego

Strunecký O, Komárek J, Johansen J et al (2013) Molecular and morphological criteria for revision of the genus *Microcoleus* (Oscillatoriales, Cyanobacteria). J Phycol 49:1167–1180

Tilden J (1910) Minnesota algae. The Myxophyceae of North America and adjacent regions including Central America, Greenland, Bermuda, The West Indies, and Hawaii. Bot Ser 8:1–328

Whitford LA, Schumaker GJ (1969) A manual of the freshwater algae in North America. N C Agric Exp Stat Techn Bull 188:1–313

Whitton BA (2005) Phylum Cyanophyta (cyanobacteria). In: John DM, Whitton BA, Brook AJ (eds) The freshwater algal flora of the British Isles. Cambridge University Press, Cambridge, pp 25–122

Whitton BA (2012a) Changing approaches to monitoring during the period of the 'Use of Algae for Monitoring Rivers' symposia. Hydrobiologia 695:7–16

Whitton BA (ed) (2012b) Ecology of cyanobacteria II: their diversity in space and time. Springer, Dordrecht

Whitton BA, Potts M (2000) The ecology of cyanobacteria: their diversity in time and space. Kluwer, Dordrecht

Chapter 3
Green Algae (Chlorophyta and Streptophyta) in Rivers

Alison R. Sherwood

Abstract The green algae represent one of the most diverse and abundant algal lineages in river systems around the world, and their evolutionary diversification led to two major lineages, the Chlorophyta and Streptophyta (with the latter including the land plants). Macroscopic and microscopic forms of green algae are common in streams, as are those living on hard substrata, epiphytically on aquatic plants or other algae, and a few free-floating forms. This chapter treats the most common genera from stream habitats, with an emphasis on the benthic forms, the macroscopic taxa, and those that are widespread in distribution. Basic descriptions of 42 genera are provided, along with illustrations and information on the habitat and phylogeny of each genus, where known.

Keywords Algae • Benthic • Biodiversity • Charophytes • Chlorophyta • Green algae • River • Streptophyta • Viridiplantae

Introduction

Phylogenetic Scope and Features of the Green Algae

The green algae (Divisions Chlorophyta and Streptophyta) comprise the green lineage, or Viridiplantae, which also includes the land plants, and represents one of the major evolutionary lines of oxygenic photosynthetic organisms. The green algae include organisms that have successfully established in marine, freshwater, and some terrestrial environments and which have led to the evolution of land plants and the subsequent development of terrestrial ecosystems. Approximately

A.R. Sherwood (✉)
Department of Botany, University of Hawaii, 3190 Maile Way, Honolulu, HI 96822, USA
e-mail: asherwoo@hawaii.edu

O. Necchi Jr. (ed.), *River Algae*, DOI 10.1007/978-3-319-31984-1_3

14,000 species of green algae have been described to date, and an estimated 21,000 species may exist (Guiry 2012). The green algae are distinguished from the other algal lineages by having chlorophyll *a* and *b* within a double membrane-bound chloroplast, an organelle that resulted from a primary endosymbiotic event. The primary photosynthetic storage product is true starch, which is stored in the chloroplast. The green algal cell wall is most commonly composed of cellulose, although a number of variations on this theme are known (Graham et al. 2009).

Current research supports an early divergence of the green algae into two main clades, whose ancestor was likely a flagellated green alga (Leliaert et al. 2012). The Chlorophyta (one of the two clades) consists of a series of early diverging prasinophyte lineages as well as the "core" chlorophytes—the classes Ulvophyceae, Trebouxiophyceae, Chlorodendrophyceae, and Chlorophyceae. The second main clade includes both the land plants and their closest green algal relatives, the charophyte green algae from which they have evolved. Green algae in stream ecosystems are represented by both of these main clades, and belong predominantly to the classes Ulvophyceae, Trebouxiophyceae, and Chlorophyceae in the Chlorophyta, and classes Charophyceae, Coleochaetophyceae, Klebsormidiophyceae, and Zygnematophyceae of the Streptophyta.

Features of Taxonomic Importance

The higher level taxonomic scheme adopted in this chapter is based on a combination of ultrastructural characters—i.e., cell division and flagellar apparatus patterns—and molecular data analyses—i.e., phylogenetic reconstruction based on nuclear, chloroplast, and mitochondrial markers—as well as, more recently, genomic analyses (Pröschold and Leliaert 2007; Friedl and Rybalka 2012; Leliaert et al. 2012). However, the features of greatest importance for genus- and species-level identification are visible at the light microscope level and pertain to cellular characteristics and the organization of the thallus. The thallus construction can be discerned by eye, with a dissecting microscope, compound light microscope, or some combination of those. Of critical importance is whether the alga is single-celled or multicellular, whether it is flagellated or nonflagellated (generally for single-celled forms), colonial or free-living, and if multicellular, whether it is filamentous, parenchymatous, or coenocytic/siphonous in organization. Taxa inhabiting streams are more often than not attached, and few unicellular flagellated forms are common. Multicellular thalli may be branched or unbranched. The number, shape, and position of the chloroplast are diagnostic for many taxa, as is the presence/absence of pyrenoids and their number and position in the chloroplast (Prescott 1951; Hall and McCourt 2015; John and Rindi 2015; Nakada and Nozaki 2015; Shubert and Gärtner 2015; John et al. 2011).

Collecting and Preserving Samples

Large growths of benthic green algae or floating mats can be easily seen unaided in stream systems, but many forms are sufficiently small that additional help is needed—this can be accomplished by using a view box, constructed of plastic or plexiglass sides with a glass bottom (Sheath and Vis 2015). Benthic macroalgal individuals, colonies, and mats are most easily collected using long-handled forceps (for robust taxa) or suction devices such as a turkey baster (for more delicate taxa). Microscopic periphyton can be collected by removing the entire community with a toothbrush or scraping with a razor blade, knife, or the edge of a metal spoon. Epiphytic taxa (e.g., those attached to macrophytes and some species of *Coleochaete*) are best sampled by collecting the substratum along with the alga of interest, which allows closer examination of the material at the microscopic level in the laboratory (Lowe and LaLiberte 2007). How the collections are further processed depends on the scope of analysis techniques to be used. For example, if only light microscope examination is anticipated, samples should be placed into vials or small bags and either immediately fixed (2.5 % $CaCO_3$-buffered glutaraldehyde is an excellent fixative because it retains the color of the algae for some years, but samples stored in this fixative must be refrigerated; Sheath and Vis 2015) or kept cool in a small amount of stream water for 1–2 days before fixation. Samples for molecular characterization (not covered further in this chapter) should be examined and cleaned, if necessary, while viewing with a dissecting microscope prior to DNA extraction (samples can also be either frozen or desiccated in silica gel if not immediately extracted). Artificial surfaces such as plastic or glass bottles can also be excellent substrata for attached stream algae and are often a good source of unialgal material and yield easily removed samples.

Classification Followed in this Chapter

Stream-inhabiting green algae of the Phyla Chlorophyta and Streptophyta are treated in this chapter, with the former represented by three classes (Chlorophyceae, Trebouxiophyceae, and Ulvophyceae) and the latter by four (Charophyceae, Coleochaetophyceae, Klebsormidiophyceae, and Zygnematophyceae). Of the classes within the Chlorophyta, the Chlorophyceae contains the greatest number of genera in stream systems, while within the Streptophyta the Zygnematophyceae is treated in most detail. Genera are arranged alphabetically within each order. Taxonomy, for the most part, follows that presented in AlgaeBase as of the time of this writing (Guiry and Guiry 2015).

Given the large number of green algal genera known from streams, not all could be treated in this chapter. Thus, emphasis is given to the more common taxa, the larger green algae, and the attached forms, and the reader is referred to some of the excellent recent and classic literature on freshwater green algae for additional information on some of the less widespread and less commonly encountered taxa (e.g., Prescott 1951; Wehr et al. 2015; John et al. 2011). Some examples of recent literature applying molecular methodologies to green algal systematics are cited under the descriptions of individual genera.

Distribution of Green Algae in Streams

Most stream-inhabiting green algae, by virtue of living in unidirectional flow, are attached to hard surfaces, including rocks, plant material (aquatic vegetation as well as decaying trees and other plant material that has fallen into the stream), animals (e.g., turtles and mollusk shells), and artificial surfaces (e.g., glass, plastic, or other man-made materials). Occasionally, algal growth can be seen on sand or mud surfaces, but these are typically too unstructured for prolific algal growth in streams, and these surfaces also tend to be rare habitat in flowing waters. The ecology of green algae in stream habitats is covered elsewhere in the literature (e.g., Allan 1995; Stevenson et al. 1996; Dodds 2002; Wehr et al. 2015), and the reader is referred to these sources for detailed information on the seasonality of algal communities as well as the effects of abiotic factors (such as light, temperature, nutrients, and flow) and biotic factors (such as grazing and competition) on the diversity and biomass of green algae.

Taxonomic key to the green algae of streams

1a	Alga microscopic	2
1b	Alga forming a macroscopic thallus	10
2a	Alga attached to benthic substratum, or living epiphytically or endophytically in algal or plant tissue	3
2b	Alga free floating	7
3a	Thallus composed of solitary cells or clusters of cells, or parenchyma/pseudoparenchyma	4
3b	Thallus filamentous, branched or unbranched	6
4a	At least some cells of the thallus with basally sheathed hairs	5
4b	Cells usually solitary, grouped in a mucilaginous matrix (occasionally in short filaments)	*Cylindrocystis*
5a	Thallus composed of individual cells or clusters of cells connected by long, gelatinous tubes	*Chaetosphaeridium*
5b	Thallus pseudoparenchymatous or parenchymatous, but usually with long, basally sheathed hairs	*Coleochaete* (in part)
6a	Thallus epiphytic, formed of prostrate and creeping filaments that can be branched or unbranched	*Aphanochaete*
6b	Thallus epi- or endophytic, composed of irregularly branched filaments	*Chaetonema*
7a	Alga unicellular, composed of needle-like cells with sharply tapered apices	*Closteriopsis*

(continued)

7b	Alga colonial, composed of multiple cells	8
8a	Thallus a sphaerical or irregularly shaped colony composed of radiating clusters of four cells	*Dictyosphaerium*
8b	Thallus not composed of radiating clusters of cells	9
9a	Thallus composed of a colony of cells in a monostromatic layer that is circular or oval in shape	*Pediastrum*
9b	Thallus composed of a colony of cells (4–32) laterally joined in a linear or alternating series	*Scenedesmus*
10a	Thallus filamentous, although may be plant-like in form	14
10b	Thallus sheet-like, parenchymatous or a gelatinous or reticulate colony (or otherwise), but not fundamentally filamentous	11
11a	Thallus sheet like and with cells in regular groupings	*Prasiola*
11b	Thallus not sheet like	12
12a	Thallus composed of sphaerical or oval-shaped cells embedded in a gelatinous matrix	*Tetraspora*
12b	Thallus parenchymatous or composed of a net-like colony of cells	13
13a	Thallus linear, elongated and "filament-like," uniseriate at the base but becoming parenchymatous above	*Schizomeris*
13b	Thallus composed of a net-like mesh of five- and six-sided polygons of cells	*Hydrodictyon*
14a	Thallus filamentous (may be plant like) and typically branched	15
14b	Thallus filamentous and typically unbranched	29
15a	Branched thallus consisting of both an upright and prostrate system of uniseriate filaments, appearing cushion-like macroscopically	*Gongrosira*
15b	Thallus not appearing cushion-like macroscopically	16
16a	Thallus large and "plant like" (not flimsy), with main axis and branches clearly visible without magnification (the "stoneworts")	17
16b	Thallus clearly "alga like" (soft, flexible, not stiff), with thallus construction not clearly visible without magnification	19
17a	Branches with cortication (more rarely with partial or complete lack of cortication), with undivided branchlets, and with bract cells at nodes	*Chara*
17b	Branches usually lacking cortication, and with branchlets divided	18
18a	Branchlets forked at least once, with resulting unicellular rays of equal length	*Nitella*
18b	Branchlets forked into multicellular rays of unequal length, resulting in bushy or "untidy" appearance	*Tolypella*
19a	Cells of thallus typically broader at the anterior end or cylindrical in shape, with "apical rings" visible at the ends of some cells	20
19b	Cells of thallus broader at the anterior end or not, but lacking "apical rings" at ends of cells	21
20a	Thallus heterotrichous	*Oedocladium*
20b	Thallus not heterotrichous, consisting only of upright filaments	*Bulbochaete*
21a	Cells with a single parietal chloroplast	22
21b	Cells with one or more reticulate parietal chloroplasts	26
22a	Cells of branches markedly smaller than those of the main axis	23
22b	Cells of branches slightly, but not markedly, smaller than those of the main axis	24

(continued)

23a	Secondary branching irregular	*Cloniophora*
23b	Secondary branches borne oppositely, alternately or in whorls, in clusters or tufts	*Draparnaldia*
24a	Chloroplasts lacking pyrenoids	*Microthamnion*
24b	Chloroplasts with pyrenoids	25
25a	Thallus macroscopically appears as a tuft or mat	*Stigeoclonium*
25b	Thallus macroscopically appears globose, arbuscular or tubercular, often enveloped in mucilage	*Chaetophora*
26a	Filaments with solitary, paired or short chains of akinetes	*Pithophora*
26b	Filaments lacking akinetes	27
27a	Angle of branching typically approaching 90°, filaments stiff and irregularly branched	*Aegagropila*
27b	Angle of branching typically much less than 90°	28
28a	Thallus heterotrichous with a prostrate layer of coalescing filaments and an upright portion of stiff filaments	*Arnoldiella*
28b	Thallus consisting of sparsely to abundantly branched uniseriate filaments, not heterotrichous	*Cladophora*
29a	Asexual, or if sexually reproducing, not by conjugation	30
29b	Sexual reproduction via conjugation	36
30a	Thallus consisting of unbranched filaments with one or more reticulate chloroplasts	31
30b	Thallus consisting of unbranched filaments with a band-shaped or plate-like chloroplast, but not reticulate	33
31a	Cells of thallus typically broader at the anterior end, with "apical rings" visible at the ends of some cells	*Oedogonium*
31b	Cell ends lacking "apical rings"	32
32a	Cells of filaments constructed of "H-walls," having a bipartite structure, chloroplasts lacking pyrenoids	*Microspora*
32b	Cells of filaments not having a bipartite structure, pyrenoids present in chloroplasts	*Rhizoclonium*
33a	Filaments surrounded by a thick, mucilaginous sheath	34
33b	Filaments not surrounded by a mucilaginous sheath	35
34a	Cells of filaments separated from one another but in a linear arrangement, cells equidistant or in pairs, but not forming "H-shaped" sections	*Geminella*
34b	Cells of filaments separated from one another but in a linear arrangement, cells equidistant or in pairs, forming "H-shaped" sections	*Binuclearia*
35a	Unbranched, uniseriate filaments attached by a basal cell	*Ulothrix*
35b	Unbranched, uniseriate filaments lacking apical/basal differentiation	*Klebsormidium*
36a	Cell wall lacking a median constriction	37
36b	Cell wall with a median constriction, thallus usually surrounded by a thick mucilaginous sheath	41
37a	Chloroplast nearly ribbon like and straight to spiraled, parietal, 1–16 per cell	38
37b	Chloroplast plate like or stellate	39

(continued)

38a	Chloroplast spiraling more than one-half a turn per cell, distinct conjugation tube formed during sexual reproduction	*Spirogyra*
38b	Chloroplast straight or nearly so, not spiraling more than one-half a turn per cell, no conjugation tube formed during sexual reproduction	*Sirogonium*
39a	Chloroplast(s) plate like	40
39b	Chloroplasts stellate, two per cell	*Zygnema*
40a	One or two axial, plate-like chloroplasts per cell, with one to several pyrenoids	*Mougeotia*
40b	One axial, plate-like chloroplast per cell, lacking pyrenoids	*Mougeotiopsis*
41a	Thallus with or without a thick mucilaginous sheath, median constriction shallow but discernable	*Desmidium*
41b	Thallus almost always with a thick mucilaginous sheath, median constriction not discernable or only slightly so	*Hyalotheca*

Descriptions of Common Genera of Green Algae in Rivers

Phylum Chlorophyta

Class Chlorophyceae: Order Chaetophorales

Aphanochaete A. Braun (Fig. 3.1a)

Thalli grow epiphytically on submerged aquatic plants and filamentous algae. Uniseriate filaments of *Aphanochaete* can be branched or unbranched, but prostrate and creeping. Cells are cylindrical or inflated, with a parietal, disk-shaped chloroplast and one to several pyrenoids. Upper surfaces of cells sometimes possess setae, one to several per cell, which can be swollen at the base. Either quadriflagellate zoospores or aplanospores can be asexually produced, while sexual reproduction is oogamous.

Remarks: *Aphanochaete* is cosmopolitan in distribution and most commonly found in hard or eutrophic waters. Caisová et al. (2011, 2013) used molecular data to characterize several strains of *Aphanochaete*, which were recovered as monophyletic toward the base of the Chaetophorales, but with high divergence from one another.

Chaetonema Nowakowski (Fig. 3.1b)

Chaetonema grows on or in the mucilaginous sheath of some freshwater macroalgae, including *Batrachospermum*, *Chaetophora*, *Draparnaldia*, and *Tetraspora*. Filaments are irregularly branched, with some cells forming short side branches arising at right angles to the main filament and others attenuating to a hair. Main

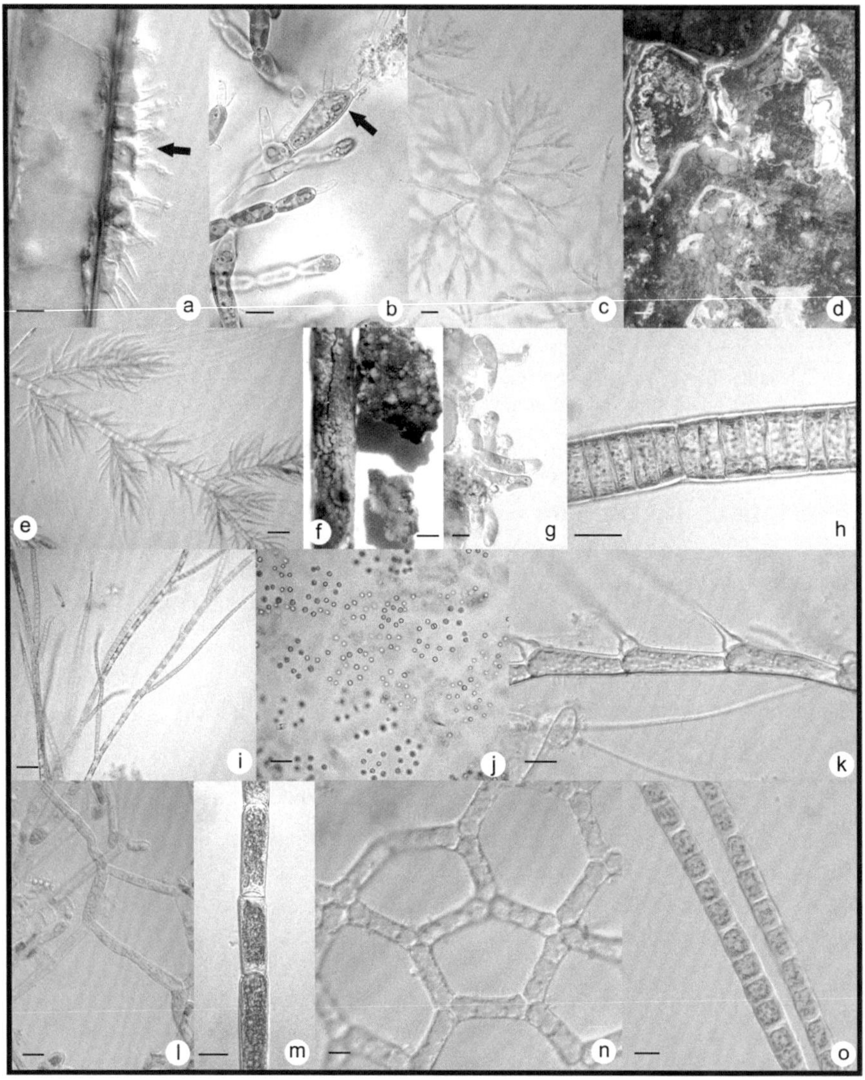

Fig. 3.1 (**a**) *Aphanochaete*, growing on a larger filamentous alga. (**b**) *Chaetonema*, growing on the freshwater red alga *Batrachospermum*. (**c, d**) *Chaetophora*: (**c**) filaments; (**d**) macroscopic colonies in a stream. (**e**) *Draparnaldia*. (**f, g**) *Gongrosira*: (**f**) Macroscopic colonies; (**g**) filaments. (**h**) *Schizomeris*. (**i**) *Stigeoclonium*. (**j**) *Tetraspora*. (**k**) *Bulbochaete*. (**l**) *Oedocladium*. (**m**) *Oedogonium*. (**n**) *Hydrodictyon*. (**o**) *Microspora*. Scale bars: (**f**) = 2 cm; (**d**) = 1 cm; (**e**) = 100 μm; (**h**) = 75 μm; (**i, j, m, n**) = 30 μm; (**c, k, l, o**) = 20 μm; (**a, b, g**) = 10 μm. Image authors: (**a**) W. Bourland; (**b**) C. Carter; (**f**) D. John; (**c, e, k, l, n,** and **o**) Y. Tsukii; (**g**) P. York

cells of the thallus are cylindrical in shape, with a parietal, plate-like chloroplast and one to two pyrenoids. Asexual reproduction occurs by production of two quadriflagellate zoospores per cell, and sexual reproduction is oogamous, with male gametes biflagellate and formed in groups of 8.

Chaetophora Schrank (Fig. 3.1c, d)

A number of thallus shapes are known for the macroscopic genus *Chaetophora*, including globose, arbuscular, or tubercular, and the thallus can be enveloped in soft or firm mucilage. Thalli consist of uniseriate filaments and possess an underdeveloped prostrate system and a highly branched erect system. Filaments terminate in a blunt point or an elongated, multicellular hair. Cells contain a single, parietal chloroplast, and one to several pyrenoids. Motile cells are produced at the tips of branches. Asexual reproduction occurs via production of quadriflagellate zoospores; sexual reproduction is isogamous and occurs following the production of biflagellate gametes.

Remarks: *Chaetophora* occurs on a wide variety of substrata and is common in fast-flowing streams. Abundance peaks during the colder months of the year, at least in temperate areas. *Chaetophora* has been demonstrated to be a polyphyletic genus in need of much further systematic work (Caisová et al. 2011, 2013).

Draparnaldia Bory de Saint-Vincent (Fig. 3.1e)

Thalli are embedded in a soft, mucilaginous matrix. The erect system is uniseriate and consists of barrel-shaped or cylindrical cells on the main axis; secondary branches in clusters or tufts with much smaller cells. Secondary branches are borne oppositely, alternatively or in whorls, and each terminates in a blunt cell or a multicellular hair. Cells of the main axis possess a single large, band-shaped parietal chloroplast that can vary in degree of entirety of its edges; cells of secondary branches have a single laminate parietal chloroplast with one to three pyrenoids. Reproductive cells are produced in the secondary branches. Asexual reproduction occurs by aplanospores and biflagellate zoospores; sexual reproduction is isogamous with quadriflagellate gametes.

Remarks: *Draparnaldia* is common in cool, clean waters and is frequently encountered in higher elevation streams and springs. Like *Stigeoclonium*, members of the genus can exhibit substantial morphological plasticity, and molecular phylogenetic assessment of the genus in the context of other members of the Chaetophorales is needed. The limited molecular data thus far suggest that the genus is monophyletic (e.g., Caisová et al. 2011).

Gongrosira Kützing (Fig. 3.1f, g)

Thalli composed of both a prostrate (uniseriate) and erect system of filaments, appearing cushion-like macroscopically. There can be considerable variation in the prostrate system, but it is usually pseudoparenchymatous in mature specimens. The erect system consists of short branches terminating in blunt tips. Cells are cylindrical in shape or sometimes clavate and can have thick and lamellated walls. Cells possess a single parietal chloroplast and one or more pyrenoids. Asexual reproduction can occur by aplanospores, zoospores, or akinetes; sexual reproduction is unknown.

Remarks: *Gongrosira* can occur on a variety of substrata.

Schizomeris Kützing (Fig. 3.1h)

Schizomeris is an unbranched, tube-like green alga with a macroscopic growth form. The alga is typically uniseriate at the base, but becoming multiseriate, such that a solid parenchyma is formed above the base. Thalli can be constricted at multiple points along their length. Cell shape tends to be elongated toward the base of the thallus but quadrate in the upper portions. Cells possess a single band-like, parietal chloroplast that encircles much of the cell. Thalli reproduce through the production of zoospores, which are quadriflagellate, and tend to be produced in the more mature sections of the thallus; sexual reproduction is unknown.

Remarks: *Schizomeris* is typically collected from slow-flowing regions of streams, but may occasionally be found near waterfalls. The genus is widespread, but not commonly reported. A single species is recognized in the genus (*S. leibleinii* Kützing), and recent molecular analyses of collections from widespread locations support the monotypic status of this genus (Caisová et al. 2011). Several studies have found phylogenetic evidence for *Schizomeris* as sister to the remainder of the order Chaetophorales (Buchheim et al. 2011; Tippery et al. 2012). It has also been reported that *Schizomeris* and *Stigeoclonium* chloroplast genomes share a number of features that support a close phylogenetic relationship between the genera (Brouard et al. 2011).

Stigeoclonium Kützing (Fig. 3.1i)

Thalli are heterotrichous with prostrate and erect systems that are developed to various degrees, depending on the species and the conditions under which they grow. The filamentous thalli are uniseriate, and upright portions can be branched alternately, oppositely, or dichotomously, or irregularly (rarely). Apices of the upright filaments are attenuated, narrowly obtuse, or with a multicellular hyaline hair; prostrate sections of the thallus are typically rhizoidal or sometimes pseudoparenchymatous. Cells of *Stigeoclonium* are either cylindrical or inflated in shape and can be either thin or thick walled. Each cell contains one parietal chloroplast and one or more pyrenoids. Asexual reproduction via quadriflagellate zoospores can be by micro- or macrozoospores; sexual reproduction occurs via production of isogametes that are bi- or quadriflagellate.

Remarks: *Stigeoclonium* is a common component of stream algal floras and appears as tufts or mats on a variety of surfaces. Members of the genus are renowned for being polymorphic and identification based on morphology of the upright thallus portion is unreliable. Culture-based analyses suggest that far fewer species should be recognized than are indicated, and molecular analyses illustrate that much more work needs to be completed on members of *Stigeoclonium*, within the context of analyses of the order Chaetophorales, before the taxonomy can be resolved. Caisová et al. (2011) recovered the genus as polyphyletic in their molecular analyses. As mentioned for *Schizomeris*, *Stigeoclonium* and that genus share a number of features of their chloroplast genomes that support their close phylogenetic relationship (Brouard et al. 2011).

Class Chlorophyceae: Order Chlamydomonadales

Tetraspora Link ex Desvaux (Fig. 3.1j)

Tetraspora is a colonial macroalga consisting of spherical or oval-shaped cells embedded in a gelatinous matrix; cells may be grouped in 2's or 4's, especially at the periphery. Colonies are spherical, elongated, or amorphous in shape. Each cell possesses a cup-shaped chloroplast with a single pyrenoid and a pair of anterior pseudoflagella that project toward the colony edge. Asexual reproduction occurs by zoospores, akinetes, and colony fragmentation; sexual reproduction via isogamous gametes has been reported for two species.

Remarks: *Tetraspora* is typically attached in stream systems and is a common and widespread genus.

Class Chlorophyceae: Order Oedogoniales

Bulbochaete C. Agardh (Fig. 3.1k)

Bulbochaete consists of uniseriate filaments that are unilaterally branched and usually attached by holdfast cells. Cells of the filaments are typically broader at the anterior end and many bear a colorless, elongated terminal hair cell with a swollen base (one of the most distinctive features of the genus). Cells each contain a parietal, reticulate chloroplast with several pyrenoids. Asexual reproduction occurs by akinetes, fragmentation of filaments or by zoospores with a distinctive, apical ring of flagella; sexual reproduction is oogamous, with female cells dividing to form the oogonium and two supporting (suffultory) cells, with dwarf male filaments produced in some species (nannandrous) and others directly producing antheridia. Cell division like for *Oedogonium* and resulting in "apical rings."

Remarks: *Bulbochaete* is a common genus that is most often found growing epiphytically and usually in slower-flowing waters. Much research remains to be completed to clarify the systematics of the Oedogoniales using molecular data; published studies thus far suggest that *Bulbochaete* is indeed a distinct genus from the other members of the order (Buchheim et al. 2001; Mei et al. 2007).

Oedocladium Stahl (Fig. 3.1l)

Oedocladium is a heterotrichous alga with branching on all sides of the filament. Cells are cylindrical in shape and have a single, parietal, reticulate chloroplast, and several pyrenoids. Cell division occurs most commonly in the apical cells and is of the same type as reported for *Bulbochaete* and *Oedogonium*. Asexual reproduction by multiflagellate zoospores, akinetes or fragmentation of the filament; sexual reproduction as for *Oedogonium*.

Remarks: *Oedocladium* can be either terrestrial or freshwater; thalli in the former have colorless and hyaline cells growing subterraneously. Molecular analyses suggest that *Oedocladium* may not be a distinct genus from *Oedogonium* (Mei et al. 2007), but further research is necessary for confirmation.

Oedogonium Link ex Hirn (Fig. 3.1m)

Oedogonium consists of unbranched filaments of cylindrical cells (which may be slightly broader at the anterior end), and are usually attached. Cells are characterized by one-to-many ring-like caps directly adjacent to the cross-wall, and possess a reticulate and parietal chloroplast that can have one or more pyrenoids. Asexual reproduction occurs by the production of zoospores (one per cell, and distinctive in having an apical ring of flagella), aplanospores, akinetes, and fragmentation of filaments; sexual reproduction is oogamous with female cells forming oogonia after division to separate the supporting (or suffultory) cell, and males as short, disk-like antheridia. Sexual reproduction is termed either nannandrous (with dwarf males—short antheridium-producing filaments attached near oogonia) or macrandrous (lacking dwarf males).

Remarks: *Oedogonium* is a very common green alga that can be attached to various substrata, including macrophytic vegetation, or free-living. The ~250 species are usually discerned based on characters pertaining to sexual reproduction, which can be rare. Molecular research thus far suggests a separation of the dioecious nannandrous taxa from the monoecious taxa (Mei et al. 2007).

Class Chlorophyceae: Order Sphaeropleales

Hydrodictyon Roth (Fig. 3.1n)

Thallus consists of a macroscopic, coenobial reticulate mesh of five- or six-sided polygons with each cell attached to two others at its end walls. Cells are cylindrical or slightly inflated in shape. Cells are multinucleate and possess a single chloroplast that is parietal and plate like with a single pyrenoid in young cells, and reticulate with multiple pyrenoids in older cells, surrounding a central vacuole. Asexual reproduction occurs via production of biflagellated zoospores that align and elongate to form a new colony (autocolony formation); sexual reproduction occurs via isogamous, biflagellate gametes that unite to form a zygote; the zygote loses its flagella and becomes a zygospore, which undergoes meiosis to form a zoospore and eventually the polyhedral stage of the alga.

Remarks: *Hydrodictyon* is a common and sometimes abundant alga that can grow to nuisance levels and is distributed in tropical and temperate regions. Four species of *Hydrodictyon* are currently recognized; molecular data representing three of these supports their recognition as distinct species (Buchheim et al. 2005; McManus and Lewis 2011).

Microspora Thuret (Fig. 3.1o)

Plants either floating or attached basally, thallus filamentous and uniseriate. Filaments constructed of "H" walls; having a bipartite structure that can typically be discerned with the light microscope. Cell shape varies from cylindrical to quadrate and sometimes slightly inflated, cell walls can be slightly lamellate. Each cell with a single chloroplast, lacking pyrenoids, which can be finely or coarsely reticulate. Asexual reproduction by biflagellate zoospores, aplanospores, and akinetes; sexual reproduction via biflagellate isogametes.

Remarks: *Microspora* is a widely distributed alga in freshwater habitats, with some species abundant in low pH environments. The genus can be easily confused with the tribophyte genus *Tribonema* (see Chap. 7) which is similar in structure but distinguished by having two or more discoid chloroplasts and lacking true starch.

Pediastrum Meyen (Fig. 3.2a)

Thallus a coenobium, composed of 4–64 (–128) cells, arranged in a monostromatic circular to oval layer. Cell shape highly variable; interior cells typically polyhedral with four or more sides, peripheral cells similar to those of the interior or with one to two lobes or processes. Cells of the colony can be completely contiguous with each other, or with gaps. Cell walls can be either smooth or ornamented. The multinucleate cells possess a single, parietal chloroplast, with one or more pyrenoids. Asexual reproduction via coenobium production from biflagellate zoospores, or by thick-walled resting spores; sexual reproduction infrequently reported but by isogamous, biflagellate gametes.

Remarks: *Pediastrum* is planktonic, but is a widespread genus and can occur in streams. Recent molecular work has demonstrated that *Pediastrum* is composed of a series of evolutionary lineages (Buchheim et al. 2005; McManus and Lewis 2011), which are beginning to be recognized at the generic level based on molecular and morphometric analyses (*Monactinus*, *Parapediastrum*, *Pseudopediastrum*, *Sorastrum*, and *Stauridium*) (Buchheim et al. 2005; McManus et al. 2011).

Scenedesmus Meyen (Fig. 3.2b)

Thallus a coenobium, composed of 4–32 elongated cells laterally joined in a linear or alternating series. Cell shape varies from ellipsoid to ovoid, crescent-shaped or elongated and tapering. Cell walls of *Scenedesmus* are smooth and lacking in spines. Cells possess a single parietal chloroplast and usually one pyrenoid. Asexual reproduction occurs via autospore production (2–32 per sporangium and usually arranged as a single coenobium); sexual reproduction presumably rare, but occurs via isogamous biflagellate gametes.

Remarks: *Scenedesmus* is planktonic, but is a widespread genus and can occur in streams. Previously recognized subgenera (*Desmodesmus* and *Scenedesmus*) were demonstrated to represent monophyletic lineages with relatively high divergence,

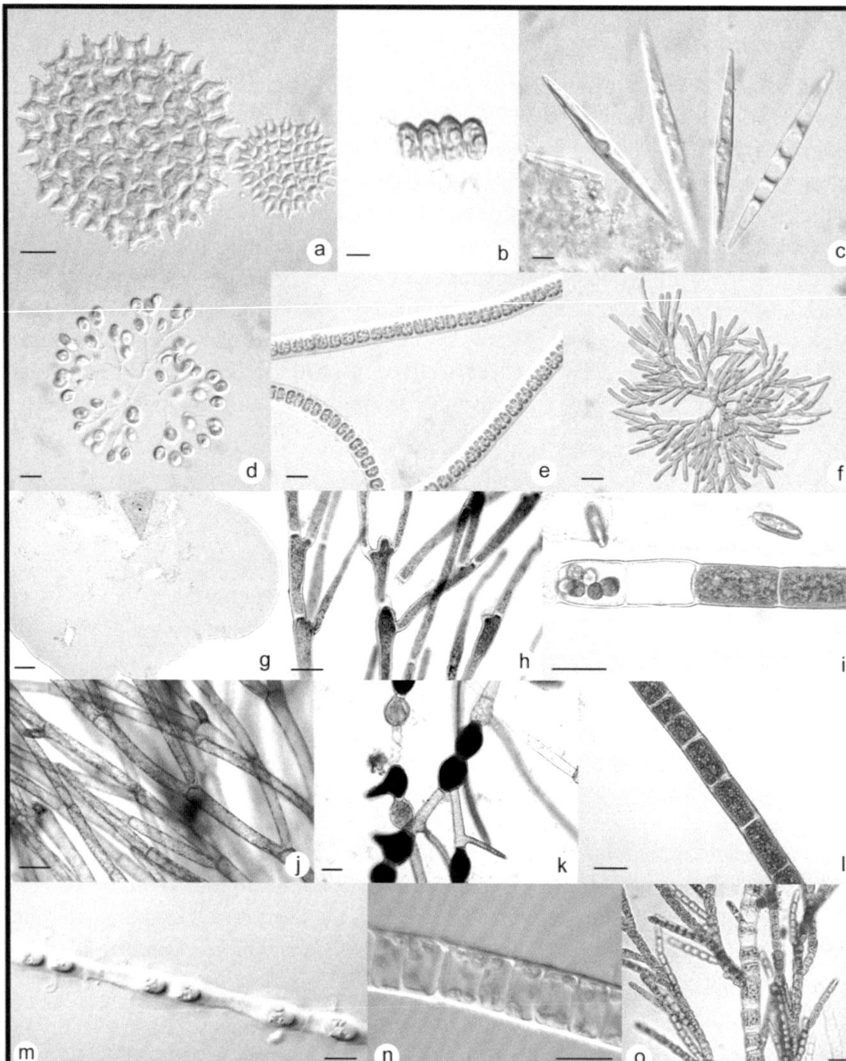

Fig. 3.2 (**a**) *Pediastrum*. (**b**) *Scenedesmus*. (**c**) *Closteriopsis*. (**d**) *Dictyosphaerium*. (**e**) *Geminella*. (**f**) *Microthamnion*. (**g**) *Prasiola*. (**h**) *Aegagropila*. (**i**) *Arnoldiella*. (**j**) *Cladophora*. (**k**) *Pithophora*. (**l**) *Rhizoclonium*. (**m**) *Binuclearia*. (**n**) *Ulothrix*. (**o**) *Cloniophora*. Scale bars: (**h**) = 150 μm; (**g**, **j**) = 100 μm; (**k**, **l**) = 80 μm; (**i**) = 60 μm; (**a**) = 25 μm; (**o**) = 20 μm; (**e**, **f**, **n**) = 15 μm; (**d**, **m**) = 10 μm; (**b**, **c**) = 8 μm. Image authors: (**a**, **c**–**f**, **n**) Y. Tsukii; (**g**) F. Rindi; (**h**) I. Bárbara; (**j**) C. Carter; (**k**) P. Škaloud; (**m**) P. York

and were proposed as separate genera (An et al. 1999). Paraphyly of *Scenedemus* (Hegewald and Wolf 2003) and the discovery of high molecular diversity within morphospecies (Vanormelingen et al. 2007) further illustrate the need for additional systematic research on this, and other related genera of coenobial green algae.

Class Trebouxiophyceae: Order Chlorellales

Closteriopsis Lemmermann (Fig. 3.2c)

Unicellular, needle-like algae with sharply tapered apices and lacking a mucilaginous envelope. Cells can be straight or curved. Cells possess a single chloroplast that can be parietal or axial in position, and ribbon-like, band-shaped or spirally twisted in shape, with multiple pyrenoids in a series. Asexual reproduction by autospores; sexual reproduction unknown.

Remarks: *Closteriopsis* is planktonic and widespread in distribution. Three species of *Closteriopsis* are currently recognized, and molecular data (nuclear 18S rDNA and chloroplast 16S rDNA) indicated the affinity of this genus with other members of the Chlorellales within the Trebouxiophyceae, rather than with the Chlorophyceae (Ustinova et al. 2001).

Dictyosphaerium Nägeli (Fig. 3.2d)

Thallus consists of unattached, spherical or irregularly shaped colonies that are formed by a series of clusters of four cells attached by cell wall material, and encased in a common mucilaginous envelope. Clusters of cells are attached via thin stalks and radiate from the center of the colony. Cells range in shape from spherical to ovoid, ellipsoid or cylindrical. Cells with one to two parietal, cup-shaped chloroplasts, each with a single pyrenoid. Asexual reproduction occurs via autospore production; sexual reproduction rarely reported but oogamous, with an elongated and biflagellate male gamete.

Remarks: *Dictyosphaerium* is planktonic or present in the metaphyton of freshwaters, and is very widely distributed. Paraphyly of the genus *Dictyosphaerium* was recently demonstrated (Krienitz et al. 2010), and several additional genera have been described in an effort to recognize monophyletic lineages (e.g., *Hindakia*—Bock et al. 2010; *Mucidosphaerium*—Bock et al. 2011).

Geminella Turpin (Fig. 3.2e)

Thallus consists of attached or free-floating uniseriate filaments (or, more accurately, pseudofilaments), with cells separated from one another but in a linear arrangement and surrounded by a thick mucilaginous sheath. Cells equidistant or in pairs, cylindrical, inflated, ellipsoidal, or ovoid in shape, each with a parietal band-shaped or laminate chloroplast (often restricted to the central section of the cell) and with a single pyrenoid. Asexual reproduction via fragmentation of filaments or thick-walled akinetes; sexual reproduction unknown.

Remarks: widespread in distribution and known from a variety of freshwater habitats; especially common in low pH habitats. DNA sequence analyses demonstrated a clear alliance of the genus with the Trebouxiophyceae rather than the Chlorophyceae, where it was previously classified (Durako 2007; Mikhailyuk et al. 2008).

Class Trebouxiophyceae: Order Microthamniales

Microthamnion Nägeli (Fig. 3.2f)

Thallus microscopic and consisting of an erect system of branched, uniseriate filaments that terminate in obtuse cells, attached by a mucilage pad. Branching pattern is characteristic of the genus with branches arising toward the distal end of cells. Cells are cylindrical in shape, with a single parietal chloroplast lacking pyrenoids. Asexual reproduction occurs via biflagellate zoospores; sexual reproduction and reproduction via akinetes and aplanospores are unconfirmed.

Remarks: *Microthamnion* occurs on a variety of benthic substrata.

Class Trebouxiophyceae: Order Prasiolales

Prasiola Meneghini (Fig. 3.2g)

Prasiola is a large and morphologically complex trebouxiophycean green alga that inhabits marine, freshwater and terrestrial environments. The freshwater species are found exclusively in clear, fast-flowing streams, often at high elevation. The thalli of the freshwater species are leafy, monostromic blades (rarely distromatic) that can be lobed or ruffled along their margins (typically 1–20 cm in length), occasionally ribbon like. The genus is readily recognizable by the cell arrangements of the thalli—cells are arranged in square or rectangular blocks of four or more; each cell has a single, stellate chloroplast with a central pyrenoid.

Remarks: freshwater species of *Prasiola* reproduce either by aplanospores, which are evident in thickened regions of the thallus, or by oogamous sexual reproduction in which the edges of the thallus develop patches of cells that produce either biflagellate male gametes or nonflagellated egg cells. Stream-inhabiting species of *Prasiola* are known from the Canadian Arctic to the tropics. Much research remains to be completed on the phylogeny of *Prasiola*, and especially the freshwater species (Rindi et al. 2007).

Class Ulvophyceae: Order Cladophorales

Aegagropila Kützing (Fig. 3.2h)

Thalli macroscopic and stiff, composed of irregularly branched filaments. Branching frequently lateral in angle and subterminal in origin, and more than one branch can originate from a cell. Rhizoidal branches sometimes produced from basal portions of cells. Cell division is primarily apical but sometimes intercalary. Numerous chloroplasts per cell form a parietal network, and contain pyrenoids. Asexual reproduction occurs through thallus fragmentation; sexual reproduction unknown. Remarks: *Aegagropila* is a widespread green alga that can form hemispherical clumps or dense mats of filaments; the range of morphological variation within the genus corresponds to only 0.1–0.5 % nucleotide variation for the nuclear ITS region (Boedeker et al. 2010). *Aegagropila* has been variously recognized as a distinct genus and as part of *Cladophora*, but 18S rDNA data established its position as a distinct entity (Hanyuda et al. 2002).

Arnoldiella V.V. Mill emend. Boedeker (=*Basicladia* W.E. Hoffmann and Tilden) (Fig. 3.2i)

Thallus heterotrichous with a prostrate layer of coalescing filaments and an upright portion of compact and rigid filaments. Degree of branching varies for the upright thallus portion, but becoming more abundant in apical regions. Branches arising subterminally or with a pseudodichotomous pattern. Cells becoming shorter and wider in the apical regions of the thallus. Numerous chloroplasts per cell form a parietal network, and contain pyrenoids. Asexual reproduction by flagellated zoospores that are produced in cells at the ends of filaments, and akinetes; sexual reproduction unknown.

Remarks: *Arnoldiella* is found growing on the shells of turtles or bivalves, or on *Cladophora* or hard substrata. *Arnoldiella* was recently emended and most species of *Basicladia* were transferred to this genus, based on a combination of molecular phylogenetic analyses and transmission electron microscopy (Boedeker et al. 2012).

Cladophora Kützing (Fig. 3.2j)

Thalli macroscopic, consisting of sparsely to abundantly branched uniseriate filaments, attached by a discoid or rhizoidal holdfast, or free-living, with primarily apical but sometimes intercalary cell divisions. Cells cylindrical or somewhat inflated, tending to be longer in sparsely branched forms. Branching irregular or pseudodichotomous. Chloroplasts parietal and reticulate with numerous pyrenoids. Asexual reproduction via bi- or quadriflagellate zoospores, rarely by akinetes; sexual reproduction by isogamous biflagellate gametes.

Remarks: *Cladophora* is largely a marine genus, but the freshwater species *C. glomerata* is one of the most cosmopolitan freshwater algal species worldwide. Members of the genus are recognized as "ecosystem engineers" for their role in hosting functional microflora (Zulkifly et al. 2013). Macroscopic growth forms vary widely, from small attached thalli, to large streaming mats. Some degree of tolerance to increased salinity has been demonstrated for *C. glomerata* specimens from Japan (Hayakawa et al. 2012).

Pithophora Wittrock (Fig. 3.2k)

Thallus macroscopic, consisting of uniseriate branched filaments that are free floating. Filaments freely branching, with branches arising subterminally at an almost right angle. Branches can be solitary or opposite. Filaments differ from *Cladophora* in having large akinetes (solitary, in pairs or in short chains) in terminal and intercalary positions, and in cells being longer at the ends of branches than at the base. Cells cylindrical, usually thick walled, each with a parietal reticulate chloroplast and multiple pyrenoids. Asexual reproduction via thallus fragmentation and production of akinetes; sexual reproduction unknown.

Remarks: *Pithophora* is widespread in the tropics and subtropics, and has recently been shown to consist of a single, broadly distributed species (*P. roettleri*) (Boedeker et al. 2012).

Rhizoclonium Kützing (Fig. 3.2l)

Thallus composed of uniseriate, mostly unbranched filaments (occasionally bearing short, rhizoidal branches). Each cell possesses a parietal reticulate chloroplast with several pyrenoids. Cells cylindrical, at least twice as long as broad, sometimes with thick and lamellate walls. Asexual reproduction occurs via thallus fragmentation, akinetes, or the production of biflagellate zoospores; sexual reproduction unknown in freshwater forms.

Remarks: *Rhizoclonium* is a widespread genus in freshwaters and frequently occurs in streams. Molecular phylogenetic work indicates that the genus is polyphyletic and in need of taxonomic revision (Hanyuda et al. 2002; Ichihara et al. 2013).

Class Ulvophyceae: Order Ulotrichales

Binuclearia Wittrock (Fig. 3.2m)

Thallus composed of uniseriate and unbranched filaments, surrounded by a thick mucilaginous sheath, attached or free living. Cells cylindrical to ellipsoid in shape and with rounded apices. Cells commonly in pairs following cytokinesis, and with their own mucilage envelope, separating more widely as they mature; older cross-walls becoming thick and lamellated and daughter cells giving rise to "H"-shaped sections. Each cell with a parietal, band-shaped chloroplast with a single, difficult to discern, pyrenoid. Asexual reproduction by fragmentation, akinetes, aplanospores, or quadriflagellate zoospores; sexual reproduction unknown.

Remarks: *Binuclearia* is widespread in low pH, soft waters.

Ulothrix Kützing (Fig. 3.2n)

Thallus composed of unbranched uniseriate filaments that are attached by a basal cell (which can be rhizoidal). Cells cylindrical or bead shaped, walls thin in young plants and thick in older plants. Each cell with a single parietal band-shaped chloroplast that can mostly or completely encircle the cell, and with a single pyrenoid. Asexual reproduction by fragmentation, akinetes, aplanospores, or quadriflagellate zoospores; sexual reproduction by isogamous biflagellate gametes. The multicellular filamentous gametophyte stage alternates with a unicellular sporophyte known as the *Codiolum* stage.

Remarks: *Ulothrix* is a widespread genus in marine and brackish waters and is occasionally reported from flowing water habitats. Although a few *Ulothrix* sequences have been generated (e.g., Hayden and Waaland 2002; O'Kelly et al. 2004), the genus awaits a thorough molecular systematic investigation.

Class Ulvophyceae: Order Ulvales

Cloniophora Tiffany (Fig. 3.2o)

Thallus composed of uniseriate filaments and characterized by the cells of the secondary branches being very short in contrast to those of the primary branches, which tend to be much longer than broad. Branching irregular, and the thallus attached by downwardly growing rhizoids. Apical cells of secondary branches conical or bluntly rounded. Cells cylindrical, inflated or capitate, each with a single parietal chloroplast and one or more pyrenoids. Asexual reproduction by biflagellate zoospores; sexual reproduction unknown.

Remarks: *Cloniophora* is generally considered a tropical to subtropical genus, where it is common in fast-flowing streams, although its distribution does include some temperate areas. Until recently, the genus was considered to be a member of the Chaetophorales; however, both molecular and morphological analyses have supported its transfer to the Ulvales (Carlile et al. 2011).

Phylum Streptophyta

Class Charophyceae: Order Charales

Chara Linnaeus (Fig. 3.3a)

Thallus macroscopic, usually lime-encrusted, composed of branched uprights and attached by branched, colorless rhizoids. The primary axis is indeterminate, formed through division of the basal part of the apical cell, and is composed of alternating node and internode cells. Nodes bear a whorl of determinate branchlets, 6–16 in

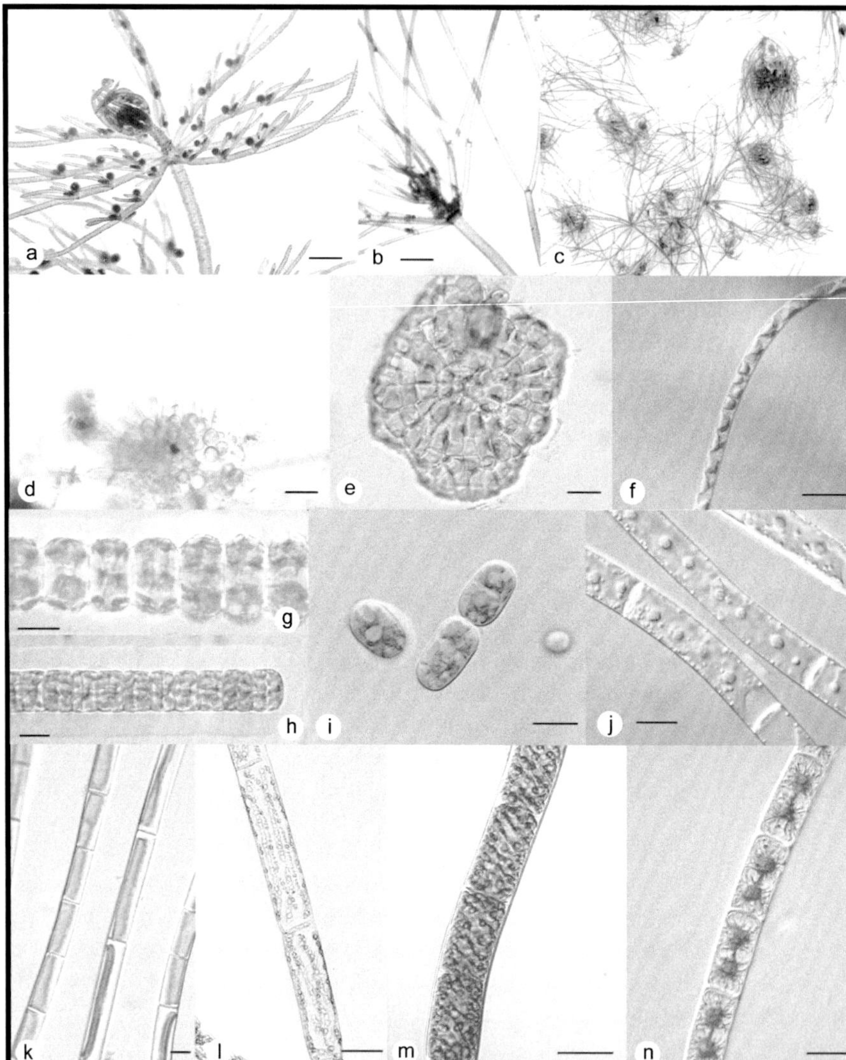

Fig. 3.3 (**a**) *Chara.* (**b**) *Nitella.* (**c**) *Tolypella.* (**d**) *Chaetosphaeridium.* (**e**) *Coleochaete.* (**f**) *Klebsormidium.* (**g**) *Desmidium.* (**h**) *Hyalotheca.* (**i**) *Cylindrocystis.* (**j**) *Mougeotia.* (**k**) *Mougeotiopsis.* (**l**) *Sirogonium.* (**m**) *Spirogyra.* (**n**) *Zygnema.* Scale bars: (**a–c**) = 1 cm; (**e, l**) = 30 µm; (**d, h, k, m**) = 25 µm; (**f, i, n**) = 20 µm; (**g, j**) = 15 µm. Image authors: (**a, c**) C. Carter; (**b**) R. Sheath; (**d–g, i, j, n**) Y. Tsukii; (**h**) P. Škaloud; (**k**) J. Brand

number, unbranched, and can be corticate or ecorticate. One or two tiers of stipulodes are present at the base of each node, as are rings of unicellular bract cells at nodes. Internodal cells can elongate to several centimeters in length and are corticate or ecorticate, and can bear single or clustered spinous cells. Asexual reproduction by vegetative propagation of rhizoids, development of buried nodal cells, or by bulbils; sexual reproduction oogamous with gametes produced in multicellular antheridia and oogonia at branchlet nodes.

Remarks: *Chara* can be abundant in hard and alkaline waters, most commonly in lakes and ponds, but is occasionally found in flowing waters. Some sequence data have been generated for members of the genus (e.g., Sanders et al. 2003; Kato et al. 2008), but this species-rich genus awaits thorough molecular systematic revision.

Nitella C. Agardh (Fig. 3.3b)

Thallus macroscopic, usually not lime-encrusted or only lightly so, composed of regularly and symmetrically branched uprights and attached by branched, colorless rhizoids. The primary axis forms through division of the basal part of the apical cell, and is composed of alternating node and internode cells, with the internodes being greatly elongated in comparison to the nodes. Nodes bear a whorl of four or more determinate ecorticate branchlets; branchlets with 3–4 orders of di- or trichotomous branching and acuminate tips. Stipulodes absent. Indeterminate axes up to two or more per node, forming in branchlet axils. Internodal cells can elongate to several centimeters in length and are ecorticate. Asexual reproduction by vegetative propagation; sexual reproduction oogamous with oogonia produced at the forks of branchlets and antheridia produced terminally.

Remarks: *Nitella* can be abundant in softer water areas and lower pH habitats than *Chara*. The phylogeny of *Nitella* based on molecular systematics was investigated by Sakayama et al. (2005, 2006).

Tolypella (A. Braun) A. Braun (Fig. 3.3c)

Thallus macroscopic and not usually heavily lime-encrusted, similar in overall construction to *Nitella*, but with longer branches that are less symmetrical and forming coarse clusters (birds' nests) of branchlets. Branches are ecorticate and lacking spines. Asexual reproduction by fragmentation; sexual reproduction oogamous with oogonia and antheridia in clusters, commonly in the basal parts of branchlets (oogonia with a corona of ten cells in two tiers).

Remarks: *Tolypella* is widespread and relatively common in slow-flowing streams and hard water habitats.

Class Coleochaetophyceae: Order Chaetosphaeridiales

Chaetosphaeridium Klebahn (Fig. 3.3d)

Thallus microscopic, growing epiphytically, endophytically, or planktonically, composed of individual cells (that can appear to be growing in a gelatinous colony) or clusters of cells connected in a series by long, gelatinous tubes. Cells globose or flask-shaped, bearing a single sheathed hair. Cells with one to two parietal, plate-like chloroplasts with a pyrenoid. Asexual reproduction by biflagellate zoospores; sexual reproduction oogamous with large oogonia and smaller, biflagellate male gametes.

Remarks: *Chaetosphaeridium* is widespread and is distributed from the tropics to polar regions. The genus is recognized as a distinct lineage with high DNA sequence divergence compared with other members of the class (Cimino and Delwiche 2002).

Class Coleochaetophyceae: Order Coleochaetales

Coleochaete Brébisson (Fig. 3.3e)

Thallus can exhibit a number of different morphologies, including forms with well-developed prostrate and erect branched systems, only prostrate branched forms, pseudoparenchymatous forms consisting of adherent branched filaments, and parenchymatous forms. Cells commonly with long, basally sheathed hairs. Each cell with a single parietal and plate-like chloroplast with one to two conspicuous pyrenoids. Asexual reproduction by biflagellate zoospores; sexual reproduction oogamous with flask-shaped oogonia and biflagellate male gametes.

Remarks: *Coleochaete* is widespread and commonly grows as an epiphyte on aquatic vegetation. Species with a flattened, discoidal morphology have been demonstrated to be not monophyletic (Delwiche et al. 2002).

Class Klebsormidiophyceae: Order Klebsormidiales

Klebsormidium P.C. Silva, K. Mattox, and W. Blackwell (Fig. 3.3f)

Thallus composed of uniseriate (occasionally biseriate or with cells as doublets) and unbranched filaments that lack apical/basal differentiation. Cells cylindrical or barrel shaped, usually with smooth walls. Each cell possesses a single parietal and plate-like chloroplast of variable shape that encircles at least half of the cell, and with a single pyrenoid. Some species produce "H" walls; having cell walls of bipartite structure. Asexual reproduction occurs by aplanospore or biflagellate zoospore production; sexual reproduction unconfirmed.

Remarks: *Klebsormidium* is an extremely common genus of green algae in freshwater habitats, especially in temperate regions, although tropical records are known. Molecular phylogenetic investigations of the genus have been used to, in part, describe new species from stream habitats (Novis 2006), provide further support for described morphospecies and demonstrate affinity of some strains with other classes of green algae (Sluiman et al. 2008), demonstrate the lack of congruence between morphology and genetics for many members of the genus (Rindi et al. 2008), and illustrate an unexpected amount of evolutionary divergence within the genus (Rindi et al. 2011).

Class Zygnematophyceae: Order Desmidiales

Desmidium C. Agardh ex Ralfs (Fig. 3.3g)

Thalli composed of long, twisted filaments of cells, often with a thick mucilaginous sheath. Cells rectangular in shape, with a shallow (usually) median constriction; elliptical or with three to five angles in apical view. Each semicell with a single chloroplast (stellate in end view) with one to two pyrenoids. Asexual reproduction by cell division; sexual reproduction via conjugation.

Remarks: *Desmidium* is a widespread genus that is typically found in oligotropic, low pH waters. *Desmidium* is closely related to the next taxon covered, *Hyalotheca* (McCourt et al. 2000; Hall et al. 2008a), and recent molecular analyses suggest that the genus is polyphyletic (Gontcharov 2008; Hall et al. 2008b).

Hyalotheca Ehrenberg ex Ralfs (Fig. 3.3h)

Thallus composed of long filaments of cells that are almost always surrounded by a thick mucilaginous sheath. Cells subcylindrical, only slightly constricted or not at all. One chloroplast per semicell, axial in position, stellate in end view, with a central pyrenoid. Asexual reproduction by cell division or aplanospores; sexual reproduction by conjugation (known more rarely).

Remarks: *Hyalotheca* includes several cosmopolitan species that are known from oligotropic and low pH waters. The genus, thus far, appears to be monophyletic based on molecular analyses (Hall et al. 2008a, b).

Class Zygnematophyceae: Order Zygnematales

Cylindrocystis Meneghini ex De Bary (Fig. 3.3i)

Cells usually solitary, occasionally filamentous, or as groups of cells in a gelatinous matrix. Cells short and cylindrical with broadly rounded apices, with smooth walls. Cells with two axial, stellate chloroplasts, each with a central pyrenoid. Chloroplasts can have radiating extensions that are extended to the outer edges of the cell. Asexual reproduction by cell division; sexual reproduction by homothallic or heterothallic conjugation.

Remarks: *Cylindrocystis* is a widespread genus that is most commonly found in low pH waters. The genus was demonstrated to be not monophyletic based on molecular analyses (Gontcharov and Melkonian 2010).

Mougeotia C. Agardh (Fig. 3.3j)

Thallus composed of uniseriate, unbranched filaments of cylindrical cells that are much longer than broad, sometimes with anchoring rhizoids. Cells with one (or rarely two) axial plate-like chloroplasts almost as long as the cell, with pyrenoids usually in a linear series. Asexual reproduction by aplanospores and parthenospores; sexual reproduction by scaliform conjugation (involving cells of two different filaments), or, more rarely, lateral conjugation (involving adjacent cells of a single filament). Remarks: *Mougeotia* is widespread in freshwater habitats, and approximately 138 species are recognized worldwide. Molecular analyses thus far indicate that the genus is monophyletic (Hall et al. 2008a).

Mougeotiopsis Palla (Fig. 3.3k)

Thallus composed of uniseriate, unbranched filaments of cylindrical cells that are somewhat longer than broad. Cells with one axial plate-like (or slightly curved) chloroplast with a thickened and granulated margin, chloroplasts almost as long as the cell, and lacking pyrenoids. Vegetatively very similar to *Mougeotia* except for the features of the chloroplast. Sexual reproduction by isogamous scaliform conjugation. Remarks: *Mougeotiopsis* is much more rarely reported than *Mougeotia*; known from freshwater habitats of low pH.

Sirogonium Kützing (Fig. 3.3l)

Thallus composed of uniseriate filaments of cells that are two to five times as long as broad, and with plane end walls (adjacent cell walls of neighboring cells are flat). Mucilaginous sheath much reduced compared with *Spirogyra* and hence filaments do not feel slippery to the touch. Cells contain two to ten parietal, ribbon-like chloroplasts that are either straight and lying parallel to the long axis of the cell, or spiraled no more than half of a turn from one end of the cell to the other. Asexual reproduction by akinetes; sexual reproduction by scaliform conjugation, but lacking conjugation tubes. Remarks: *Sirogonium* is widespread in freshwater habitats, although not commonly reported. *Sirogonium* is likely not a distinct genus from *Spirogyra* based on molecular phylogenetic investigations (Drummond et al. 2005; Hall et al. 2008a; Chen et al. 2012; Stancheva et al. 2013).

Spirogyra Link (Fig. 3.3m)

Spirogyra is characterized by unbranched, uniseriate filaments containing 1–16 chloroplasts, and with a mucilaginous sheath that results in the filaments feeling slippery to the touch, filaments sometimes attached by rhizoids. Cells may be one to six times as long as broad. Chloroplasts distinctive, ribbon-like in shape and parietal in position, and spiraled through the cell. Degree of spiraling depends on the species, and ranges from less than one to several complete spirals in a cell. Each chloroplast contains several pyrenoids. Nucleus large and positioned centrally, suspended by cytoplasmic strands from the edges of the cell. Adjacent walls of neighboring cells can be either flat (i.e., plane) or folded (i.e., replicate). Species-level identification is highly dependent on characters associated with the process of sexual reproduction, which occurs by either lateral (involving adjacent cells of a single filament) or scaliform (involving cells of two different filaments) conjugation; mature zygospores are also necessary for species-level identification.

Remarks: polyploidy events may have contributed to the taxonomic diversity within the genus (McCourt and Hoshaw 1990), which is substantial (almost 400 species are currently recognized). *Spirogyra* is an extremely common and recognizable genus of green algae that can, under slow-flowing and nutrient-rich conditions, become abundant and mat-forming. Molecular analyses have highlighted substantial variation within the genus that does not always correlate with vegetative morphology, suggesting that the genus *Sirogonium* is not distinct from *Spirogyra* (Drummond et al. 2005; Hall et al. 2008a), and indicating large amounts of cryptic diversity within the genus (Chen et al. 2012; Stancheva et al. 2013).

Zygnema C. Agardh (Fig. 3.3n)

Filaments of *Zygnema* are recognizable by their uniseriate and unbranched nature, and by the presence of two axial stellate chloroplasts in each cell, each with a central pyrenoid. A central nucleus is present between the two chloroplasts. Filaments can be attached in flowing water habitats by rhizoidal extensions. As for *Spirogyra*, species-level identification is highly dependent on characters associated with the process of sexual reproduction, which can occur by lateral (involving adjacent cells of a single filament; rare for *Zygnema*) or scaliform conjugation (involving cells of two different filaments; more common for *Zygnema*); mature zygospores are also necessary for species-level identification. Aplanospores and akinetes are occasionally formed.

Remarks: like *Spirogyra*, the genus is widely distributed and can form large growths. Approximately 120 species are known. Mesospore color was recently found to correlate with molecular clades recovered in analyses of the *cox*3 and *rbc*L gene regions (Stancheva et al. 2012); these same analyses also suggested that *Zygogonium* is likely not distinct from *Zygnema*. Sequence data for the *psb*A gene have also been recently published for a few species (Kim et al. 2012).

Acknowledgments Collection and characterization of some of the green algae illustrated in this chapter was supported by a grant from the U.S. National Science Foundation (DEB-0841734). The following people are thanked for allowing the use of their green algal images: Ignacio Bárbara, William Bourland, Jerry Brand, Chris Carter, David John, Fabio Rindi, Robert Sheath, Pavel Škaloud, Yuuji Tsukii, and Peter York. Many images were originally located on the Protist, UTEX, Natural History Museum and AlgaeBase websites.

References

Allan JD (1995) Stream ecology: structure and function of running waters. Chapman and Hall, London

An SS, Friedl T, Hegewald E (1999) Phylogenetic relationships of *Scenedesmus* and *Scenedesmus*-like coccoid green algae as inferred from ITS-2 rDNA sequence comparisons. Plant Biol 1:418–428

Bock C, Pröschold T, Krienitz L (2010) Two new *Dictyosphaerium*-morphotype lineages of the Chlorellaceae (Trebouxiophyceae): *Heynigia* gen. nov. and *Hindakia* gen. nov. Eur J Phycol 46:267–277

Bock C, Pröschold T, Krienitz L (2011) Updating the genus *Dictyosphaerium* and description of *Mucidosphaerium* gen. nov. (Trebouxiophyceae) based on morphological and molecular data. J Phycol 47:638–652

Boedeker C, Eggert A, Immer A et al (2010) Biogeography of *Aegagropila linnaei* (Cladophorophyceae, Chlorophyta): a widespread freshwater alga with low effective dispersal potential shows a glacial imprint in its distribution. J Biogeogr 37:1491–1503

Boedeker C, O'Kelly CJ, Star W et al (2012) Molecular phylogeny and taxonomy of the Aegagropila clade (Cladophorales, Ulvophyceae), including the description of Aegagropilopsis gen. nov. and Pseudocladophora gen. nov. J Phycol 48:808–825

Brouard J-S, Otis C, Lemieux C et al (2011) The chloroplast genome of the green alga *Schizomeris leibleinii* (Chlorophyceae) provides evidence for bidirectional DNA replication from a single origin in the Chaetophorales. Genome Biol Evol 3:505–515

Buchheim MA, Michalopulos EA, Buchheim JA (2001) Phylogeny of the Chlorophyceae with special reference to the Sphaeropleales: a study of 18S and 26S rDNA data. J Phycol 37:819–835

Buchheim M, Buchheim J, Carlson T et al (2005) Phylogeny of the Hydrodictaceae (Chlorophyceae): inferences from rDNA data. J Phycol 41:1039–1054

Buchheim MA, Sutherland DM, Schleicher T et al (2011) Phylogeny of the Oedogoniales, Chaetophorales and Chaetopeltidales (Chlorophyceae): inferences from sequence-structure analysis of ITS2. Ann Bot 107:109–116

Caisová L, Marin B, Sausen N et al (2011) Polyphyly of *Chaetophora* and *Stigeoclonium* within the Chaetophorales (Chlorophyceae), revealed by sequence comparisons of nuclear-encoded SSU rRNA genes. J Phycol 47:164–177

Caisová L, Marin B, Melkonian M (2013) A consensus secondary structure of ITS2 in the Chlorophyta identified by phylogenetic reconstruction. Protist 164:482–496

Carlile AL, O'Kelly CJ, Sherwood AR (2011) The green algal genus Cloniophora represents a novel lineage in the Ulvales: a proposal for Cloniophoraceae fam. nov. J Phycol 47: 1379–1387

Chen C, Barfuss MHJ, Pröschold T et al (2012) Hidden genetic diversity in the green alga *Spirogya* (Zygnematophyceae, Streptophyta). BMC Evol Biol 12:77

Cimino MT, Delwiche CF (2002) Molecular and morphological data identify a cryptic species complex in endophytic members of the genus *Coleochaete* Bréb. (Charophyta: Coleochaetaceae). J Phycol 38:213–1221

Delwiche CF, Karol KG, Cimino MT et al (2002) Phylogeny of the genus Coleochaete (Coleochaetales, Charophyta) and related taxa inferred by analysis of the chloroplast gene rbcL. J Phycol 38:394–403

Dodds WK (2002) Freshwater ecology: concepts and environmental applications. Academic Press, San Diego

Drummond CS, Hall J, Karol KG et al (2005) Phylogeny of *Spirogyra* and *Sirogonium* (Zygnematophyceae) based on *rbc*L sequence data. J Phycol 41:1055–1064

Durako MR (2007) A reassessment of *Geminella* (Chlorophyta) based upon photosynthetic pigments, DNA sequence analysis and electron microscopy. MS thesis, University of North Carolina, Wilmington

Friedl T, Rybalka N (2012) Systematics of the green algae: a brief introduction to the current status. Prog Bot 73:259–280

Gontcharov AA (2008) Phylogeny and classification of Zygnematophyceae (Streptophyta): current state of affairs. Fottea 8:87–104

Gontcharov AA, Melkonian M (2010) Molecular phylogeny and revision of the genus *Netrium* (Zygnematophyceae, Streptophyta): *Nucleotaenium* gen. nov. J Phycol 46:346–362

Graham LE, Graham JM, Wilcox LW (2009) Algae, 2nd edn. Benjamin Cummings, San Francisco

Guiry MD (2012) How many species of algae are there? J Phycol 48:1057–1063

Guiry MD, Guiry GM (2015) AlgaeBase. World-wide electronic publication, National University of Ireland, Galway, http://www.algaebase.org. Accessed 15 Dec 2015

Hall JD, McCourt RM (2015) Conjugating green algae including desmids. In: Wehr JD, Sheath RG, Kociolek JP (eds) Freshwater algae of North America. Ecology and classification. Academic, San Diego, pp 429–457

Hall JD, Karol KG, McCourt RM et al (2008a) Phylogeny of the conjugating green algae based on chloroplast and mitochondrial nucleotide sequence data. J Phycol 44:467–477

Hall JD, McCourt RM, Delwiche CF (2008b) Patterns of cell division in the filamentous Desmidiaceae, close green algal relatives of land plants. Am J Bot 95:643–654

Hanyuda T, Wakana I, Arai S et al (2002) Phylogenetic relationships within Cladophorales (Ulvophyceae, Chlorophyta) inferred from 18S rRNA gene sequences, with special reference to *Aegagropila linnaei*. J Phycol 38:564–571

Hayakawa Y, Ogawa T, Yoshikawa S et al (2012) Genetic and ecophysiological diversity of *Cladophora* (Cladophorales, Ulvophyceae) in various salinity regimes. Phycol Res 60:86–97

Hayden HS, Waaland JR (2002) Phylogenetic systematics of the Ulvaceae (Ulvales, Ulvophyceae) using chloroplast and nuclear DNA sequences. J Phycol 38:1200–1212

Hegewald E, Wolf M (2003) Phylogenetic relationships of *Scenedesmus* and *Acutodesmus* (Chlorophyta, Chlorophyceae) as inferred from 18S rDNA and ITS-2 sequence comparisons. Plant Syst Evol 241:185–191

Ichihara K, Shimada S, Miyaji K (2013) Systematics of *Rhizoclonium*-like algae (Cladophorales, Chlorophyta) from Japanese brackish waters, based on molecular phylogenetic and morphological analyses. Phycologia 52:398–410

John DM, Rindi F (2015) Filamentous (non-conjugating) and plant-like green algae. In: Wehr JD, Sheath RG, Kociolek JP (eds) Freshwater algae of North America. Ecology and classification. Academic Press, San Diego, pp 375–427

John DM, Whitton BA, Brook AJ (2011) The freshwater algal flora of the British Isles: an identification guide to freshwater and terrestrial algae, 2nd edn. Cambridge University Press, Cambridge

Kato S, Sakayama H, Sano S et al (2008) Morphological variation and intraspecific phylogeny of the ubiquitous species *Chara braunii* (Charales, Charophyceae) in Japan. Phycologia 47:191–202

Kim J-H, Boo SM, Kim YH (2012) Morphology and plastid *psb*A phylogeny of *Zygnema* (Zygnemataceae, Chlorophyta) from Korea: *Z. insigne* and *Z. leiospermum*. Algae 27:225–234

Krienitz L, Bock C, Luo W et al (2010) Polyphyletic origin of the *Dictyosphaerium* morphotype within Chlorellaceae (Trebouxiophyceae). J Phycol 46:559–563

Leliaert F, Smith DR, Moreau H et al (2012) Phylogeny and molecular evolution of the green algae. Crit Rev Plant Sci 31:1–46

Lowe RL, LaLiberte GD (2007) Benthic stream algae: distribution and structure. In: Hauer FR, Lamberti GA (eds) Methods in stream ecology, 2nd edn. Academic, Amsterdam, pp 327–356

McCourt RM, Hoshaw RW (1990) Non-correspondence of breeding groups, morphology, and monophyletic groups in *Spirogyra* (Zygnemataceae: Chlorophyta) and the application of species concepts. Syst Bot 15:69–78

McCourt RM, Karol KG, Bell J et al (2000) Phylogeny of the conjugating green algae (Zygnemophyceae) based on *rbc*L sequences. J Phycol 36:747–758

McManus HA, Lewis LA (2011) Molecular phylogenetic relationships in the freshwater family Hydrodictyaceae (Sphaeropleales, Chlorophyceae), with an emphasis on *Pediastrum duplex*. J Phycol 47:152–163

McManus HA, Lewis LA, Schultz ET (2011) Distinguishing multiple lineages of *Pediastrum duplex* with morphometrics and a proposal for *Lacunastrum* gen. nov. J Phycol 47:123–130

Mei H, Luo W, Liu GX et al (2007) Phylogeny of Oedogoniales (Chlorophyceae, Chlorophyta) inferred from 18S rDNA sequences with emphasis on the relationships in the genus *Oedogonium* based on ITS-2 sequences. Plant Syst Evol 265:179–191

Mikhailyuk TI, Sluiman HJ, Massalski A et al (2008) New streptophyte green algae from terrestrial habitats and an assessment of the genus *Interfilum* (Klebsormidiophyceae, Streptophyta). J Phycol 44:1586–1603

Nakada T, Nozaki H (2015) Flagellate green algae. In: Wehr JD, Sheath RG, Kociolek JP (eds) Freshwater algae of North America. Ecology and classification. Academic Press, San Diego, pp 265–313

Novis P (2006) Taxonomy of *Klebsormidium* (Klebsormidiales, Charophyceae) in New Zealand streams and the significance of low-pH habitats. Phycologia 45:293–301

OKelly CJ, Wysor B, Bellows WK (2004) *Collinsiella* (Ulvophyceae, Chlorophyta) and other ulotrichalean taxa with shell-boring sporophytes form a monophyletic clade. Phycologia 43:41–49

Prescott GW (1951) Algae of the western Great Lakes area, with an illustrated key to the genera of desmids and freshwater diatoms. Wm. C Brown, Dubuque

Pröschold T, Leliaert F (2007) Systematics of the green algae: conflict of classic and modern approaches. In: Brodie J, Lewis J (eds) Unravelling the algae: the past, present, and future of algal systematics. CRC Press (Taylor and Francis Group), Boca Raton, pp 123–153

Rindi F, McIvor L, Sherwood AR et al (2007) Molecular phylogeny of the green algal order Prasiolales (Trebouxiophyceae, Chlorophyta). J Phycol 43:811–822

Rindi F, Guiry MD, López-Bautista J (2008) Distribution, morphology, and phylogeny of *Klebsormidium* (Klebsormidiales, Charophyceae) in urban environments in Europe. J Phycol 44:1529–1540

Rindi F, Mikhailyuk TI, Sluiman HJ et al (2011) Phylogenetic relationships in *Interfilum* and *Klebsormidium* (Klebsormidiophyceae, Streptophyta). Mol Phylogenet Evol 58:218–231

Sakayama H, Miyaji K, Nagumo T et al (2005) Taxonomic re-examination of 17 species of *Nitella* subgenus *Tieffallenia* (Charales, Charophyceae) based on internal morphology of the oospore wall and multiple DNA marker sequences. J Phycol 41:195–211

Sakayama H, Arai S, Nozaki H et al (2006) Morphology, molecular phylogeny and taxonomy of *Nitella comptonii* (Charales, Characeae). Phycologia 45:417–421

Sanders ER, Karol KG, McCourt RM (2003) Occurrence of *matK* in a *trnK* Group II intron in charophyte green algae and phylogeny of the Characeae. Am J Bot 90:628–633

Sheath RG, Vis ML (2015) Red algae. In: Wehr JD, Sheath RG, Kociolek JP (eds) Freshwater algae of North America. Ecology and classification. Academic Press, San Diego, pp 237–264

Shubert LE, Gärtner G (2015) Nonmotile coccoid and colonial green algae. In: Wehr JD, Sheath RG, Kociolek JP (eds) Freshwater algae of North America. Ecology and classification. Academic Press, San Diego, pp 315–373

Sluiman HJ, Guihal C, Mudimu O (2008) Assessing phylogenetic affinities and species delimitations in Klebsormidiales (Streptophyta): nuclear-encoded rDNA phylogenies and ITS secondary structure models in *Klebsormidium*, *Hormidiella*, and *Entransia*. J Phycol 44:183–195

Stancheva R, Hall JD, McCourt RM, Sheath RG (2012) Systematics of the genus *Zygnema* (Zygnematophyceae, Charophyta) from Californian watersheds. J Phycol 48:409–422

Stancheva R, Hall JD, McCourt RM et al (2013) Identity and phylogenetic placement of *Spirogyra* species (Zygnematophyceae, Charophyta) from California streams and elsewhere. J Phycol 49:588–607

Stevenson RJ, Bothwell ML, Lowe RL (1996) Algal ecology: freshwater benthic ecosystems. Academic Press, San Diego

Tippery NP, Fučiková K, Lewis PO et al (2012) Probing the monophyly of the Sphaeropleales (Chlorophyceae) using data from five genes. J Phycol 48:1482–1493
Ustinova I, Krienitz L, Huss VAR (2001) *Closteriopsis acicularis* (G.M. Smith) Belcher et Swale is a fusiform alga closely related to *Chlorella kessleri* Fott et Nováková (Chlorophyta, Trebouxiophyceae). Eur J Phycol 36:341–351
Vanormelingen P, Hegewald E, Braband A et al (2007) The systematics of a small spineless *Desmodesmus* species, *D. costato-granulatus* (Sphaeropleales, Chlorophyceae), based on ITS2 rDNA sequence analyses and cell wall morphology. J Phycol 43:378–396
Wehr JD, Sheath RG, Kociolek JP (2015) Freshwater algae of North America, ecology and classification. Academic Press, San Diego
Zulkifly SB, Graham JM, Young EB et al (2013) The genus *Cladophora* Kützing (Ulvophyceae) as a globally distributed ecological engineer. J Phycol 49:1–17

Chapter 4
Red Algae (Rhodophyta) in Rivers

Orlando Necchi Jr.

Abstract Freshwater red algae are presently classified into five classes based on recent molecular phylogenetic and supporting morphological analyses: Bangiophyceae, Compsopogonophyceae, Florideophyceae, Porphyridiophyceae, and Stylonematophyceae. Red algae are well represented in river ecosystems and many freshwater members occur exclusively in lotic habitats. These algae are usually important constituents of stream floras, either in terms of abundance or distribution from local scale to biomes. This chapter treats all the genera from river habitats, with an emphasis on the benthic forms and macroscopic taxa. Basic descriptions of 26 genera are provided, along with illustrations and information on the habitat.

Keywords Algae • Benthic • Biodiversity • Genus • Red algae • Rhodophyta • River

Introduction

Phylogenetic Relationships and Features of the Red Algae

The higher taxonomic categories of red algae were recently revised by Yoon et al. (2006) using an analysis of the PSI P700 chl *a* apoprotein A1 (psaA) and *rbc*L coding regions river-inhabiting rhodophytes are now placed in five classes based on this phylogeny and related morphological features: the Bangiophyceae, Compsopogonophyceae, Florideophyceae, Porphyridiophyceae, and Stylonematophyceae. Each of these classes is briefly described below.

O. Necchi Jr. (✉)
Department of Zoology and Botany, São Paulo State University,
Rua Cristóvão Colombo, 2265, São José do Rio Preto, SP 15054-000, Brazil
e-mail: orlando@ibilce.unesp.br

O. Necchi Jr. (ed.), *River Algae*, DOI 10.1007/978-3-319-31984-1_4

Bangiophyceae

These organisms typically have a heteromorphic life history with an alternating macroscopic gametophyte and microscopic sporophyte, the "Conchocelis" stage (Müller et al. 2010). The Bangiophyceae almost exclusively consist of species attached to substrata, such as rough boulders in intertidal and upper subtidal zones in both temperate and tropical oceans. One order, the Bangiales, extends into rivers.

Compsopogonophyceae

The class is differentiated from other bangiophyte groupings by having the combination of monosporangia and spermatangia cleaved by curved walls from vegetative cells, Golgi body–ER associations, encircling outer thylakoids in the chloroplast and a biphasic life history where known (Saunders and Hommersand 2004). All the members of the Compsopogonophyceae are benthic, either being epilithic or epiphytic. Most are intertidal to upper subtidal, but some genera can extend into freshwater streams, such as *Boldia* and *Compsopogon* of the order Compsopogonales.

Porphyridiophyceae

This class consists of unicellular red algae with a single branched or stellate plastid with or without pyrenoid; Golgi association with mitochondria and ER; cells with floridoside as a low molecular weight carbohydrate; reproduction by cell division. These algae are found on moist soils but the genus *Kyliniella* of the order Porphyridiales has been collected in streams from Europe and North America (Kumano 2002; Sheath and Vis 2015).

Stylonematophyceae

This group has unicellular, pseudofilamentous, or filamentous red algae; various plastid morphologies with or without pyrenoid; Golgi association with mitochondria and ER; reproduction by cell division or monospores (Yoon et al. 2006). Taxa within the Stylonematophyceae are generally rare and often overlooked in floras, as they typically occur as epiphytes on larger algae or macrophytes or as small colonies in members of the order Stylonematales (Sheath and Vis 2015).

Florideophyceae

The florideophytes are characterized as a separate group of red algae by molecular data and two reproductive features, the tetrasporophyte and the carposporophyte (Graham et al. 2009). There are more than 20 orders distinguished by molecular, reproductive, and structural features, including pit plugs. In river environments there are seven orders: Acrochaetiales, Balbianales, Batrachopermales, Ceramiales, Gigartinales, Hildenbrandiales, and Thoreales.

Sample Collection and Preservation

Information on procedures, equipment, and tools for collection and preservation of freshwater red algae can be found in Entwisle et al. (1997), Lowe and Laliberte (2007), Eloranta and Kwandrans (2012), and Sheath and Vis (2015). We will describe hereafter the most relevant aspects for a nonexperienced collector. The vast majority of red algae representatives are macroalgae and thus we will concentrate on macroscopic forms. Considering that freshwater red algae are usually not very frequent and abundant in drainage basins (Sheath and Vis 2015), it is often necessary thoroughly search stream segments, as well as to take into account the best season for collecting. One should be keep in mind to avoid rainy season in tropical regions (due to high flow and turbidity) and summer in temperate areas (due to heavy canopy cover by surrounding vegetation). Usually stream segments of 20–50 m length should be examined, using a view box with glass bottom, for a representative sampling. Tools as long forceps, razor blades, spatulas, and vacuum pipettes are useful for successfully removing the specimens from the variable substrata (rock beds, boulders, stones, logs, tree roots, macrophytes, and other macroalgae). Equipment for measuring selected stream environmental variables usually include, as portable single meters or multiple probes: thermometer, pH meter, conductivity meter, turbidity meter, and flowmeter.

Specimens are best observed shortly after collection in a living state. Brief examination during collection with a portable field microscope are useful in some cases (e.g., more than one taxon present or rough identification). For morphological purposes, they should be fixed in 2.5 % $CaCO_3$-buffered, histological-grade glutaraldehyde, and kept in a dark and cool environment. Other fixatives include 4 % $CaCO_3$-formaldehyde, Transeau, FAA, or Lugol's, but might cause more distortions in morphology and pigmentation. For molecular purposes, specimens can be previously cleaned, if necessary, with a dissecting microscope, and then frozen or dried in silica gel desiccant. Drying of herbarium specimens are desirable for usual archival procedures but result in considerable morphological damage, in some cases not allowing observation of essential features. The correct preservation is essential for observation of the vegetative and reproductive diagnostic characters required for a proper identification, as indicated in the key and descriptions below.

Distribution of Red Algae in Rivers

Red algae are primarily marine, with less than 3 % of the over 6500 species occurring in freshwater habitats (Guiry and Guiry 2015; Sheath and Vis 2015). Most of the inland rhodophytes are restricted to streams and rivers, though some species can be collected in lakes, on moist soil, in sloth hair, and on cave walls. In major surveys, 18–65 % of river reaches contained red algae (Sheath and Hambrook 1990; Necchi et al. 1999 and references therein). The maximum number of species collected per river segment ranges from three to six (Sheath and Hambrook 1990; Necchi et al. 1999).

Most riverine red algae occur in moderate flow regimes (mean 29–57 cm s^{-1}) (Sheath and Hambrook 1990; Necchi et al. 1999). Moderate flow enhances primary productivity, pigment levels, respiration rates, and nutrient uptake rates as well as causing washout of loosely attached competitors. Flowing waters also replenish nutrients and gases and reduce boundary layers and the zone of depletion around the thallus. Some freshwater red algae, such as *Hildenbrandia*, *Lemanea*, and *Paralemanea*, are common in stream reaches with high current velocities (>1 m s^{-1}).

The light regime, which includes changes in intensity, quality, and photoperiod, is one of the major factors affecting the distribution and seasonality of freshwater red algae (Sheath and Hambrook 1990). Radiation affects algal growth by photosynthesis, processes indirectly related to photosynthesis, and those processes unrelated to photosynthesis. In deciduous forested regions, the surrounding tree canopy can reduce light penetration in a shaded reach by 90–99 %, and freshwater rhodophytes tend to disappear during periods of peak shading, such as *Sheathia americana* in a headwater stream (Sheath and Vis 2015). In addition, stream inhabiting rhodophytes exhibit photosynthetic characteristics of shade-adapted algae: occurrence of photoinhibition, low values of the saturation (<225 μmol m^{-2} s^{-1}) and compensation irradiances (<20 μmol m^{-2} s^{-1}), and relatively high values of the effective quantum yield of photosystem II ($\Delta F = F'_{m}\,0.45$) (Necchi 2005). Localized distribution in a drainage basin will also be affected by the light regime, and red algae tend to be most abundant in mid-order reaches (Sheath and Hambrook 1990).

Temperature influences the latitude, elevation, and drainage basin distribution patterns of freshwater red algae (Sheath and Hambrook 1990). In North America, the greatest species diversity has been observed in temperate and tropical latitudes. Certain taxa are most commonly distributed in tropical and subtropical regions, such as *Compsopogon* and *Kumanoa*. In contrast, the genus *Lemanea* is widespread in boreal and alpine habitats. In temperate regions, most red algae are restricted to elevations less than 900 m above sea level, and they exhibit their maximum biomass, growth, and reproduction from late fall to early spring. This seasonality is related to temperature but probably more so to light and nutrient availability.

In general, red algae occur in river parts that have low to moderate nutrient levels and are good indicators of relatively little disturbed sites. Flow replenishment and reduction of the boundary layer are factors accounting for this distribution, and many freshwater red algae can produce colorless hair cells that can increase the surface area for nutrient uptake (Sheath and Hambrook 1990). A number of taxa are considered to be indicators of good water quality (Eloranta and Kwandrans 2004). The interaction between pH and the form of inorganic carbon greatly influences the distribution of freshwater red algae. Rates of photosynthesis in response to pH showed that several species of freshwater red algae are able to use inorganic carbon as carbon dioxide or bicarbonate (Necchi and Zucchi 2001). Many species most frequently occur between pH 6.0 and 7.5, a range in which both forms are reasonably plentiful. The majority of these species is restricted to freshwater habitats and can be classified as specialists in terms of habitat (Sheath and Vis 2015). Competition for suitable substrata can occur among species of freshwater red algae or with other benthic algae in streams and rivers. Forms growing within and near the boundary layer, such as the "Chantransia" stages of the Batrachospermales, Thoreales, and Balbianiales, *Audouinella* and *Hildenbrandia*, are components of the stream epilithic community and compete with a complex periphyton association usually dominated by diatoms. The semierect forms compete with macrophytes, such as bryophytes, as well as other species of macroalgae.

Taxonomic key to the genera of red algae in rivers

1a	Thalli unicellular, colonial, or pseudofilamentous	2
1b	Thalli filamentous, filliform, sac like, or pseudoparenchimatous	4
2a	Cells solitary or joined	*Chroothece*
2a	Cells forming pseudofilaments	3
3a	Pseudofilaments with subspherical to elliptical cells	*Chroodactylon*
3a	Pseudofilaments with discoid or cubic cells	*Kyliniella*
4a	Thalli consisting of a monostromatic sac	*Boldia*
4b	Thalli filamentous, filiform, or pseudoparenchymatous	5
5a	Thalli filiform, unbranched, consisting of unbranched cylinders	*Bangia*
5a	Thalli filamentous or pseudoparenchymatous	6
6a	Thalli uniseriate or multiseriate	7
6a	Thalli entirely multiseriate	13
7a	Thalli entirely uniseriate	8
7b	Thalli uniseriate in young parts and corticated when mature	10
8a	Spermatangia unstalked, arising on branch tips	*Audouinella*
8b	Spermatangia stalked, arising on specialized colorless branches	9
9a	Erect thalli consisting of elongate cylindrical cells (10–15 times longer than wide)	*Balbiania*
9b	Erect thalli consisting of large barrel-shaped axial and determinate lateral branches with smaller cells	*Rhododraparnaldia*
10a	Cortex formed by polygonal and irregularly arranged cells	*Compsopogon*
10a	Cortex formed by cylindrical and regularly arranged cells	11
11a	Thalli consisting of flat blade with a distinctive midrib	*Caloglossa*
11b	Thalli with cylindrical axes, polysiphonous	12
12a	Trichoblasts (hair cells) present, stichidia (inflated structure with tetrasporangia) lacking	*Polysiphonia*
12b	Trichoblasts lacking, stichidia on terminal branches	*Bostrychia*
13a	Thalli multiaxial, cord like (Thoreales)	14
13b	Thalli uniaxial, compact, or hollow	15
14a	Outer region formed by densely arranged filaments; sporangia located at the thallus surface	*Nemalionopsis*
14b	Outer region formed by loosely arranged filaments; sporangia located in the inner part of the thallus	*Thorea*

15a	Axial cell surrounded by tightly aggregated layers of medullary and outer cortical cells	*Sterrocladia*
15b	Axial cells surrounded by loosely aggregated layers of medullary and outer cortical cells	16
16a	Thalli cartilaginous, rarely mucilaginous, pseudoparenchymatous	17
16b	Thalli mucilaginous, not pseudoparenchymatous	22
17a	Thalli with beaded appearance (with nodes and internodes), sometimes only when young	18
17b	Thalli lacking beaded appearance	21
18a	Thalli hollow, with central axis not connected to the outer cortex	19
18b	Thalli compact, with central axis connected to the outer filaments	20
19a	Central axis surrounded by rhizoidal filaments; spermatangia formed in rings at nodal regions	*Paralemanea*
19b	Central axis lacking rhizoidal filaments; spermatangia formed in patches at nodal regions	*Lemanea*
20a	Outer portion formed by densely arranged filaments; secondary fascicles abundant, covering the internode	*Nothocladus*
20b	Outer cortex formed by loosely arranged filaments, forming an outer cortex; secondary fascicles lacking	*Tuomeya*
21a	Thalli stalked, hollow, with axial cells surrounded by rhizoidal filaments not connected to the outer cortex	*Petrohua*
21b	Thalli unstalked, compact, with axial cells surrounded by medullary filaments, connected to the outer cortex	*Psilosiphon*
22a	Carposporophytes indefinite, diffuse	*Sirodotia*
22b	Carposporophytes definite, spherical, or semispherical	23
23a	Carpogonial branches helically twisted (more than one turn) or less often curved, carpogonia asymmetric	*Kumanoa*
23b	Carpogonial branches straight or less often curved (less than one turn), carpogonia symmetric	24
24a	Axis heterocorticated (rhizoidal filaments with bulbous cells)	*Sheathia*
24b	Axis homocorticated (rhizoidal filaments without bulbous cells)	*Batrachospermum*

Descriptions of Genera of Red Algae in Rivers

Phylum Rhodophyta

Class Stylonematophyceae—Order Stylonematales: Genera Chroodactylon, Chroothece

Chroodactylon Hansgirg (Fig. 4.1a)

Pseudofilaments with false branching or unbranched, composed of subspherical to elliptical cells, encircled in a broad, gelatinous sheath with an irregular and uniseriate arrangement. Cells are 6–28 μm in length and 3–17 μm in diameter. Each cell contains one blue-green, stellate, axial chloroplast, with a prominent pyrenoid. Reproduction by monosporangia and fragmentation. Sexual reproduction unknown.

Remarks: two species were described from freshwater (Kumano 2002; Sheath and Vis 2015) and occur usually as epiphyte on macroalgae (e.g., *Cladophora*), largely in hardwaters with specific conductivity of 170–540 and pH of 7.8–8.5.

Chroothece Hansgirg (Fig. 4.1b, c)

Cells solitary or joined pole to pole, immersed in a firm and stratified gelatinous matrix, the basal pole extending into a lamelated stalk; cells elipsoidal to cylindrical, 30–46 μm in length and 20–30 μm in diameter. Each cell contains one blue-green, stellate, axial chloroplast, with a prominent pyrenoid. Reproduction by cell division and possibly by colony fragmentation. Sexual reproduction unknown.

Remarks: few species have been described from freshwater and are usually rare components of streams in North America and Europe (Sheath and Sherwood 2002; Sheath and Vis 2015).

Class Porphyridiophyceae—Order Porphyridiales: Genus Kyliniella

Kyliniella Skuja (Fig. 4.1d)

Pseudofilaments unbranched, arising from a discoid and pseudoparenchymatous base. Cells discoid or cubic, 5–17.5 μm in length and 15–19 μm in diameter, length/diameter 0.5–1.0, contiguous or separate, arranged within a broad gelatinous matrix. Rhizoidal outgrowths occur at points of contact among cells. Each cell contains several blue-green, parietal, discoid chloroplasts, without a pyrenoid. Vegetative reproduction by fragmentation and asexual reproduction by monosporangia shed by release from the sheath. Presumptive sexual reproduction with small colorless spermatia and large, pigmented carpogonia with tubular projections. Postfertilization structures unknown.

Remarks: only one species (*K. latvica*) has been described in the genus (Kumano 2002; Sheath and Vis 2015) and occur usually as epiphyte on macroalgae or macrophytes in softwater streams in North America and Europe.

Class Compsopogonophyceae—Order Compsopogonales: Genera Boldia, Compsopogon

Boldia Herndon (Fig. 4.1e, f)

Thalli consisting of pink to reddish brown, rarely olive green, monostromatic sac, or tube, occurring singly or in clusters arising from a microscopic disc or an irregular aggregation of cells. The disc or aggregation produces a cushion-like structure that

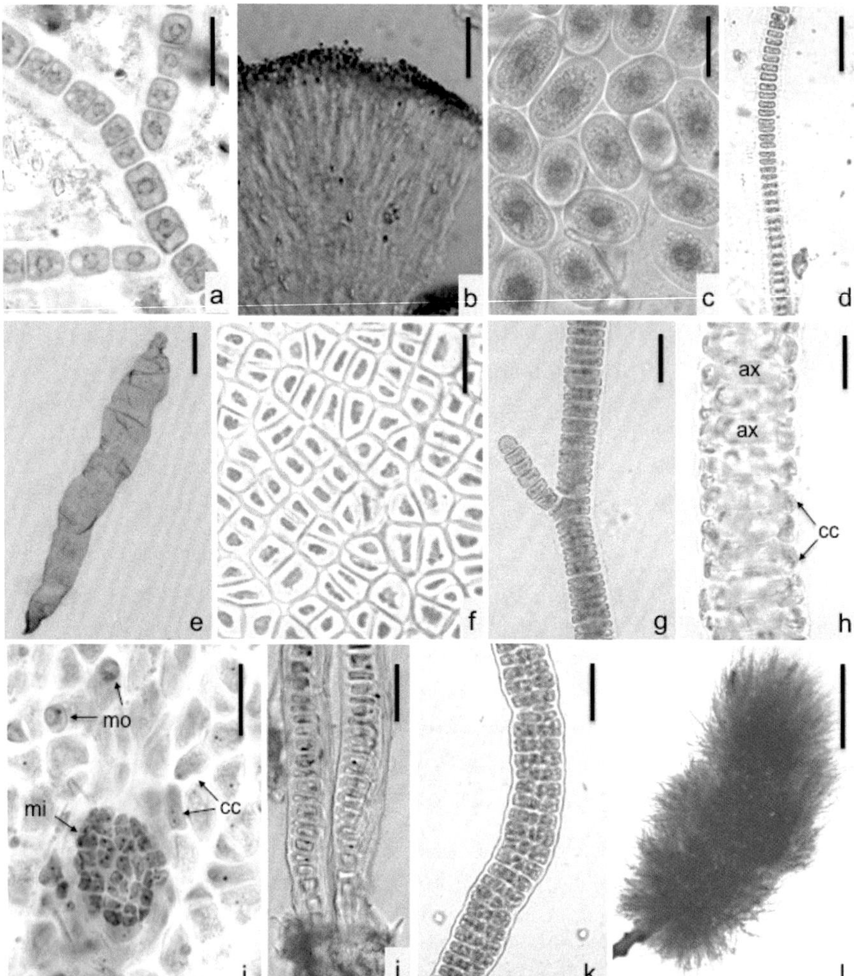

Fig. 4.1 (**a**) *Chroodactylon*. (**b** and **c**) *Chroothece*: (**b**) general view of the gelatinous matrix; (**c**) detail of cells. (**d**) *Kyliniella*. (**e** and **f**) *Boldia*: (**e**) general view of a dried thallus; (**f**) cells in surface view. (**g–i**) *Compsopogon*: (**g**) uniseriate filament; (**h**) corticated filament with axial cells (ax) and cortical cells (cc); (**i**) surface view with cortical cells (cc), monosporangia (mo) and microsporangia (mi). (**j** and **k**) *Bangia*: (**j**) basal part with uniseriate filaments; (**k**) multiseriate filament. (**l**) *Audouinella*: tuft epiphytic on *Lemanea*. Scale bars: (**e**) = 10 mm; (**l**) = 1 mm; (**b**) = 250 μm; (**h**, **i**) = 50 μm; (**a**, **c**, **d**, **f**, **g**, **k**) = 25 μm; (**j**) = 10 μm. Image authors: (**a**, **i**) C. Carter; (**b**, **d**, **k**) M. Aboal; (**c**) I. Chapuis; (**f**) Sheath and Vis (2015, Elsevier); (**j**) I. Bárbara

functions as a perennial holdfast. Vegetative cells are isodiametric or rectangular in surface view, containing several peripheral, ribbon-like chloroplasts, and a large central vacuole. Secondary filaments arise as outgrowths from vegetative cells, elongating between and above vegetative cells. Reproduction by monosporangia, formed from vegetative cells, 5–9 μm in diameter.

Remarks: a single species (*B. erythrosiphon*) occurs in scattered streams in eastern North America from Alabama to Ontario and Quebec (Sheath and Vis 2015). In the southern range, it is often reported as epizoic on snails, whereas in the north it is largely epilithic.

Compsopogon Montagne (Fig. 4.1g–i)

Thalli consisting of branched, bluish to violet-green, uniseriate filaments with mature portions corticated. Cortical filaments consisting of a cortex with one to several layers, produced by vertical division of axial cells. Thalli up to 20–50 cm long and 250–2000 μm in diameter, may be free floating or benthic. Axial cells large and evident by slight constrictions in mature portions, usually barrel shaped. Axial cells may break down, leaving a hollow cylinder. Attached thalli fixed by rhizoids or discoid base, formed by outgrowths of lower cortical cells. Cortical cells 7–22 × 10–48 μm, polygonals, containing several peripheral, discoid chloroplasts. Reproduction by fragmentation or monosporangia, which are formed by oblique divisions of cortical or uncorticated axial cells. Monospore divides into creeping, branched filament; a central cell eventually elongates vertically and divides to form erect stage. Sexual reproduction not confirmed; microsporangia reported may represent spermatia but not confirmed.

Remarks: a recent study involving molecular data in a global sampling (Necchi et al. 2013) revealed a low genetic diversity in the genus, with a single species recognized (*C. caeruleus*) within the monotypic family Compsopogonaceae. The genus is reported under a wide range of environmental variables, ranging from hard to softwaters and clean to moderately polluted water bodies, mainly from tropical and subtropical regions, but there are also few reports from temperate habitats.

Class Bangiophyceae—Order Bangiales: Genus Bangia

Bangia Lyngbye (Fig. 4.1j, k)

Thalli filamentous, filiform, red or dark-purple, consisting of unbranched cylinders of cells embedded in a firm gelatinous matrix. Attachment by down-growing rhizoids, usually in dense clumps. Basal region uniseriate becoming multiseriate in upper portions. Cells spherical or cubic, containing a large, axial stellate chloroplast with prominent pyrenoid; cell number and filament length correlated in uniaxial filaments. Asexual reproduction by monosporangia produced in apical packets. Sexual reproduction not observed in freshwater habitats.

Remarks: according to the recent generic revision by Sutherland et al. (2011), *Bangia* is a distinct genus among the filamentous Bangiales, occurring exclusively in freshwaters and with only one species (*B. atropurpurea*). It is mostly found in hardwater lakes (as the Great Lakes in North America, Sheath and Vis 2015) but there are some reports from streams in North America and Europe.

Class Florideophyceae—Order Acrochaetiales: Genus Audouinella

Audouinella Bory (Figs. 4.11 and 4.2a)

Thalli filamentous, uniaxial, reddish, growing typically in dense tufts. Basal portion composed of rhizoidal outgrowth, simple or parenchymatous disc. Erect filaments consisting of cylindrical cells with unilateral, opposite, or alternate branches. Cells contain reddish, parietal, ribbon-like chloroplasts. Cell diameter 6–26 µm. Asexual reproduction by monosporangia formed at branch tips, obovoidal to spherical. Sexual reproduction by colorless spermatangia arising in clusters in branch tips; carpogonia have a cylindrical base and thin trichogyne. Carposporophytes are spherical to subspherical, consisting of a compact mass of short gonimoblast filaments; carposporangia are obovoid to subspherical. Tetrasporangia are cruciate, formed at branch tips. Asexual reproduction by monosporangia, obovoidal, elliptical, spherical, or subspherical, arising from short branches on gametophytes or tetrasporophytes.

Remarks: species of *Audouinella* can be misidentified, by the morphological resemblance, with the acrochaetioid “Chantransia” stage of members of the freshwater orders Batrachospermales, Balbianiales, and Thoreales. Presence of sexual reproductive structures in true species of *Audouinella* (Necchi et al. 1993a) is the unequivocal diagnostic character. However, these structures are not often found and other characters are helpful to distinguish these two entities. Freshwater members of *Audouinella* are usually reddish (Necchi et al. 1993a), whereas the “Chantransia” stages are bluish (Necchi et al. 1993b; Zucchi and Necchi 2003). Molecular evidence was used to prove that two widely distributed bluish species (*A. macrospora* and *A. pygmaea*) are “Chantransia” stages (Necchi and Oliveira 2011). Six reddish species have been recognized as good species (*A. eugenea*, *A. hermannii*, *A. huastecana*, *A. meiospora*, *A. scopulata*, and *A. tenella*), whereas others remain uncertain. The genus is widespread in streams, ranging from Alaska to Costa Rica in North America (Sheath and Vis 2015), Brazil in South America (Necchi and Zucchi 1995), and all over Europe (Eloranta et al. 2011). *Audouinella hermannii*, the most common species in North America, tends to occur in cool waters (11 °C), with low ion content (specific conductivity 104 µS cm^{-1}) and mildly alkaline pH (7.5). In contrast, *A. eugenea* tended to occur in warm (23 °C), mildly alkaline waters (pH 7.6) with high ion content (530 µS cm^{-1}) in Mexico and Brazil (Carmona and Necchi 2001a).

Class Florideophyceae—Order Balbianiales: Genera Balbiania, Rhododraparnaldia

Balbiania Sirodot (Fig. 4.2b, c)

Thalli microscopic, composed of red-purple, uniseriate filaments, epiphytic on thalli of *Batrachospermum*. Filaments with a prostrate and an erect part; prostrate part composed of creeping filaments with short cylindrical or barrel-shaped cells, 1–4 times longer than large; erect part formed by elongate cylindrical cells, 10–15

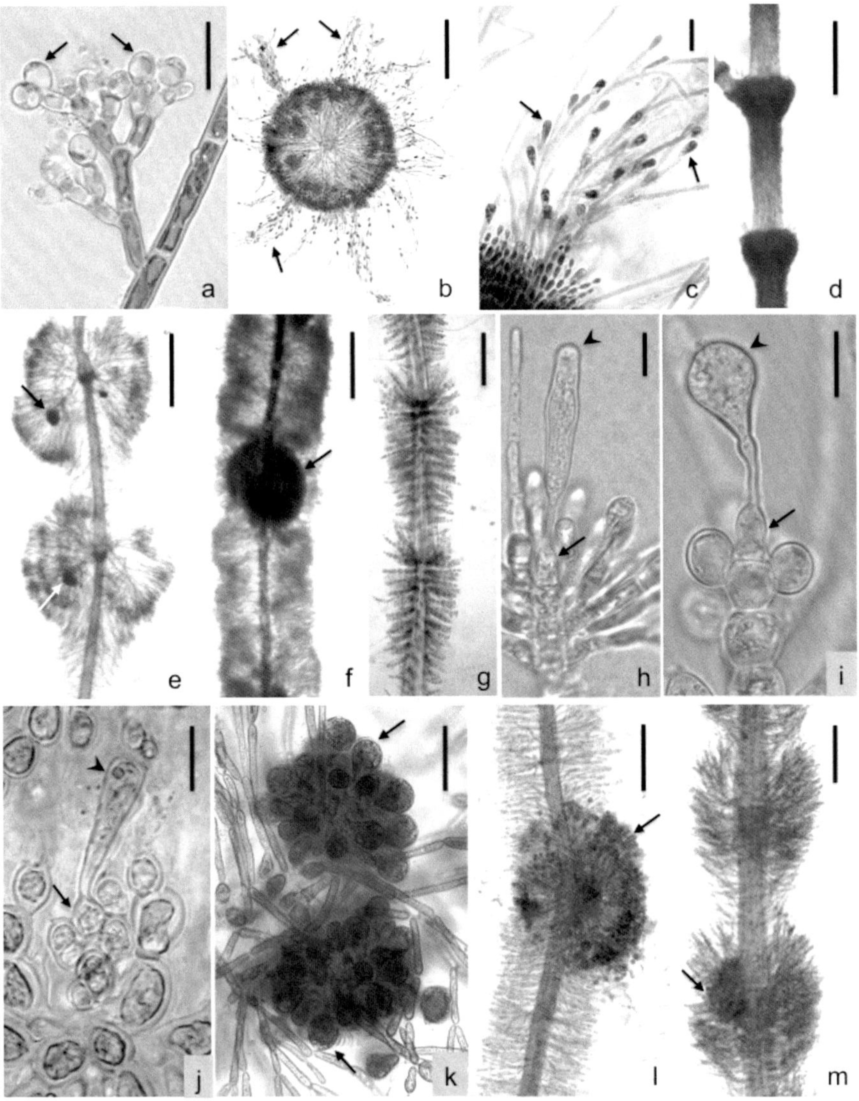

Fig. 4.2 (**a**) *Audouinella*: filament with monosporangia; (**b**, **c**) *Balbiania*; (**b**) thalli (*arrows*) epiphytic on *Batrachospermum*; (**c**) filaments with monosporangia (*arrows*); (**d–k**) *Batrachospermum*: (**d–g**) whorls and carposporophytes (*arrows*); (**h–j**) carpogonia (*arrows*) and trichogynes (*arrowheads*); (**k**) carposporophytes (*arrows*). (**l**, **m**) *Kumanoa*: whorls and carposporophytes (*arrows*). Scale bars: (**e**) = 500 μm; (**b**, **d**, **f**, **l**, **m**) = 250 μm; (**g**) = 100 μm; (**k**) = 50 μm; (**a**, **c**) = 25 μm; (**h–j**) = 10 μm. Image authors: (**a–c**) C. Carter

times longer than wide. Asexual reproduction by obovoidal or subspherical monosporangia, arising terminally on short lateral branches. Spermatangia formed in clusters, stalked, terminal on short branchlets. Carpogonia terminal on two to three celled branches, ovoidal with a long and thin trichogyne. Carposporophytes are spherical to subspherical, consisting of a compact mass of short gonimoblast filaments; carposporangia are obovoid or subspherical.

Remarks: two species were described in the genus (*B. investiens* and *B. meiospora*) but the latter is recognized within the genus *Audouinella* (see comments above), as it does not develop the specialized spermatangial branch characteristic of the genus. *Balbiania investiens* is known from several countries in Europe (Eloranta et al. 2011). Reports of this species from Australia (Skinner and Entwisle 2001) require further investigation (Entwisle pers. comm.).

Rhododraparnaldia Sheath, Whittick, and Cole

Thalli composed of crimson, uniseriate filaments, composed of large barrel-shaped axial cells (17.3–30.1 μm in diameter) and determinate lateral branches with smaller cells (4.3–8.5 μm in diameter). Spermatangia are complex, formed at the tips of colorless stalks, consisting of two to three cells: the central cell is spherical, like a typical spermatangium and the outer one or two cells are more cylindrical. Carpogonium is borne on an undifferentiated branch and has a swollen, cylindrical base and thin trichogyne. Carposporophyte spherical, consisting of a compact mass of short gonimoblast filaments. Carposporangia spherical, 5.5–8.0 μm in diameter. Carpospores germinate into a "Chantransia" stage with composed of cylindrical cells 16–38 μm in length, 5–7 μm in diameter.

Remarks: the single species (*R. oregonica*) combines characteristics of both the Acrochaetiales (carpogonium shape) and Batrachospermales (thallus with main axes and determinate laterals; presence of a "Chantransia" stage). It is known only from two mountain streams in Oregon (Sheath et al. 1994); the type locality has moderate current velocity (35–61 cm s^{-1}), temperatures 8–11 °C, pH 8.3, and specific conductivity 30 μS cm^{-1}.

Class Florideophyceae—Order Batrachospermales: Genera Batrachospermum, Kumanoa, Lemanea, Nothocladus, Paralemanea, Petrohua, Psilosiphon, Sheathia, Sirodotia, Tuomeya

Batrachospermum Roth (Fig. 4.2d–k)

Thalli poorly to abundantly mucilaginous, delicate or rigid, with beaded appearance, uniaxial, formed by a large axial cell surrounded by rhizoidal filaments with cylindrical to barrel-shaped cells; pericentral cells in variable number around the nodes, producing from the upper side one to five primary fascicles of limited growth and from the lower side rhizoidal filaments growing attached to axial cells and forming one to four layers; branching irregular or pseudodichotomous. Whorls well developed or reduced, usually distinct, but sometimes becoming confluent and indistinct in older parts. Primary fascicles straight or curved with 2–35 cell-storeys; branching di- or trichotomous; fascicle cells variable in shape, generally distinct between proximal and distal parts. Secondary fascicles usually abundant, covering the entire internode and as long as primary fascicles, less often few, sparse covering part of the internode or lacking. Monosporangia occur in few species. Monoecious, less often dioecious or polyoecious. Spermatangia usually spherical to obovoid, terminal or subterminal on primary or secondary fascicles, rarely on involucral filaments. Carpogonial branches ranging from well differentiated to little modified from the fascicle cells, straight, composed of 1–22 cells; it usually arises from pericentral cells, less often from fascicle cells, carpogonial branches, or rhizoidal filaments; involucral filaments short to long, one to ten cell-storeys. Carpogonia symmetric, straight, inserted on the center of the carpogonial branch distal cell; trichogynes variable in shape, unstalked, or stalked. Carposporophytes one to two per whorl, dense or loose, spherical or hemispherical, lower or higher than the whorl radius; gonimoblast filaments with one to ten cell-storeys; cells cylindrical, elliptical, or barrel shaped; carposporangia variable in shape; usually of one type, but two types (small and large) and bisporangia rarely observed; propagules observed in one species.

Remarks: *Batrachospermum* is morphologically the most diverse genus in the Batrachospermales and due to this wide variation in vegetative and reproductive features it is usually split in sections. The genus has been demonstrated to be paraphyletic in phylogenetic analyses (Entwisle et al. 2009), resulting in a recent trend of raising monophyletic sections to the genus rank, such as *Kumanoa*, for the former sections *Contorta* and *Hybrida* (Entwisle et al. 2009) and *Sheathia*, former section *Helminthoidea* (Salomaki et al. 2014). The genus has been reported in a wide range of environmental conditions (see Kumano 2002; Sheath and Vis 2015; and references therein).

Key to the sections currently recognized within the genus *Batrachospermum*

1a	Carposporophyte lacking; "Chantransia" stage developing directly from the fertilized carpogonium onto the gametophyte	*Acarposporophytum*
1b	Carposporophyte present; "Chantransia" stage developing from the germination of the carposporangia, free living	2
2a	Whorls reduced; carposporophytes inserted on the axis, forming wart-like structures on the nodes	*Setacea*
2b	Whorls well developed; carposporophytes inserted on the axis or stalked	3
3a	Carpogonial branches short; carposporophytes inserted on the axis	4
3b	Carpogonial branches long; carposporophytes stalked and distributed at different positions within the whorl	6
4a	Carposporophytes with segmented propagules	*Gonimopropagulum*
4a	Carposporophytes with regular, nonsegemented carposporangia	5
5a	Carposporophytes dense with one type of gonimoblast filaments (upright) radiating from the fertilized carpogonia	*Virescentia*
5b	Carposporophytes dense or loose, sometimes abortive, with two types of gonimoblast filaments (upright and prostrate), radiating from the fertilized carpogonia or spreading on the rhizoidal filaments	*Turfosa*

6a	Carpogonial branches little differentiated from the fascicles	*Batrachospermum*
6b	Carpogonial branches well differentiated from the fascicles	7
7a	Carposporangia large (≥30 μm long and 20 μm in diameter)	*Macrospora*
7b	Carposporangia small (≤25 μm long and 15 μm in diameter)	*Aristata*

Kumanoa Entwisle, Vis, Chiasson, Necchi, and Sherwood (Figs. 4.2l, m and 4.3a–c)

Thalli poorly, moderately, or abundantly mucilaginous, delicate or rigid, usually with evident beaded appearance; branching irregular or pseudodichotomous. Whorls well developed or reduced, usually distinct, but sometimes becoming confluent and indistinct in older parts. Pericentral cells in variable number around the nodes, each forming one to four primary fascicles of limited growth and rhizoidal filaments growing attached to axial cells forming one to four layers; rhizoidal filaments homocorticated (with cylindrical cells only); primary fascicles straight or curved with 3–20 cell-storeys; branching di- or trichotomous; fascicle cells variable in shape. Secondary fascicles usually abundant, covering the entire internode and as long as primary fascicles, less often few, sparse covering part of the internode or lacking. Monosporangia occur in few species. Monoecious, rarely dioecious. Spermatangia usually spherical to obovoid, terminal or subterminal on primary or secondary fascicles, rarely on involucral filaments or specialized branches. Carpogonial branches helically twisted (more than one turn), less often curved or slightly curved (less than one turn), usually arising from pericentral cells, less often from fascicle cells, carpogonial branches or rhizoidal filaments, composed of 2–26 disc- or barrel-shaped cells; involucral filaments short, one to five cell-storeys. Carpogonia asymmetric, twisted or straight, inserted off-center on the carpogonial branch distal cell; trichogynes variable in shape, unstalked, or stalked. Carposporophytes one to two per whorl, dense or loose, hemispherical, rarely spherical, lower or higher than the whorl radius; gonimoblast filaments with three to ten cell-storeys; cells cylindrical, elliptical, or barrel shaped; carposporangia variable in shape; usually of one type, but two types (small and large) and bisporangia rarely observed.

Remarks: *Kumanoa* is the most species rich genus in the Batrachospermales, with 35 species recognized worldwide by Necchi and Vis (2012); at least three new species have been described or reinstated recently. The genus occurs mainly in tropical and subtropical regions of the world.

Lemanea Bory (Fig. 4.3d–g)

Thalli cartilaginous, rigid, pseudoparenchymatous, with beaded appearance, uniaxial, hollow, formed by a long axial cell lacking rhizoidal filaments not connected to the outer cortex; branching irregular and sparse. Pericentral cells consisting of

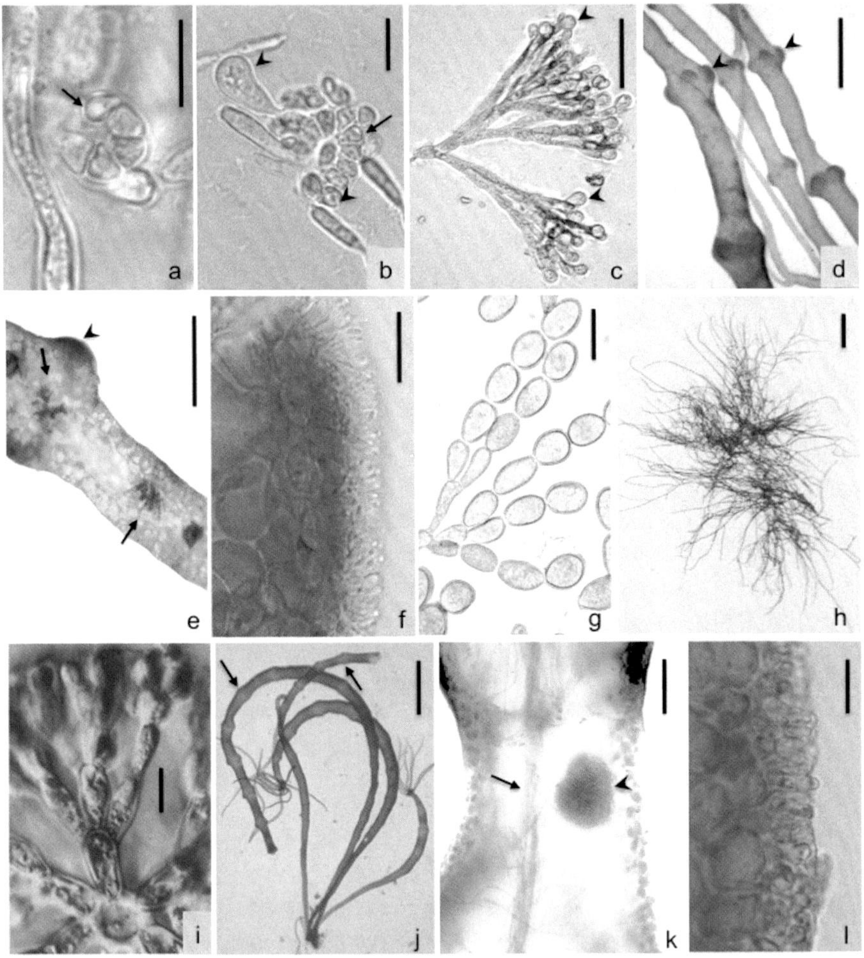

Fig. 4.3 (**a**) *Kumanoa*: (**a**, **b**) carpogonial branches (*arrows*) and trichogyne (*arrowhead*); (**c**) gonimoblast filaments with carposporangia (*arrowheads*). (**d–g**) *Lemanea*: (**d**) nodes with spermatangia in patches (*arrowheads*); (**e**) mid portion showing carposporophytes (*arrows*) and spermatangia in patches (*arrowhead*); (**f**) detail of spermatangia and cortical cells; (**g**) gonimoblast filaments with carposporangia. (**h**, **i**) *Nothocladus*: (**h**) general view of a dried thallus; (**i**) detail of primary fascicles. (**j–l**) *Paralemanea*: (**j**) general view of thalli with spermatangia in rings (*arrows*); (**k**) mid portion showing rhizoidal filaments (*arrows*) and carposporophyte (*arrowhead*); (**l**) detail of spermatangia and cortical cells. Scale bars: (**h–j**) = 1 cm; (**d**) = 500 μm; (**e**) = 250 μm; (**k**) = 100 μm; (**c**, **f**, **l**) = 50 μm; (**a**, **b**, **g**) = 10 μm. Image authors: (**d–g**) C. Carter; (**h**) Museum of New Zealand; (**i**) T. Entwisle; (**j–l**) J. Carmona

T- or L-shaped ray cells closely applied to the outer cortex. Outer cortex formed by two to four cell layers with closely appressed cells; the inner layers formed by large and colorless cells; the outer layer formed by small and pigmented cells. Monoecious. Spermatangia formed in patches at the nodes. Carpogonial branches internal, straight, short, composed of three to four cells. Carpogonia symmetric, inserted on the center of the carpogonial branch distal cell; trichogynes thin, protruding beyond the outer cortex layer. Carposporophytes microscopic, forming dense spherical masses; all cells of the gonimoblast filaments produce carposporangia; carposporangia ellipsoid to subspherical, formed in chains within the central cavity.

Remarks: *Lemanea* is widespread in lotic systems of temperate and boreal regions of Europe (Eloranta et al. 2011) and North America (Vis and Sheath 1992). Kumano (2002) recognized 11 species in the world, whereas two new species were described later from China (Shi 2006) and one from India (Ganesan et al. 2015), totaling 14 species in the genus.

Nothocladus Skuja (Fig. 4.3h, i)

Thalli cartilaginous or mucilaginous, rigid, more or less pseudoparenchymatous, with beaded appearance when young, but cylindrical or barrel shaped when mature, uniaxial; branching irregular and abundant. Whorls reduced, distinct on the apices, but becoming confluent and indistinct in older parts. Pericentral cells in variable number around the nodes, each forming one to three primary fascicles of limited growth and rhizoidal filaments growing attached to large axial cells, forming several layers; primary fascicles straight with four to nine cell-storeys; branching di- or trichotomous; fascicle cells variable in shape; elongate distal cells loosely arranged. Secondary fascicles abundant, covering the entire internode and as long as primary fascicles. Monoecious. Spermatangia obovoid to spherical, terminal on primary and secondary fascicles or on specialized branches arising from pericentral cells. Carpogonial branches curved or twisted (more than one turn), arising from pericentral cells, proximal fascicle cells, or pericentral cells, composed of 4–14 disc- or barrel-shaped cells; involucral filaments short, one to three cell-storeys. Carpogonia asymmetric, inserted off-center on the carpogonial branch distal cell; trichogynes club shaped or cylindrical, unstalked. Carposporophytes indefinite in shape; gonimoblast filaments of unlimited growth, spreading among fascicles; carposporangia obovoid to ellipsoidal.

Remarks: *Nothocladus* has three species, two occurring in southeastern Australia and one endemic to Madagascar (Entwisle and Foard 2007).

Paralemanea (Silva) Vis and Sheath (Fig. 4.3j–l)

Thalli cartilaginous, rigid, pseudoparenchymatous, with beaded appearance, uniaxial, hollow, formed by a long axial cell surrounded by dense rhizoidal filaments not connected the outer cortex; branching irregular and sparse. Pericentral cells consisting of simple ray cells not abutting the outer cortex. Outer cortex formed by two to four cell layers with closely appressed cells; the inner layers formed by large and colorless cells; the outer layer formed by small and pigmented cells. Monoecious.

Spermatangia formed in rings at the nodes. Carpogonial branches internal, straight, short to long, composed of three to tend cells. Carpogonia symmetric, inserted on the center of the carpogonial branch distal cell; trichogynes thin, protruding beyond the outer cortex layer. Carposporophytes forming dense spherical masses; all cells of the gonimoblast filaments produce carposporangia; carposporangia ellipsoid to subspherical, formed in chains within the central cavity.

Remarks: *Paralemanea* is widespread in lotic systems of temperate, subtropical, and alpine regions of Europe (Eloranta et al. 2011) and North America (Vis and Sheath 1992). An accurate estimation on the number of species in the genus is presently difficult, as Kumano (2002) recognized four species in the world, but additional species have been accepted or reinstated more recently for some regions, e.g., China (Shi 2006) and Central Europe (Eloranta et al. 2011).

Petrohua G.W. Saunders

Thalli cartilaginous, rigid, pseudoparenchymatous, stalked, lacking beaded appearance, uniaxial, hollow, formed by a long axial cell surrounded by rhizoidal filaments not connected to the outer cortex; branching irregular and rare. Pericentral cells consisting of simple ray cells not abutting the outer cortex. Outer cortex formed by two to three cell layers of densely arranged cells; the inner layers formed by large and colorless cells; the outer layer formed by small and pigmented cells. Monoecious? Putative spermatangia formed in patches on thallus surface. Putative carpogonial branches internal, composed of a cluster of small cells, arising from elongated cells of the cortical cells. Carpogonia asymmetric, inserted on the center of the carpogonial branch distal cell; putative trichogynes club shaped, projecting toward the outer cortex layer. Carposporophytes microscopic, forming spherical masses; gonimoblast filaments arising from clusters of small cells; all cells of the gonimoblast filaments apparently produce carposporangia; carposporangia formed in chains within the central cavity, carposporangia released by an ostiole, formed by modified outer cortical cells.

Remarks: *Petrohua bernabei* is the single *Petrohua* (Saunders, G.W.)species in the genus, known only from the type locality in Chile, South America (Vis et al. 2007).

Psilosiphon Entwisle (Fig. 4.4a)

Thalli cartilaginous, rigid, pseudoparenchymatous, unstalked, lacking beaded appearance, uniaxial, compact, formed by a long axial cell surrounded by medullary filaments connected to the outer cortex; branching irregular and sparse. Four pericentral cells arising from axial cells. Medullary filaments occupying the area between the axis and the cortex, forming an open tangle mass; medullary cells with numerous chloroplasts. Outer cortex formed by cortical filaments, radially branched, compact, and cohesive; cortical filaments arising laterally and adventitiously from

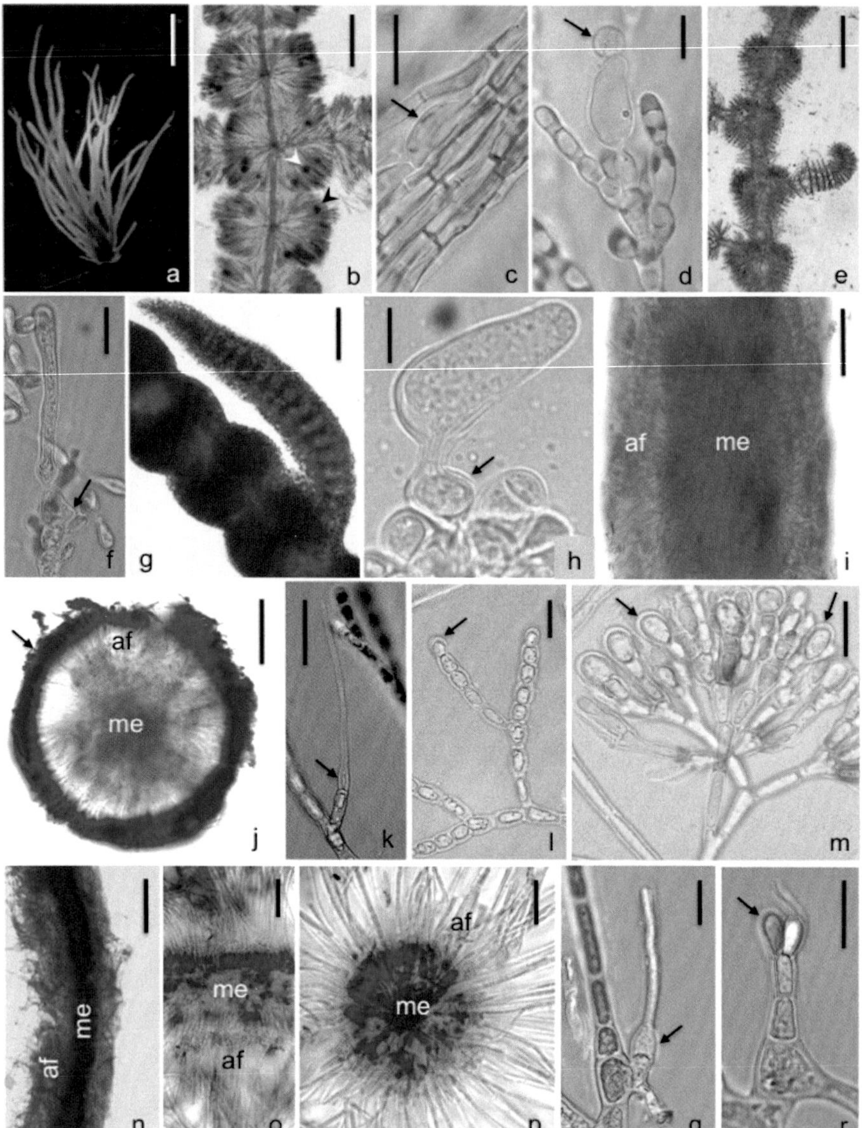

Fig. 4.4 (**a**) *Psilosiphon*: general view of thalli. (**b–d**) *Sheathia*: (**b**) whorls with carposporophytes (*arrowheads*); (**c**) rhizoidal filaments heterocorticated with bulbous cells (*arrow*); (**d**) carpogonium with attached spermatium (*arrow*); (**e, f**) *Sirodotia*: whorls; (**f**) carpogonia with protuberance (*arrow*). (**g, h**) *Tuomeya*: (**g**) whorls; (**h**) carpogonia (*arrow*) and trichogyne. (**i–m**) *Nemalionopsis*: (**i**) mid part showing medulla (me) and assimilatory filaments (af); (**j**) cross section showing medulla (me) and assimilatory filaments (af); (**k**) carpogonia (*arrow*) and trichogyne; (**l**) assimilatory filaments with spermatangia (*arrow*); (**m**) gonimoblast filaments with carposporangia (*arrows*). (**n–r**) *Thorea*: (**n**) mid part showing medulla (me) and assimilatory filaments (af); (**o**) longitudinal section showing (me) and assimilatory filaments (af); (**p**) cross section showing medulla (me) and assimilatory filaments (af); (**q**) carpogonia (*arrow*) and young trichogyne; (**r**) short assimilatory filaments with spermatangia (*arrow*). Scale bars: (**a**) = 5 mm; (**b, e, g, i, j, n**) = 250 μm; (**c, l, m**) = 25 μm; (**d, h, q, r**) = 10 μm. Image authors: (**a**) T. Entwisle; (**b–d**) C. Carter; (**g, h**) Sheath and Vis (2015, Elsevier)

medullary cells. Sexual reproduction poorly known. Putative spermatangia scattered on thallus surface. Apparently asexual sporangia arising from apical cells of cortical filaments. Adventitious plantlets produced vegetatively, apparently from indeterminate growth of some cortical filaments.

Remarks: there is one species only, *Psilosiphon scoparium*, from southeastern Australia and northern New Zealand, restricted to rocky streams in or around waterfalls (Entwisle and Foard 2007).

Sheathia Salomaki and Vis (Fig. 4.4b–d)

Thalli moderately or abundantly mucilaginous, delicate, usually with evident beaded appearance; branching irregular. Whorls well developed, usually distinct, but sometimes becoming confluent and indistinct in older parts. Pericentral cells in variable number around the nodes, each forming two to four primary fascicles of limited growth and rhizoidal filaments growing attached to axial cells forming one to three layers; rhizoidal filaments usually heterocorticated (with bulbous and cylindrical cells) but with regular cortication in one species; primary fascicles straight with 6–23 cell-storeys; branching di- or trichotomous; fascicle cells variable in shape. Secondary fascicles usually abundant, covering the entire internode and as long as primary fascicles, less often sparse covering part of the internode or lacking. Monoecious or dioecious. Spermatangia usually spherical to obovoid, terminal, or subterminal on primary or secondary fascicles, rarely on involucral filaments. Carpogonial branches straight, usually arising from fascicle cells, rarely from pericentral cells; involucral filaments undifferentiated from fascicle cells; long, one to five cell-storeys. Carpogonia symmetric, inserted on the center of the carpogonial branch distal cell; trichogynes club shaped or lanceolate, unstalked. Carposporophytes 1 to several per whorl, dense, spherical, or subspherical, stalked, distributed at variable positions within the whorl or exerted; gonimoblast filaments with two to five cell-storeys; cells cylindrical; carposporangia obovoid to spherical.

Remarks: *Sheathia* was recently proposed as a genus, resulting from raising of the former section *Helminthoidea* to genus (Salomaki et al. 2014), with eight species recognized worldwide. The genus occurs mainly in temperate regions of Europe and North America.

Sirodotia Kylin (Fig. 4.4e, f)

Thalli moderately or abundantly mucilaginous, delicate, with evident nodes and internodes; branching irregular and abundant. Whorls well developed or reduced, obconic or pear shaped, distinct, but sometimes becoming confluent and indistinct in older parts. Pericentral cells in variable number around the nodes, each forming one to three primary fascicles of limited growth and rhizoidal filaments growing attached to axial cells forming one to three layers; primary fascicles straight or curved with 5–13 cell-storeys; branching di- or trichotomous; fascicle cells variable in shape. Secondary fascicles usually abundant, covering the entire internode and as long as primary fascicles. Monoecious, dioecious or polyoecious. Spermatangia usually spherical to obovoid, terminal or subterminal on primary or secondary fascicles, less often on involucral filaments or specialized branches. Carpogonial branches straight or slightly curved, usually arising from pericentral cells, less often from fascicle cells, carpogonial branches or rhizoidal filaments, composed of two to nine disc- or barrel-shaped cells; involucral filaments short, one to four cell-storeys. Carpogonia asymmetric, inserted off-center on the carpogonial branch distal cell with a protuberance on one side; trichogynes variable in shape, unstalked or stalked. Carposporophytes indefinite in shape, extending along the internodes; gonimoblast filaments of two types: prostrate filaments creeping among primary and secondary fascicles and rhizoidal filaments; erect filaments arising from prostrate filaments, producing carposporangia; carposporangia variable in shape.

Remarks: *Sirodotia* is widespread in tropical and subtropical regions of North (Necchi et al. 1993c) and South America (Necchi 1991), but one species (*S. suecica*) is commonly found in temperate or boreal regions of North America (Necchi et al. 1993c) and Europe (Eloranta et al. 2011). Kumano (2002) recognized six species in the genus, but other (probably new) species have been reported in recent studies using molecular data (Lam et al. 2012; Paiano and Necchi 2013).

Tuomeya W.H. Harvey (Fig. 4.4g, h)

Thalli cartilaginous, rigid, pseudoparenchymatous, with beaded appearance when young, but cylindrical when mature, uniaxial; branching irregular and abundant. Whorls reduced, distinct on the apices, but becoming confluent and indistinct in older parts. Pericentral cells in variable number around the nodes, each forming one to three primary fascicles of limited growth and rhizoidal filaments growing attached to large axial cells, forming two to three layers; primary fascicles straight with four to seven cell-storeys; branching di- or trichotomous; fascicle cells variable in shape; small distal cells densely arranged, forming an outer cortex with two to three layers. Secondary fascicles lacking. A cavity formed between two whorls limited by the inner cortex and the axis with rhizoidal filaments. Monoecious. Spermatangia usually elongate obovoid or ellipsoidal, terminal on primary fascicles. Carpogonial branches curved or slightly twisted (less than one turn), arising from pericentral cells, composed of four to seven disc- or barrel-shaped cells; involucral filaments short, one to three cell-storeys. Carpogonia asymmetric, inserted off-center on the carpogonial branch distal cell; trichogynes club-shaped, unstalked. Carposporophytes consisting of a globular and dense mass; gonimoblast filaments radially spreading from the fertilized carpogonia, with three to five storeys of cylindrical cells; carposporangia obovoid to ellipsoid.

Remarks: *Tuomeya* is a monotypic genus, with the single species *T. americana* occurring in eastern North America, although there are some unsubstantiated reports from Europe (Eloranta et al. 2011) and Asia (Shi 2006).

Class Florideophyceae—Order Thoreales: Genera Nemalionopsis and Thorea

Nemalionopsis Skuja (Fig. 4.4i–m)

Thalli mucilaginous, cord-like, multiaxial, composed of a central medullary region and an outer layer of lateral filaments of limited growth; branching irregular and abundant. Medullary region composed of interwoven branched and twisted, colorless or yellowish filaments with cylindrical short to elongate cells, occupying half to two thirds of the thallus diameter. Cortical region consisting of densely arranged assimilatory filaments, in an outer whorl at a right angle to the main axis, disposed along the entire thallus, with cylindrical or barrel-shaped cells containing chloroplasts. Assimilatory filaments densely arranged at the distal portion composed of cylindrical, elliptical or barrel-shaped cells. Spermatangia arising from terminal or subterminal cells of assimilatory filaments at the thallus surface, in clusters, pairs or single; spermatangia can regenerate from old empty spermatangial cells. Carpogonia bottle shaped or ovoid, inserted directly on the basal cell of assimilatory filaments or on short carpogonial branches with cylindrical or barrel-shaped cells; trichogynes elongate and filiform, reaching half to two thirds of the outer layer. Gonimoblast filaments spreading among the assimilatory filaments, radially arranged in the distal portion, composed of cylindrical cells. Carposporangia in clusters, near the thallus surface, larger than spermatangia. Monosporangia arising from the long assimilatory filaments, similarly to spermatangia and carposporangia.

Remarks: two species are widely recognized in the genus (*N. shawii* and *N. tortuosa*) occurring mostly in warm and hard water from Asia and North America. Reproductive structures usually described as monosporangia are possibly spermatangia or carposporangia, as observed in the genus *Thorea* (Necchi et al. 2016). Carpogonia, spermatangia and carposporangia were recently reported in *N. shawii* from Indonesia and Nepal (Johnston et al. 2014; Necchi et al. 2016).

Thorea Bory (Fig. 4.4n–r)

Thalli mucilaginous, cord-like, multiaxial, composed of a central medullary region and an outer layer of lateral filaments of limited growth; branching irregular and abundant. Medullary region composed of interwoven, colorless filaments, occupying one to two thirds of the thallus diameter. Outer region consisting of loosely arranged assimilatory filaments, disposed along the entire thallus, with cylindrical or less often barrel-shaped cells containing chloroplasts; assimilatory filaments are of two types: short and restricted to the proximal portion of the outer layer; long and reaching the thallus surface. Monoecious or dioecious. Spermatangia arising on the basal part of the outer region, from the short assimilatory filaments. Carpogonia consisting of short carpogonial branches, with few cells, arising from the short assimilatory filaments; trichogynes elongate and filiform, reaching half to two thirds of the outer layer. Gonimoblast filaments spreading among the short assimilatory filaments, arising directly from the fertilized carpogonia. Carposporangia in clusters, arising from short, erect gonimoblast filaments. Monosporangia arising from the short assimilatory filaments, similarly to spermatangia and carposporangia.

Remarks: more than ten species are currently recognized in the genus, based on morphological (Kumano 2002) or molecular data (Vis, pers. comm.). Species occur predominantly in warm and hard waters in tropical or subtropical regions of most continents. Reproductive structures usually described as monosporangia by some authors are possibly spermatangia or carposporangia, as noted by Necchi (1987) and Necchi and Zucchi (1997). However, the coexistence of asexual monosporangia with sexual reproductive structures (carpogonia and spermatangia) and carposporangia has also been confirmed in some species (Carmona and Necchi 2001b).

Class Florideophyceae—Order Hildenbrandiales: Genus Hildenbrandia

Hildenbrandia Nardo (Fig. 4.5a–e)

Thalli crustose, bright red, firmly attached to the substratum, with a smooth surface or wart-like protuberances, composed of a basal layer and erect filaments. Basal layer thin and formed by branched filaments that are densely aggregated laterally. Erect filaments formed by cubic or short-cylindrical cells. Sexual reproduction and tetrasporangia are unknown in freshwaters. Asexual reproduction by gemmae, consisting of a small and dense packet of filaments, that release and germinate into new crusts. Vegetative reproduction by fragments of older thalli.

Remarks: two species are usually recognized in freshwater habitats, with *H. angolensis* distributed in Africa, Australia, North and South America, and *H. rivularis* in Europe and Asia (Eloranta et al. 2011; Sheath and Vis 2015; Entwisle pers. comm.). The two species are usually reported in shaded and fast-flowing streams.

Class Florideophyceae—Order Ceramiales: Genera Bostrychia, Caloglossa and Polysiphonia

Bostrychia Montagne (Fig. 4.5f–i)

Thalli dark reddish, pseudodichotomously or dichotomously branched, corticated or uncorticated, attached to substrata by haptera (specialized rhizoidal branches). Vegetative branching bilateral; apical branches are usually curved. Axis composed of polysiphonous axes and branches consisting of axial cells having two to six whorls of four to eight periaxial cells. Tetrasporangia are the most usually reproductive structures observed in freshwater collections, consisting of inflated, multichambered structures termed stichidia, on the upper portion of the thalli. Carpogonia and cystocarps described in one species.

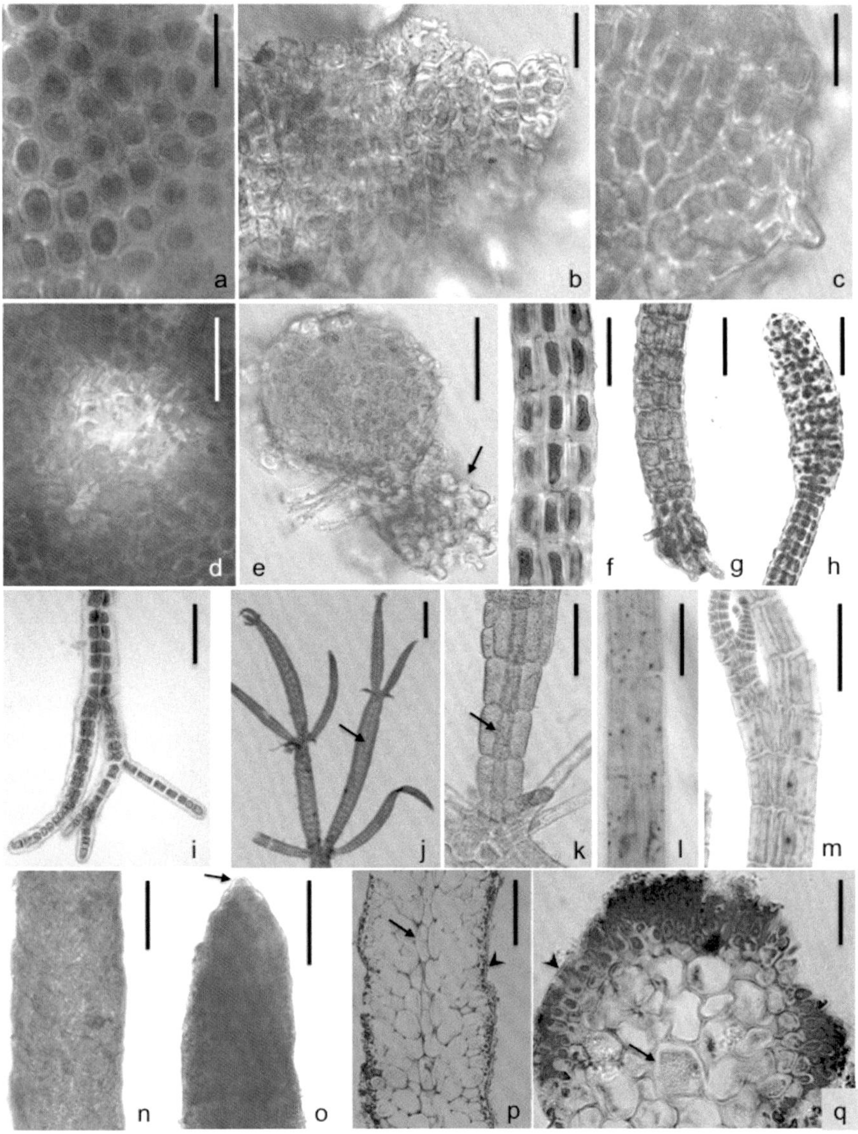

Fig. 4.5 (**a–e**) *Hildenbrandia*: (**a**) surface view of the crust; (**b**) erect filaments; (**c**) marginal part of the crust; (**d**) surface view of the crust with a gap formed by release of a gemma; (**e**) released gemma with germinating filaments (*arrow*). (**f–i**) *Bostrychia*: (**f**) uncorticated part with polysiphonous axis; (**g**) hapteron; (**h**) apical part with stichidium; (**i**) uniseriate terminal branches. (**j, k**) *Caloglossa*: leafy branches with prominent midribs (*arrow*); (**k**) young leafy branch with prominent midrib (*arrow*). (**l, m**) *Polysiphonia*: (**l**) mid part with polysiphonous axis; (**m**) apical part with polysiphonous and uniseriate axes. (**n–q**) *Sterrocladia*: (**n**) mid portion showing cortical cells; (**o**) apical portion showing apical cell (*arrow*); (**p**) longitudinal section showing axial cells (*arrow*) and cortical cells (*arrowhead*); (**r**) cross section showing axial cell (*arrow*) and cortical cells (*arrowhead*). Scale bars: (**j**) = 500 μm; (**h**) = 250 μm; (**f, g, i, k–p**) = 100 μm; (**d, q**) = 50 μm; (**e**) = 25 μm; (**a–c**) = 10 μm. Image authors: (**j**) M. Kamiya; (**k**) J. West

Remarks: six species have been reported in freshwater habitats (rivers, streams, and ponds; Kumano 1979, 2002). D'Lacoste and Ganesan (1987) described carpogonia, cystocarps and stichidia in *B. moritziana* from a freshwater population in Venezuela.

Lotic populations tend to occur in streams or rivers with relatively high temperatures (17–26 °C), neutral to alkaline (pH 7.0–8.4) and a wide range in specific conductivity (38–440 μS cm^{-1}) (Sheath et al. 1993; Necchi et al. 1999).

Caloglossa (Harvey) Martens (Fig. 4.5j, k)

Thalli reddish, dichotomously, rarely trichotomously branches, consisting of flat, articulate, narrow leafy segments; a prominent midrib is evident and is composed of a broad axial cell, surrounded by four periaxial cells, appearing like three rows of cylindrical or barrel-shaped cells in surface view. Outer portions of the leafy segments monostromatic, with oblique series of hexagonal cells. Rhizoids and new leafy segments arise at constrictions, either from the midrib area or peripheral layer of cells. Freshwater populations vegetative; reproductive structures unknown.

Remarks: Five species have been reported from freshwaters (Kumano 2002), but two species (*C. leprieurii* and *C. ogasawarensis*) are found most often. Current velocities are usually moderate (33–43 cm s^{-1}), temperature high (23–24 °C), pH alkaline (7.6–8.4) and specific conductivity moderate (100–200 μS cm^{-1}) in North America (Sheath et al. 1993).

Polysiphonia Greville (Fig. 4.5l, m)

Thalli dark reddish, consisting of erect, polysiphonous filaments with a single tier of pericentral cells around the axial cell. Freshwater collections with no additional layer of cortication that could be present in few marine species. Rhizoidal branches arise from pericentral cells. Delicately branched hairs (trichoblasts) formed in upper portions of the thalli. Freshwater populations vegetative; reproductive structures unknown.

Remarks: only one species (*P. subtilissima*) has been reported from stream habitats, from warm (22–26 °C), alkaline (pH 7.6–8.8) and high conductivity (1150–2700 μS cm^{-1}) waters (Sheath et al. 1993; Lam et al. 2013).

Class Florideophyceae—Order Gigartinales: Genus Sterrocladia

Sterrocladia F. Schmitz (Fig. 4.5n–q)

Thalli pseudoparenchymatous, cylindrical and irregularly branched, with main axes and branches of similar size. Apices rounded and slightly mucronated due to the occurrence of a prominent apical cell, which undergoes divisions in early stages to form axial and pericentral cells. Uniaxial structure evident in cross section, with the presence of an axial cell surrounded by two to three inner layers of large and irregular medullary cells and one, rarely two, outer layers of small cortical cells. Medulla entire, with all cells abutting in cross section and axial cell adherent to the outer cortex. One row of cylindrical to elliptical axial cells revealed in longitudinal section, surrounded by medullary and cortical cells. Small, irregularly shaped, polygonal and densely arranged cortical cells viewed in surface view. Reproductive structures developed in nemathecia, forming wart-like protuberances on the thallus surface. Nemathecia composed of short, branched filaments, producing terminal "sporangia"; it is not clear if these structures are sporangia or spermatangia, but their large size (16–20 μm long, 15–19 μm in diameter) and presence of chloroplasts suggest they are sporangia; in this case they could be asexual monosporangia, sexually related carposporangia or young (undivided) tetrasporangia.

Remarks: *Sterrocladia* is the only completely freshwater genus belonging to the order Gigartinales (Sherwood et al. 2012) with an *incertae sedis* status at the familial level. Two species were described in the genus: the type species *S. amnica* from Guyana and French Guyana and the recently described *S. belizeana*. Both are rarely reported in these areas.

Acknowledgments I am grateful to the financial support from the Brazilian agencies FAPESP and CNPq (Brazil) for the several grant projects received along the years that allowed me a much better knowledge of the red algae. I am thankful to the following people for kindly sharing their images to be used in this chapter: Marina Aboal, Ignacio Bárbara, Javier Carmona, Chris Carter, Iara Chapuis, Tim Entwisle, Mitsunobu Kamiya and John West. The help in image editing by Cauê Necchi is greatly appreciated.

References

Carmona JJ, Necchi O Jr. (2001a) A new species and expanded distributions of freshwater *Audouinella* (Acrochaetiaceae, Rhodophyta) from Central Mexico and southeastern Brazil. Eur J Phycol 36:217–226

Carmona JJ, Necchi O Jr. (2001b) Systematics and distribution of *Thorea* (Thoreaceae, Rhodophyta) in Central Mexico and southeastern Brazil. Phycol Res 49:231–240

D'Lacoste LGV, Ganesan EK (1987) Notes on Venezuelan freshwater algae—I. Nova Hedwigia 45:263–281

Eloranta P, Kwandrans J (2004) Indicator value of freshwater red algae in running waters for water quality assessment. Oceanol Hydrobiol Stud 33:47–54

Eloranta P, Kwandrans J (2012) Illustrated guidebook to common freshwater red algae. Polish Academy of Sciences, Kraków

Eloranta P, Kwandrans J, Kusel-Fetzmann E (2011) Rhodophyta and Phaeophyceae. Süßwasserflora von Mitteleuropa, Band 7. Spectrum Akademischer, Heidelberg

Entwisle TJ, Foard HJ (2007) Batrachospermales. In: Entwisle TJ, Skinner S, Lewis SH, Foard HJ (eds) Algae of Australia: Batrachospermales, Thoreales, Oedogoniales and Zygnemaceae. ABRS, Canberra, pp 1–25

Entwisle TJ, Sonneman JA, Lewis SH (1997) Freshwater algae in Australia. Sainty and Associates, Potts Point

Entwisle TJ, Vis ML, Chiasson WB et al (2009) Systematics of the Batrachospermales (Rhodophyta)—a synthesis. J Phycol 44:704–715

Ganesan EK, West JA, Zuccarello GC, de Goër S, Rout J (2015) *Lemanea manipurensis* sp. nov. (Batrachospermales), a freshwater red algal species from North-East India. Algae 30:1–13

Graham LE, Graham JM, Wilcox LW (2009) Algae, 2nd edn. Benjamin Cummings, San Francisco

Guiry MD, Guiry GM (2015) AlgaeBase. World-wide electronic publication, National University of Ireland, Galway, http://www.algaebase.org. Accessed 7 Dec 2015

Johnston ET, Lim PE, Buhari N et al (2014) Diversity of freshwater red algae (Rhodophyta) in Malaysia and Indonesia from morphological and molecular data. Phycologia 53:329–341

Kumano S (1979) Morphological study of nine taxa of *Bostrychia* (Rhodophyta) from southwestern Japan, Hong Kong and Guam. Micronesica 15:13–33

Kumano S (2002) Freshwater red algae of the world. Biopress, Bristol

Lam D, Entwisle TJ, Eloranta P et al (2012) Circumscription of species in the genus *Sirodotia* (Batrachospermales, Rhodophyta) based on molecular and morphological data. Eur J Phycol 47:42–50

Lam D, Garcia-Fernandez ME, Aboal MS et al (2013) *Polysiphonia subtilissima* (Ceramiales, Rhodophyta) from freshwater habitats in North America and Europe is confirmed as conspecific with marine collections. Phycologia 52:156–160

Lowe RL, LaLiberte GD (2007) Benthic stream algae: distribution and structure. In: Hauer FR, Lamberti GA (eds) Methods in stream ecology, 2nd edn. Academic, Amsterdam, pp 327–356

Müller KM, Lynch MDJ, Sheath RG (2010) Bangiophytes: from one class to six; where do we go from here? Moving the bangiophytes into the genomic age. In: Chapman DJ, Seckback J (eds) Red algae in the genomic age. Springer, New York, pp 241–259

Necchi O Jr. (1987) Sexual reproduction in *Thorea* Bory (Rhodophyta, Thoreaceae). Jpn J Phycol 35(106):112

Necchi O Jr. (1991) The section *Sirodotia* of *Batrachospermum* (Rhodophyta, Batrachospermaceae) in Brazil. Algol Stud 62:17–30

Necchi O Jr. (2005) Light-related photosynthetic characteristics of freshwater Rhodophyta. Aquat Bot 82:193–209

Necchi O Jr., Oliveira MC (2011) Phylogenetic affinities of 'Chantransia' stages in members of the Batrachospermales and Thoreales (Rhodophyta). J Phycol 46:680–686

Necchi O Jr., Vis ML (2012) Monograph of the genus *Kumanoa* (Rhodophyta, Batrachospermales). J Cramer, Stuttgart

Necchi O Jr., Zucchi MR (1995) Systematics and distribution of freshwater *Audouinella* (Acrochaetiaceae, Rhodophyta) in Brazil. Eur J Phycol 30:209–218

Necchi O Jr., Zucchi MR (1997) Taxonomy and distribution of *Thorea* (Thoreaceae, Rhodophyta) in Brazil. Algol Stud 84:84–90

Necchi O Jr., Zucchi MR (2001) Photosynthetic performance of freshwater Rhodophyta in response to temperature, irradiance, pH and diurnal rhythm. Phycol Res 49:305–318

Necchi O Jr., Sheath RG, Cole KM (1993a) Systematics of freshwater *Audouinella* (Rhodophyta, Acroachaetiaceae) in North America. 1. The reddish species. Algol Stud 70:11–28

Necchi O Jr., Sheath RG, Cole KM (1993b) Systematics of freshwater *Audouinella* (Rhodophyta, Acroachaetiaceae) in North America. 2. The bluish species. Algol Stud 71:13–21

Necchi O Jr., Sheath RG, Cole KM (1993c) Distribution and systematics of the freshwater genus *Sirodotia* (Batrachospermales, Rhodophyta) in North America. J Phycol 29:236–243

Necchi O Jr., Branco CCZ, Branco LHZ (1999) Distribution of Rhodophyta in streams from São Paulo State, southeastern Brazil. Arch Hydrobiol 147:73–89

Necchi O Jr., Garcia Fo AS, Salomaki ED et al (2013) Global sampling reveals low genetic diversity within the genus, *Compsopogon* (Compsopogonales, Rhodophyta). Eur J Phycol 48:152–162

Necchi O Jr., West JA, Rai SK, Ganesan EK, Rossignolo NL, Goër SV (2016) Phylogeny and morphology of the freshwater red alga *Nemalionopsis shawii* (Rhodophyta, Thoreales) from Nepal. Phycol Res 64(1):11–18

Paiano MO, Necchi O Jr. (2013) Phylogeography of the freshwater red alga *Sirodotia* (Batrachospermales, Rhodophyta) in Brazil. Phycol Res 61:249–255
Salomaki ED, Kwandrans J, Eloranta P et al (2014) Molecular and morphological evidence for *Sheathia* gen. nov. (Batrachospermales, Rhodophyta) and three new species. J Phycol 50:526–542
Saunders GW, Hommersand MH (2004) Assessing red algal supraordinal diversity and taxonomy in the context of contemporary systematic data. Am J Bot 91:1494–1507
Sheath RG, Hambrook JA (1990) Freshwater ecology. In: Cole KM, Sheath RG (eds) Biology of the red algae. Cambridge University Press, Cambridge, pp 423–453
Sheath RG, Sherwood AR (2002) Phyllum Rhodophyta (red algae). In: John DJ, Whitton BA, Brook AJ (eds) The freshwater algal flora of the British Isles. Cambridge University Press, Cambridge, pp 123–143
Sheath RG, Vis ML (2015) Red algae. In: Wehr JD, Sheath RG, Kociolek JP (eds) Freshwater algae of North America. Ecology and classification. Academic, San Diego, pp 237–264
Sheath RG, Vis ML, Cole KM (1993) Distribution and systematics of freshwater Ceramiales (Rhodophyta) in North America. J Phycol 29:108–117
Sheath RG, Whittick A, Cole KM (1994) *Rhododraparnaldia oregonica*, a new freshwater red algal genus and species intermediate between the Acrochaetiales and the Batrachospermales. Phycologia 33:1–7
Sherwood AR, Necchi O Jr., Carlike AL et al (2012) Characterization of a novel freshwater gigartinalean red alga from Belize, with a description of *Sterrocladia belizeana* sp. nov. (Rhodophyta). Phycologia 51:627–635
Shi Z (2006) Flora algarum Sinicarum aquae dulcis, Tomus 13: Rhodophyta, Phaeophyta. Science Press, Beijing
Skinner S, Entwisle TJ (2001) Non-marine algae of Australia: 3. *Audouinella* and *Balbiania* (Rhodophyta). Telopea 9:713–723
Sutherland JE, Lindstrom SC, Nelson WA et al (2011) A new look at an ancient order: generic revision of the Bangiales (Rhodophyta). J Phycol 47:1131–1151
Vis ML, Sheath RG (1992) Systematics of the freshwater red algal family Lemaneaceae in North America. Phycologia 31:64–179
Vis ML, Harper JT, Saunders GW (2007) Large subunit rDNA and rbcL gene sequence data place *Petrohua bernabei* gen. et sp. nov. in the Batrachospermales (Rhodophyta), but do not provide further resolution among taxa in this order. Phycol Res 55:103–112
Yoon HS, Müller KM, Sheath RG, Ott FD, Bhattacharya D (2006) Defining the major lineages of red algae (Rhodophyta). J Phycol 42:482–492
Zucchi MR, Necchi O Jr. (2003) Blue-greenish acrochaetioid algae in freshwater habitats are 'Chantransia' stages of Batrachospermales *sensu lato* (Rhodophyta). Cryptog Algol 24:117–131

Chapter 5
Diatoms (Bacillariophyta) in Rivers

Ana Luiza Burliga and J. Patrick Kociolek

Abstract Photosynthetic stramenopiles (heterokonts) constitute at least 11 distinct lineages, including some of the most abundant algae, particularly diatoms. Diatoms are unicellular algae that possess a cell wall composed of silicon dioxide often highly ornamented, made up of two parts. Diatoms are very diverse and common in streams and rivers, with a significant role in primary production. The benthic diatoms, in particular, have developed successful strategies to establish and maintain their position in lotic environments. Fifty-nine most common diatom genera from lotic systems are described, including illustrations and a key to genera.

Keywords Bacillariophyceae • Diatoms • Genera • Morphology • Rivers • Taxonomy

Introduction

Background Information

Diatoms are microscopic unicellular organisms with chlorophylls *a* and *c*, and significant xanthophyll and carotenoid pigments. Diatoms are included in Stramenopiles, a diverse group that includes, for example, non-photosynthetic colorless aquatic fungi as well as brown algae (Phaeophyceae) and Synurophyceae. Stramenopiles are characterized by the presence of flagellated cells (vegetative cells, zoospores or gametes) that have two types of flagella: a flagellum with rows of stiff, tripartite hairs (which reverse the flow around the flagellum so that the cell is dragged forward rather than pushed along), and a second shorter, smooth flagellum (de Reviers 2003; Baldauf 2008).

The diatom group is frequently dominant in aquatic ecosystems and through their photosynthetic activity is responsible for significant carbon fixation globally

A.L. Burliga (✉)
Rhithron Associates, Inc., 33 Fort Missoula Road, Missoula, MT 59804, USA
e-mail: amiranda@rhithron.com

J.P. Kociolek
Museum of Natural History and Department of Ecology and Evolutionary Biology,
University of Colorado, Boulder, CO, USA

O. Necchi Jr. (ed.), *River Algae*, DOI 10.1007/978-3-319-31984-1_5

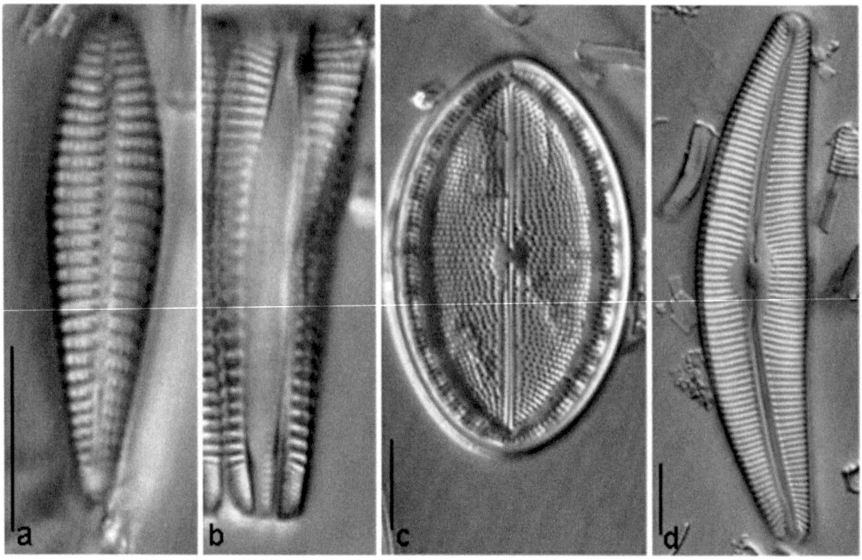

Fig. 5.1 (**A** , **b**)Pennate diatom *Rhoicosphenia*: (**a**) Valve view. (**b**) Girdle view. (**c**) Monoraphid raphe system in *Cocconeis*. (**d**) Asymmetrical biraphid system in *Cymbella*. Scale bars = 10 μm

(Falkowski et al. 2000). In streams and rivers, diatoms play significant roles in primary production (Fry and Wainright 1991; Kelly et al. 1998a), as well as silica cycling (Grady et al. 2007). Diatoms are valuable indicators of environmental conditions in rivers and streams since they respond directly and sensitively to specific physical, chemical, and biological changes (Stevenson et al. 2010; Hill et al. 2001). A review of the study of diatoms and river pollution is offered by Rimet (2012) and Lobo (see Chap. 11).

Diatoms are the most diverse groups of eukaryotic microorganisms and probably contain well in excess of 100,000 species (Mann and Droop 1996). Diatoms have the unique feature of cell walls comprised of opaline silica. These organisms remove dissolved silica and construct their cells walls from this material. The cell walls are put together in two pieces, one slightly larger than the other, much like a pillbox or Petri dish. The cell is termed a *frustule*, and each of the two pieces of the bipartite cell wall is a *valve*. The slightly larger valve is termed the *epivalve* and the slightly smaller valve is the *hypovalve* (Fig. 5.1a, b). Between the two valves a series of lightly silicified bands occur, known as *girdle bands* (the entire combination of bands is termed the *cingulum*).

Diatom Structure and Features of Taxonomic Importance

Cell walls of diatoms can be viewed using light or electron microscopy from a variety of perspectives or orientations. Much of the taxonomy of most diatom species is based on viewing the cell in *valve view*, where the face of the valve is evident

(Fig. 5.1a). In some forms, due to their structure, the most commonly viewed orientation is *girdle view* where both the side of the valve (the *mantle*) as well as the girdle bands are evident (Fig. 5.1b).

A general, commonly referenced feature of diatom valves is symmetry, and diatom classification in part, from the highest, more general levels to species-level differences have been based on this feature. A useful distinction between groups of diatoms is whether they are radial or bilaterally symmetrical in valve view. Müller's (1895) study of symmetry types within the *pennate* diatoms (diatoms generally with bilateral symmetry) is still useful today, and many classification schemes still refer to groups based on symmetry.

Since diatoms "live in glass houses" there must be ways for them to take up nutrients, release metabolic wastes, and physically keep invaders at bay. So, the cell walls of diatoms are perforated by small holes (termed *punctae* or *areolae*) that in turn may be covered by sieve plates (in a variety of forms, sometimes on the inside or outside of the cell wall) with minute holes that cover the punctae or areolae. Areolae are aligned into *striae*, and the structure and patterns of the areolae and striae help with broad- and fine-scale taxonomic designations. The holes are arranged in diverse patterns (along with symmetry contributing to the aesthetic beauty of diatoms). The structure of the valves keeps diatoms lightweight but also offers a high degree of structural integrity.

Other perforations through the glass cell walls include smaller holes arranged in higher densities than areolae where mucopolysaccharides are exuded. Generally termed "pore fields" (though by tradition the terminology for these pores may include *apical pore fields*, *ocellulimbus*, *ocellus*, *pseudocellus*). A *raphe* is a slit through the siliceous cell wall, either continuous or in two parts, on one or both valves, that is found in a particular (large) group of diatoms. The *raphid* diatoms can move over long distances quickly relative to their body size. This allows them to micro-position themselves in sediments or among substrates in more quiescent waters. Some examples of diatom raphes are shown in Fig. 5.1c, d.

Strategies in Running Waters

In aquatic environments, *flow* is the primary factor that distinguishes streams from lakes. The constant mixing in streams eliminate all vertical gradients, except for light (Lampert and Sommer 2007). In this environment, the establishment of benthic algae is the outcome of a complex series of interactions. The relative importance of proximate factors, i.e., variables directly controlling accrual and loss, such as resources (e.g., light, N, P) and stressors (e.g., pH, temperature, current velocity, toxic substances), largely reflects higher-scale environmental features (climate, geology, land use) (Biggs 1996; Stevenson 1997).

The unidirectionality of stream flow is a very powerful selective factor, to which stream organisms must be adapted. For small organisms it means that, unless they can attach themselves to a secure structure, they will be transported downstream,

Table 5.1 Diatom attachment strategies and basic characteristics

Attachment structures	Characteristics of attachment
Adhesive pads	Small EPS globular structures attaching cells to other cells or to a substratum
Adhering film/ biofilm	Surface-attached and matrix-enclosed microbial communities secreting a matrix of mucilaginous extracellular polymers (EPS)
Cell coatings	Including both cell walls (siliceous frustules with an organic component)
Fibrils	Highly organized crystalline threads developing from the valve periphery
Stalks	Unidirectional deposited, multilayered EPS structures/filament attaching cells to substrata
Tubes	Structured EPS pseudo-filaments enclosing cells

with no possibility of being carried back to their original location simply by chance (Lampert and Sommer 2007).

While many diatoms are planktonic, floating or drifting near the surface of the water, most of the diatoms considered here are benthic, growing on a variety of substrates in the flow of rivers and streams. But how do benthic algae maintain their position on substrates in lotic environments? Benthic diatoms in particular may produce extracellular polymeric substances (EPS). The EPS components encompass a wide variety of morphological forms, which range from highly crystalline, rigid fibrils to highly hydrated, mucilaginous capsules (Hoagland et al. 1993). These morphological structures are referred to as pads, stalks, tubes, adhering films, fibrils, and cell coatings (Hoagland et al. 1993; Wang et al. 2014) (Table 5.1). EPS is formed from the plasmalemma, consists primarily of carbohydrates and is extruded outside the cell (Bahulikar and Kroth 2007). The raphe and the apical pore field (of pennate diatoms) are the sites that secrete EPS (Hoagland et al. 1993). Although not documented, a combination of EPS and silica-forming fibrils may be a regular component of diatom attachment strategies (Wang et al. 2014).

In faster flowing waters, some diatoms grow adnate to the substrate using adhering films, where their cells are attached directly to stones, algal filaments or sand grains. Examples include *Cocconeis* (Fig. 5.2a), *Epithemia* and *Staurosirella*. EPS strategies are also used to maintain the integrity of colonies for cell-to-cell attachments (by mucilage pads or intercellular mucilaginous layers). Examples include *Pleurosira* and *Terpsinoe* which may form zig-zag colonies (Fig. 5.2b, c) and *Melosira* and *Achnanthes* which form straight chains (Fig. 5.2d).

Some genera such as *Ulnaria*, *Eunotia*, and *Achnanthidium* grow at the ends of pads or short peduncles, while *Gomphonema* and *Cymbella* grow at the ends of long stalks (Fig. 5.2e–h).

An excessive extracellular stalk production has been associated with *Didymosphenia geminata* blooms (Spaulding and Elwell 2007). A study by Bothwell et al. (2014) indicates that the proximate cause of blooms or more precisely, excessive extracellular stalk production is low P in the water. The stalk production in

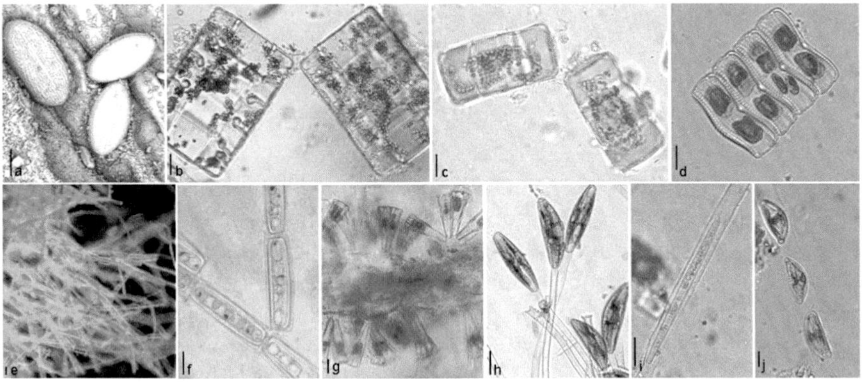

Fig. 5.2 (**a**) Examples of EPS adhering film in *Cocconeis*. (**b**, **c**) Adhesive pads at the corners between cells in: (**b**) *Terpsinoë;* (**c**) *Pleurosira*. (**d**) Cell coating in *Achnanthes*. (**e**, **f**) Adhesive pads in *Eunotia*. (**g**, **h**) stalks in: (**g**) *Gomphonema*; (**h**) *Cymbella*. (**i**, **j**) Tubes in: (**i**) *Nitzschia*; (**j**) *Encyonema*. Scale bars = 10 μm

response to very low P may be a strategy to move cells out of the benthic boundary layer and into the water column where there may be greater delivery of growth-limiting P (Bothwell et al. 2014).

In other situations, diatom colonies may contain many cells within a mucilaginous tube. In this tube, cells are arranged in a row and are capable of moving within the tube (Cox 1981). A curious incidence of algal species coexistence within diatom mucilage tubes has been described (Carr and Hergenrader 1987). Common genera using this strategy are *Nitzschia* and *Encyonema* (Fig. 5.2i, j).

Collecting and Processing Diatoms for Study

The study of diatoms has many applications, including water quality analysis. Programs utilizing diatoms in this way may be local, statewide, regional, national or across national boundaries. Application of diatoms to questions regarding the status and trends in water quality in streams and rivers (as well as other water bodies) have led to well-documented methods for the collection of samples, for both qualitative and quantitative analysis, as well as processing diatom samples to make permanent slides and the storage of cleaned material. Some of these protocols include those of Kelly et al. (1998b), Stevenson and Bahls (1999), Taylor et al. (2007), and Karthick et al. (2010). Permanent slides and "cleaned" material make an important archive for diatom work, allowing samples to be revisited for decades or centuries for reasons ranging from species descriptions, quality assurance/quality control and reassessment of population data.

Classification Followed in this Chapter

The most-used classification system for diatoms is still the one proposed by Round et al. (1990) some 25 years ago. Although this classification system does not align with our current understanding of diatom phylogeny (some of the major groups proposed and applied by Round et al. are understood to be non-monophyletic), the lack of a synthesis of the implications of our evolving understanding of diatom relationships has not resulted in major updates to the classification of Round et al. Therefore, the system we utilize here for classification is more or less aligned with the approach by Round et al. (1990), acknowledging the disconnect with our evolving understanding of diatom phylogenetic relationships.

Taxonomic key to the common diatom genera from rivers

1a	Valves radially symmetrical	2
1b	Valves bilaterally symmetrical	7
2a	Valves circular in valve view, with elongated mantle; perforations not visible with LM	3
2b	Valves some other shape in valve view, perforations evident in LM	4
3a	Valves without evident ornamentation; no interdigitating spines between valves	*Melosira*
3b	Valves with evident ornamentation; striae evident on valve mantle. Spines evident between opposing valve faces	*Aulacoseira*
4a	Valves nearly oblong to nearly circular in valve view; two (rarely) to three ocelli present on opposite sides of the valve	*Pleurosira*
4b	Valves round, hexagonal to undulate, not rounded	5
5a	Valves six-sided, with pseudocelli on three of the sides	*Hydrosera*
5b	Valves round or elongated in valve view	6
6a	Valves elongated, undulate in outline; pseudocelli at the apices	*Terpsinoe*
6b	Valves round, with two patterns of ornamentation obvious	*Cyclotella*
7a	Valves without a raphe system (sometimes a raphe may occur mostly on the valve mantle, with a short segment on the valve face)	8
7b	Valves with a raphe system	19
8a	Valves with thickened ribs (costae) on the valve face	9
8b	Valves without costae	10
9a	Valves symmetrical to the apical and transapical axes	*Diatoma*
9b	Valves symmetrical to the apical axis, asymmetrical to the transapical axis	*Meridion*
10a	Frustules with siliceous laminae on the girdle bands (septa); valves bone shaped	*Tabellaria*
10b	Frustules without septa	11
11a	Valves asymmetrical to the apical axis (except *Synedra cyclopum*)	*Hannaea*
11b	Valves symmetrical to both axis	12
12a	Two labiate processes (rimoportulae) per valve; many cells with two spines at each apex	*Ulnaria*
12b	Rimportula single, or lacking, per valve	13

(continued)

(continued)

13a	Frustules in colonies, formed valve face to valve face, connected by interconnected spines	*Fragilaria*
13b	Frustules otherwise	14
14a	Valves elongate, with very wide central sternum	*Tabularia*
14b	Valves elliptical, to rounded, if elongate without a wide central sternum	15
15a	Rimoportulae absent	16
15b	this Rimoportulae	17
16a	Striae wide, composed of lineate areolae	*Staurosirella*
16b	Striae more narrow, composed of multiseriate rows of puncta	*Punctastriata*
17a	Sternum very narrow, barely discernable	*Fragilariforma*
17b	Sternum less narrow, more easily visible	18
18a	Valves with round areolae that are regularly spaces	*Staurosira*
18b	Valves with a few, transapically elongate areolae	*Pseudostaurosira*
19a	Raphe predominantly on the valve mantle, with only the distal raphe ends present on the valve face; raphe not contained in a canal	20
19b	Raphe on the face, or if on the mantle raphe contained in a canal	21
20a	Valves asymmetrical about the apical axis, symmetrical about the transapical axis, rectangular in girdle view	*Eunotia*
20b	Valves asymmetrical about the apical and transapical axes, wedge shaped in girdle view	*Actinella*
21a	Frustules with a raphe on one valve only	22
21b	Frustules with raphe on both valves	26
22a	Valves elliptical in outline, may be flexed about the apical axis, raphe-bearing valve bordered near the margin by a hyaline area	*Cocconeis*
22b	Valves more linear, or if elliptical, not bearing a hyaline border on the raphe valve; frustules straight or flexed about the transapical axis	23
23a	Valve without raphe bearing an unornamented, horseshoe-shaped area	*Planothidium*
23b	Valves without horseshoe-shaped unornamented area	24
24a	Valves with distinctly punctate striae, in some species the striae are interrupted longitudinally on the rapheless valve; valves not flexed	*Karayevia*
24b	Valves otherwise; valves usually flexed about the transapical axis	25
25a	Valves usually highly flexed about the transapical axis; distal raphe fissures straight or turned to the same side	*Achnanthidium*
25b	Valves not usually highly flexed; distal raphe fissures straight or turned in opposite directions	*Psammothidium*
26a	Raphe branches straight, evident in the center on the valve face; valve symmetrical to both axes, or asymmetrical to either the apical or transapical axis	27
26b	Raphe not evident in the center or not straight; raphe contained in a canal	54
27a	Valves symmetrical to both the apical and transapical axis	28
27b	Valves asymmetrical to either the apical or transapical axis	44

(continued)

(continued)

28a	Valves and raphe system sigmoid	29
28b	Valves and raphe system not sigmoid	30
29a	Valves with striae forming two different patterns of ornamentation: parallel and perpendicular to the raphe system	*Gyrosigma*
29b	Valves with striae forming three different patterns of ornamentation: parallel, perpendicular and oblique to the raphe system	*Pleurosigma*
30a	Valvocopulae with rectangular chambers or locules present	*Mastogloia*
30b	Valvocopulae without locules	31
31a	Striae costate, without evident areolae comprising the striae, with raphe filiform to complex, distal raphe ends more or less bayonet shaped	*Pinnularia*
31b	Striae with more or less evident areolae comprising the striae	32
32a	Valves with a ribs bordering the raphe, ending with ribs fused to the raphe terminus	33
32b	Valves with or without margin canals, but without ribs and nothing fused to raphe terminus	34
33a	Raphe short, appearing as 'needle eyes' near the ends of the valve; striae difficult to discern	*Amphipleura*
33b	Raphe elongate, raphe fused with ribs more evident, forming a porte crayon	*Frustulia*
34a	Central area an expanded, thickened fascia reaching to the margins	*Stauroneis*
34b	Cells may have an expanded central area, but not a thickened fascia	35
35a	Longitudinal canals running near the margins; external proximal raphe ends usually deflected in opposite directions	*Neidium*
35b	Cells without longitudinal canals and proximal raphe ends not usually deflected	36
36a	Valves without evident striae, or only barely visible around the center; external distal raphe ends curved in the same direction	*Kobayasiella*
36b	Valves with evident striae	37
37a	Longitudinal hyaline area (conopeum) bordering axial area, with apical area delineated by "T"-shaped thickenings	*Sellaphora*
37b	Valves without conopeum and apical thickenings	38
38a	Central area bearing a stigma, cells mostly small and distinctly punctate	*Luticola*
38b	Central area without a stigma	39
39a	Striae interrupted at the valve terminus; central area may bear an isolated punctum	*Geissleria*
39b	Valves without interruptions to the striae at the apices	40
40a	Valves with line-like (lineolate) striae	*Navicula*
40b	Valves with areolae comprising striae more rounded or dash-like	41
41a	Dash-like striae appearing interrupted	*Brachysira*
41b	Striae otherwise	42
42a	Striae composed of rounded to elliptical areolae, not appearing interrupted	*Placoneis*
42b	Striae otherwise	43
43a	Valves asymmetrical to the transapical axis	44
43b	Valves asymmetrical to the apical axis	48

(continued)

(continued)

44a	Cells large, with multiple stigmata in the central area, footpole with apical pore field but not bisected by the raphe; septa and pseudosepta absent	*Didymosphenia*
44b	Cells otherwise	45
45a	Cells flexed about transapical plane	46
45b	Cells not flexed about transapical plane	47
46a	Septa and pseudosepta present, valves with raphe of differing sizes	*Rhoicosphenia*
46b	Septa and pseudosepta absent, valves with raphe of same sizes	*Gomphosphenia*
47a	Valves with singly-punctate striae, without a stigma or with one stigma only, longitudinal lines absent; apical pore fields differentiated from areolae	*Gomphonema*
47b	Valves with doubly punctate striae, without stigma or with four stigmoids and longitudinal lines absent and pore fields undifferentiated OR one stigma present, with longitudinal lines and pore fields differentiated from striae	*Gomphoneis*
48a	Valves with amphoroid symmetry, dorsal mantle height much higher than ventral mantle height	49
48b	Valves with dorsal mantle heights the same or slightly different	50
49a	Valves without a dorsal ledge bordering raphe; hyaline central area present on dorsal side of axial area	*Amphora*
49b	Dorsal ledge bordering raphe; hyaline central area wanting on dorsal side of axial area	*Halamphora*
50a	Valves with apical pore fields present; stigmata ventral; external distal raphe ends deflected dorsally	*Cymbella*
50b	Valves otherwise	51
51a	Valves with dorsal stigma, or stigma wanting; external distal raphe ends deflected ventrally	*Encyonema*
51b	Valves with a different combination of features	52
52a	Cells small, stigma central, with central portion of ventral margin unornamented and tumid; external distal raphe ends deflected ventrally	*Reimeria*
52b	Valves without a tumid ventral margin	53
53a	Valves larger, asymmetry distinct, usually with protracted apices	*Cymbopleura*
53b	Valves usually smaller, asymmetry indistinct	*Encyonopsis*
54a	Raphe positioned on one side of the valve	55
54b	Raphe positioned around the periphery of the valve, biarcuate in shape, or valve in the shape of a canoe	57
55a	Valve with riblike struts (fibulae) extending across the entire face of the valve	*Denticula*
55b	Valve with fibulae not extending across the valve face; usually positioned near the raphe only	56
56a	Longitudinal fold appearing as unornamented area on valve face	*Tryblionella*
56b	Longitudinal fold not present	*Nitzschia*
57a	Raphe located around the periphery of the valve	*Surirella*
57b	Valve asymmetrical to the apical axis and raphe biarcuate or valve shaped like a canoe	58
58a	Raphe biarcuate in shape	*Epithemia*
58b	Valve canoe shaped	*Rhopalodia*

Descriptions of Common Genera of Diatoms in Rivers

Phyllum Bacillariophyta

Class Coscinodiscophyceae, Subclass Coscinodiscophycidae: Order Aulacoseirales, Family Aulacoseiraceae

Aulacoseira Thwaites (Fig. 5.3a, b)

Valves round in valve view, with areolae distributed across the valve face or restricted to the margin. Valve mantle expanded, and possessing curved or straight rows of striae. Unornamented area present on the mantle, and internally this area is thickened (known as a ringleiste). Each valve is connected to an opposing valve by spines. In some cases the spines are few, elongated and they fit into grooves on the opposing valve mantle. In other taxa the species are shaped such as to interdigitate with those on the opposite valve. The connected valves can form long filaments.

Remarks: there are many species in this genus, and the ecological breadth of *Aulacoseira* is considerable. English and Potapova (2009) and Kociolek et al. (2014) described several new species recently. Species may be found in oligotrophic to eutrophic conditions, and some species occur in quite low pH environments (Camburn and Kingston 1986). Many species now in *Aulacoseira* were previously treated in the genus *Melosira*.

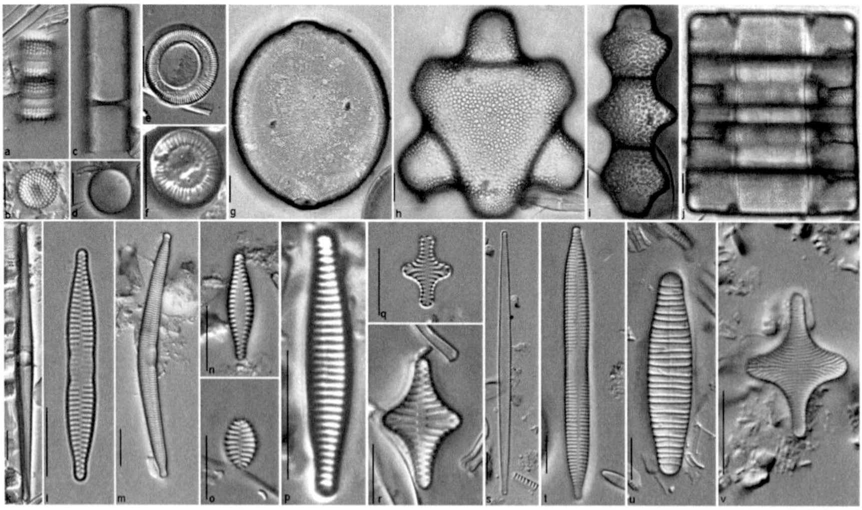

Fig. 5.3 (**A** , **b**)*Aulacoseira*. (**c**, **d**) *Melosira*. (**e**, **f**) *Cyclotella*. (**g**) *Pleurosira*. (**h**) *Hydrosera*. (**i**, **j**) *Terpsinoë*. (**k**) *Ctenophora*. (**l**) *Fragilaria*. (**m**) *Hannaea*. (**n**) *Pseudostaurosira*. (**o**) *Punctastriata*. (**p**) *Stauroforma*. (**q**) *Staurosira*. (**r**) *Staurosirella*. (**s**) *Tabularia*. (**t**) *Ulnaria*. (**u**) *Diatoma*. (**v**) *Fragilariforma*. LM, Scale bars = 10 μm

Order Melosirales, Family Melosiraceae

Melosira C.A. Agardh (Fig. 5.3c, d)

Frustules cylindrical. Cells grow end to end, forming filamentous. Valves round in valve view, with elongated valve mantles so that cells appear predominately in girdle view. Valve face and mantle with very fine puncta, not resolvable with light microscopy. No marginal spines present; colonies held together by mucilage. Colonies break up upon processing for permanent slides.

Remarks: in older literature many species assigned to this genus were removed by Simonsen (1979) and placed in the genus *Aulacoseira* G.H.K. Thwaites, a genus of mostly planktonic and/or lentic environments. The most common species of *Melosira* is *M. varians* C.A. Agardh, which is cosmopolitan and found in slower moving, usually more nutrient-rich streams.

Order Thalassiosirales, Family Thalassiosiraceae

Cyclotella Brebisson ex Kützing (Fig. 5.3e, f)

The traditional diagnosis for *Cyclotella* is that valves are mostly circular in valve view, with the valve face having two different patterns of ornamentation (Lowe 1975). Many species were placed in this broad concept of the genus, and many have been removed to other, newly-created genera such as *Discostella*, *Handmannia*, *Tertiarius*, and *Puncticulata* (Khursevich and Kociolek 2012). Remaining in *Cyclotella* are those species that have a central area without areolae, the center being formed by a solid plate of silica. Included in this “strict” concept of *Cyclotella* is *C. meneghiniana* Kützing, *C. atomus* Hustedt, and *C. gamma* Sovereign, among others.

Remarks: *Cyclotella* species may be free-floating, single cells or cells may be attached to one another by fine chitinous strands at the center of the valve. Cells may also be benthic, attached to sand grains or other substrates. Recent treatments of the genus are provided in Håkansson (2002), Tanaka (2007) and Kiss et al. (2012).

Subclass Biddulphiophycidae: Order Triceratiales, Family Triceratiaceae

Pleurosira (Meneghini) Trevisan San Leon (Fig. 5.3g)

Frustules occur in straight or zig-zag colonies, held together by mucilage pads exuded from ocelli connecting adjoining cells. Frustules are cylindrical, with valve faces round to broadly elliptical. Valves show radiating striae of unequal lengths, without hyaline ribs or distinct sectors. Two, sometimes three, ocelli occur opposite each other, located on the margin of the valve face. In the center, 0–4 rimportulae are present.

Remarks: A single species, *P. laevis* (C.G. Ehrenberg) P. Compere, occurs in many river systems, particularly wider stretches that have elevated NaCl and nutrient levels. Kociolek et al. (1983) described the morphology of this species from the USA. Karthick and Kociolek (2011) described a new species from rivers in India. In Hawaiian mountain streams this genus can be found at higher elevations.

Order Biddulphiales, Family Biddulphiaceae

Hydrosera G.C. Wallich (Fig. 5.3h)

Frustules form straight chains, formed by mucilage secreted from three pseudocelli, which occur on termini of the valve. In girdle view, frustules are quadrate to nearly cylindrical. In valve view the valve outline is hexagonal. Valves have variously shaped large areolae, giving the valve a rugose appearance. A single rimoportula is present near the center of the valve. Pseudosepta are present near the bases of the termini bearing pseudocelli.

Remarks: two species, *H. whampoensis* (Schwartz) Deby and *H. triquetra* Wallich, are reported mainly from the larger sections of tropical rivers as well as in marine environments (Simonsen 1965). The genus occurs in mountain streams on the Hawaiian Islands (Fungladda et al. 1983; Julius 2007). Qi et al. (1984) reported on the valve morphology of the genus.

Terpsinoë C.G. Ehrenberg (Fig. 5.3i, j)

Frustules form zig-zag colonies, held together by mucilage pads secreted from pseudocelli at the apices of the valve. Frustules are quadrate in girdle view, triundulate in valve view. Areolae on the valve face are rather large, giving the valve face a rugose appearance. A single rimoportula is present near the center of the valve. Pseudosepta are present internally at the points of undulation, and smaller ones may also be present. These small pseudosepta appears in shape of musical notes in girdle view of *T. musica* Ehrenberg, a species whose morphology was studied by Luttenton et al. (1986).

Remarks: usually found near the mouths of rivers, this genus is also known from higher elevations in the mountain streams of Hawaii (Fungladda et al. 1983).

Class Fragilariophyceae: Order Fragilariales, Family Fragilariaceae

Ctenophora (A. Grunow) D.M. Williams and F.E. Round (Fig. 5.3k)

Valves linear-lanceolate, with rounded apices. Central sternum narrow, straight. Striae opposite, composed of rounded to nearly rectangular areolae. Central area prominent, an expanded, thickened fascia that appears rounded to rectangular in shape, sometimes with weakly developed ("ghost") striae. A rimoportula is evident at each apex. Apical pore fields at both of the ocellimbus type.

Remarks: included in this distinctive genus *is C. pulchella* (Ralfs ex Kützing) Williams and Round, a species found in slow-flowing waters with elevated conductance.

Fragilaria H.C. Lyngbye (Fig. 5.3l)

Cells usually forming long, band-shaped colonies with valves face to face, adjoined by interdigitating spines. Valves symmetrical about both the apical and transapical axes. Valves lanceolate to elliptical. Central sternum narrow and striae parallel. Apical pore fields are of the ocellimbus type, positioned on the valve mantle at the apices. One rimportula per valve, located near the apex.

Remarks: The long history of the genus, and its interpretations, is summarized briefly by Kociolek et al. (2015). Williams and Round (1987) is an important reference to the current understanding of the genus. While some species are planktonic, others occur in the benthos of streams and rivers. Many species respond positively to increased nutrient concentrations (Patrick and Reimer 1966).

Hannaea R.M. Patrick in R.M. Patrick and C.W. Reimer(Fig. 5.3m)

Cells occurring individually, or linked together along their long axis to form arched, basket-like colonies. Valves asymmetrical about the apical axis, symmetrical about the transapical axis. The central sternum is evident, and in the center one side (the concave or ventral margin) is swollen and lacks typical striae. Labiate processes present near each capitate apex.

Remarks: *Hannaea arcus* (Ehrenberg) Patrick in Patrick and Reimer (1966) is the often-cited species, common in mountain streams in the northern hemisphere. This species was assigned to the genus *Ceratoneis* Ehrenberg prior to Patrick's (Patrick and Reimer 1966) proposal of *Hannaea* (in honor of the diatomist G. Dallas Hanna).

Pseudostaurosira D.M. Williams and F.E. Round (Fig. 5.3n)

Cells forming straight-chain colonies, valves face to face, connected by interdigitating spines. Valves with short striae, comprised of one to a few elongate areolae. Central sternum expanded. Rimoportulae wanting. Cingulum with many bands.

Remarks: *Pseudostaurosira brevistriata* (Grunow) Williams and Round is a common species in this genus, which was separated from *Fragilaria* by Williams and Round (1987). Species are commonly benthic, attached to sand grains and stones in slower flowing waters. Kociolek et al. (2014) and Morales (2001) illustrated species of this genus.

Punctastriata D.M. Williams and F.E. Round (Fig. 5.3o)

Cells single or forming short filaments, straight or branched. Valves small, elliptical, symmetrical to both axes or asymmetrical to the transapical axis. Central sternum narrow. Striae wide, alternate, formed of fine areolae in three to five rows. Striae extend onto the mantle. Marginal spines wide and positioned over the striae. Rimportulae wanting. Apical pore field present at one pole, indistinct.

Remarks: the genus includes *P. lancettula* (Schumann) Hamilton and Siver and its relatives (Siver et al. 2005; Siver and Hamilton 2011).

Stauroforma R.J. Flower, V.J. Jones and F.E. Round (Fig. 5.3p)

Cells forming straight-chain colonies, connected by marginal spines extended from the striae. Valves symmetrical to the apical and transapical axes. Striae are continuous across the valve face, interrupted by a very narrow, sometimes indistinct central sternum. Rimportulae are absent. Cingulum with a few bands; valvocopula broader than the rest of the cingulum elements.

Remarks: species assigned to this genus were separated out from *Fragilaria*, based on the structure of the striae and lack of a rimoportula. The genus is found in acid habitats, including bogs and other lentic environments as well as streams.

Staurosira C.G. Ehrenberg (Fig. 5.3q)

Cells occurring singly or in long filamentous colonies, attached by interdigitating spines. Valves elliptical to cross-shaped, with wide striae. Striae areolae are rounded (not lineate). Apical pore fields represented by a small group of rounded pores. Rimportula absent. Marginal spines evident, forked or spathulate, originating from the ribs between striae, Copulae numerous.

Remarks: cells are free-living, or attached to sand grains or other substrates, in larger streams and rivers. *Staurosira construens* Ehrenberg and *S. elliptica* (Schumann) Williams and Round are commonly reported species of this genus.

Staurosirella D.M. Williams and F.E. Round (Fig. 5.3r)

Cells single or attached in straight or zig-zag chains. Valves symmetrical to both axes, or asymmetrical to the transapical axis. Valves linear to cross shaped. Central sternum relatively wide. Striae broad, formed of lineate areolae. Marginal spines forked or branched, or absent in some taxa. Apical pore field of the ocellulimbus type, present on one apex. Rimoportulae wanting.

Remarks: included in this circumscription of *Staurosirella* are species previously assigned to *Martyana* and *Fragilaria*.

Tabularia (F.T. Kützing) D.M. Williams (Fig. 5.3s)

Cells single, or in sprays with cells attached to substrate at a common point via a common mucilage pad. Valves elongate-lanceolate, symmetrical to the apical and transapical axes, with a wide, distinct central sternum. A pore field of the ocellulimbus type occurs at each apex on the mantle.

Remarks: this genus occurs in streams near coasts, in areas of elevated salt and nutrient concentrations (Kociolek 2012). Snoeijs(1992) studied type material of the genus.

Ulnaria (F.T. Kützing) P. Compere (Fig. 5.3t)

Cells occur singly, in short, linear filaments or as a large spray of individuals attached to the substrate at a single point. Valves small to large (some species more than 300 μm long), linear to linear-lanceolate in shape. Central sternum straight, narrow. Striae usually uniseriate, opposite (rarely offset). Central area usually evident, expanded to one or both sides, and in many species containing weakly recognizable striae ("ghost striae"). A rimoportula occurs at both apices. Apical pore fields present on the mantle of both apices, structure of the ocellulimbus type.

Remarks: many species formerly assigned to *Synedra* are now referred to *Ulnaria*, including the often-reported *U. ulna* (Nitzsch) P. accomplice. Some species are long and narrow, and occur in the plankton. Others are found frequently at the lower reached of streams and rivers.

Order Tabellariales, Family Diatomaceae

Diatoma J.B.M. Bory de Saint-Vincent (Fig. 5.3u)

Frustules forming straight or zig-zag colonies, held together by mucilage exuded from apical pore fields at both poles. Upon cleaning, the colonies are not evident. Valves symmetrical to both the apical and transapical axes, with robust siliceous ribs extending partially or fully across the valve face. A single, usually obvious, rimoportule occurs near one apex. Central sternum more or less evident. While in the LM there is the appearance of striae, in fact striae densities are beyond the limits of LM resolution. Cingulum usually with many bands.

Remarks: Williams (1985) applied cladistic analysis to circumscribe the genus based on monophyly, and his revision is the most recent for this genus and *Meridion*. Kociolek and Lowe (1983) described the ultrastructure of one species from streams in Great Smoky Mountains National Park (USA). The genus can form evident filaments in cold-water streams and seeps, and many species are reported as cosmopolitan in their distributions.

Fragilariforma (J. Ralfs) D.M. Williams and F.E. Round (Fig. 5.3v)

Cells forming long, straight or zig-zag colonies. Valves elliptical to linear-lanceolate, with narrow, indistinct central sternum. Striae are composed of uniseriate rows of rounded areolae. Externally, the appear to cross the central sternum. Apical pore fields comprised of a group of pores, located at the apices and extending on to the mantle. One rimportula occurs near the apex. Marginal spines evident. Girdle bands numerous.

Meridion C.A. Agardh (Fig. 5.4a)

Frustules forming fan-shaped colonies, cells wedge shaped in girdle view. Valves symmetrical to the apical axis, asymmetrical to the transapical axis. Valves have robust costae running fully or partially across the valve face. A single rimportula is evident at the headpole. Central sternum evident. Striae are not visible in the LM, although reports of striae counts exist for taxa in the genus (e.g. Patrick and Reimer 1966). Members of the genus can form *Innenschalen* (internal valves) that are thought to function like spores (Geitler 1971). Kociolek et al. (2011) documented the differences in valve morphology between vegetative and *Innenschalen* valves in populations of the most commonly cited species, *M. circulare* (Greville) Agardh, from lotic habitats in China. Brant (2003) described a new *Meridion* species from western North Carolina (USA).

Remarks: *Meridion circulare* is commonly collected in cold water, small streams, and springs throughout the northern hemisphere and in New Zealand (Cassie 1989). There are occurrences of this species reported from South America (Montoya-Moreno et al. 2013; Oliveira and Krau 1970; Schneck et al. 2007; Ludwig and Flores 1995; Santos et al. 2011).

Order Tabellariales, Family Tabellariaceae

Tabellaria C.G. Ehrenberg ex FT Kützing (Fig. 5.4b)

Frustules forming zig-zag colonies, cells attached by evident mucilage pads at the apices. Valves symmetrical to both apical and transapical axes. Valves bone shaped, with the center and apices swollen. The central sternum is narrow, striae evident, and a single rimoportula occurs near the center of the valve. Elements of the cingulum possess septa, alternating between girdle bands present on one side or the other.

Remarks: Progress on the taxonomy of this group was made in the works by Knudson (1952, 1953) and Koppen (1975, 1978). A new species was described recently by Potapova (2011). Commonly reported species include *T. fenestrata* (Lyngbye) Kützing and *T. flocculosa* (Roth) Kützing. The former *T. binalis* (Ehrenberg) Grunow has been removed to the genus *Oxyneis* F.E. Round in F.E. Round, R.M. Crawford, and D.G. Mann (1990).

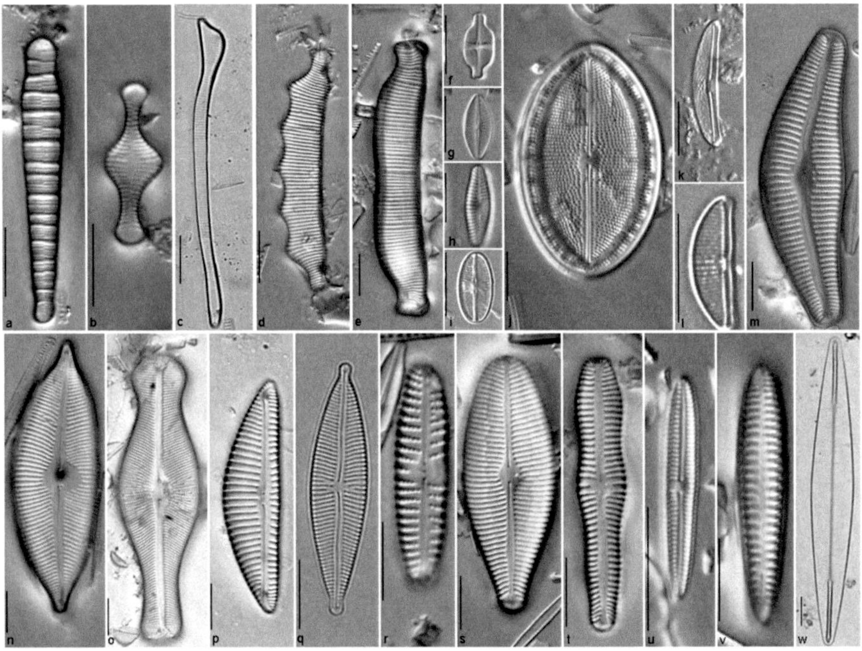

Fig. 5.4 (**a**) *Meridion.* (**b**) *Tabellaria.* (**c**) *Actinella.* (**d**, **e**) *Eunotia.* (**f**) *Achnanthidium.* (**g**) *Karayevia.* (**h**) *Planothidium.* (**i**) *Psammothidium.* (**j**) *Cocconeis.* (**k**) *Amphora.* (**l**) *Halamphora.* (**m**) *Cymbella.* (**n**) *Cymbopleura.* (**o**) *Didymosphenia.*(**p**) *Encyonema.* (**q**) *Encyonopsis.* (**r**) *Reimeria.* (**s**) *Gomphoneis.* (**t**) *Gomphonema.* (**u**) *Gomphosphenia.* (**v**) *Rhoicosphenia.* (**w**) *Amphipleura.* LM, Scale bars = 10 μm

Tabellaria is cosmopolitan in a variety of lentic and lotic environments. In more quiescent waters, the genus may form large blooms (Hustedt 1931).

Class Bacillariophyceae: Order Eunotiales, Family Eunotiaceae

Actinella Lewis (Fig. 5.4c)

Valves individual, unattached, or attached by small peduncles, cells attached by a small mucilage pad. Valves asymmetrical to the apical and transapical axes. Biraphid valves with most of the raphe branches found on the valve mantle, visible in girdle view, with the distal raphe ends curving onto the valve face only. One rimoportula present at one or both apices. Pseudoseptae present or absent. Striae extending across the valve face, interrupted near or at the ventral margin. Marginal spines usually present.

Remarks: *Actinella* is a relatively small genus, with greatest diversity found in South America (Kociolek et al. 2001; Metzeltin and Lange-Bertalot 2007). The genus reaches its greatest diversity in acidic waters.

Eunotia C.G. Ehrenberg (Fig. 5.4d, e)

Valves individual, unattached, or attached by small peduncles, forming sprays or many cells attached by a small mucilage pad, or forming long filaments attached to one another valves face to face. Valves asymmetrical to the apical axis. Biraphid valves with most of the raphe branches found on the valve mantle, visible in girdle view, with the distal raphe ends curving onto the valve face only. One to two rimportulae present. Striae extending across the valve face, interrupted near or at the ventral margin. In some species, marginal spines or ridges present.

Remarks: this is a large genus, with hundreds of taxa described (Fourtanier and Kociolek 2011). The genus may reach its greatest diversity in acidic waters, attached as epiphytes to a variety of substrates. *Eunotia* is notoriously difficult taxonomically, but several large treatises have been produced recently (Lange-Bertalot et al. 2011; Furey et al. 2011).

Order Achnanthales, Family Achnanthidiaceae

Achnanthidium F.T. Kützing (Fig. 5.4f)

Cells single, in short chains or attached to a variety of substrates via short peduncles. Valves symmetrical to the apical and transpical axis, but usually strongly flexed about the transapical plane. Valves small, linear to linear-lanceolate, small and narrow. Raphe and rapheless valves more or less similar (except see figs of *A. exiguum* var. *heterovalvum* (Krasske) Czarnecki). Raphe filiform, with slightly dilated external proximal raphe ends, and external distal ends straight to highly deflected.

Remarks: members of this genus have been treated by several individuals and groups of researchers in the last several years, including (e.g., Kobayasi 1997; Potapova and Ponader 2004; Ponader and Potapova 2007; Rimet et al. 2010; Morales et al. 2011; Wojtal et al. 2011; Péres et al. 2012). Members of this genus occur in fast-flowing waters, attached to rocks and other substrates.

Karayevia F.E. Round and L. Bukhtiyarova ex F.E. Round (Fig. 5.4g)

Cells single. Valves elliptical-lanceolate, with rounded or produced apices. The raphe valve appears different from the rapheless valve. Raphe valve with striae of differing lengths, comprised of elongated areolae. Raphe filiform. Rapheless valve with distinctly puncate striae more regular in length and comprised of rounded areolae.

Remarks: species of this genus occur attached on sand grains and in benthic habitats.

Planothidium F.E. Round and L.Bukhtiyarova (Fig. 5.4h)

Cells occur singly. Valves elliptical-lanceolate, with rounded to narrowly protracted apices. Raphe valve with filiform raphe, external proximal raphe ends dilated. Striae appear costate, comprised of fine areolae in several rows (beyond the limits of LM resolution). The rapheless valve may have striae features similar to or quite dissimilar to the raphe valve. In addition, a distinct, horse-shoe-shaped structure (termed a cavum) that is a small, internal pocket is present on one side of the central area.

Remarks: *Planothidium* species occur in fast to moderately flowing waters. A commonly-reported species, *P. lanceolatum*, is widely distributed. A similar genus, *Gliwiczia*, which has a cavum on both the raphe and rapheless valves was described from Lake Baikal, Russia (Kulikovskiy et al. 2013). *Planothidium* occurs in streams and rivers that outflow from lakes in North America.

Psammothidium L. Bukhtiyarova and F.E. Round (Fig. 5.4i)

Valves symmetrical to the apical and transapical axes. Frustules are flexed about the apical and transapical axes. The raphe valve is convex and the rapheless valve is concave, a condition opposite most other monoraphid diatoms. Valves are small, elliptical to elliptical-lanceolate, with apices rounded. The striae pattern is usually similar on both valves.

Remarks: *Psammothidium* is a relatively large genus with many taxa found in acidic waters (Flower and Jones 1989; Bukhtiyarova and Round 1996; Potapova 2012).

Order Achnanthales, Family Cocconeidaceae

Cocconeis C.G. Ehrenberg (Fig. 5.4j)

Valves elliptical to nearly circular, heterovalvate, flat or flexed apical the apical axis. Striae often are uniseriate, but multiseriate striae and loculate areolae are present in some taxa. The raphe valve has a hyaline area near the edge of the valve. Raphe filiform, straight, with dilated external proximal raphe ends. External distal raphe ends are straight. The axial area and central sternum of the rapheless valve appear similar. Valvocopulae are closed, and may have fimbriate projections.

Remarks: *Cocconeis* species are common in fast flowing waters, attached to filamentous green algae, as well as aquatic plants, wood and rocks. Two commonly reported species are *C. pediculus* Ehrenberg and *C. placentula* Ehrenberg, the latter having several named varieties.

Order Naviculales, Family Catenulaceae

Amphora C.G. Ehrenberg ex F.T. Kützing (Fig. 5.4k)

Valves are asymmetrical to the apical axis and symmetrical to transapical axis. On the dorsal margin, the valve mantle is taller than the ventral margin. Striae on the dorsal margin are usually interrupted by hyaline areas. Shortened striae occur on the ventral margin, and depending upon the situation of the valve, may be difficult to observe. The raphe is moderately to strongly eccentric, straight, sigmoid, or arched. Stigmata and apical pore fields are wanting.

Levkov (2009) has recently monographed this genus, and *Halamphora*, for species occurring in Europe. Stepanek and Kociolek (2012) and Kociolek et al. (2014) described new species from the western USA. *Amphora* species occur in benthic communities, and may be attached to a variety of substrates.

Halamphora (P.T. Cleve) Z. Levkov (Fig. 5.4l)

Valves asymmetrical to the apical axis, with the valve mantle height on the dorsal margin taller than the ventral margin. Raphe eccentric, filiform, with external proximal ends deflected dorsally. A raphe ledge is present on the ventral side in *Halamphora*. In addition to these features, the fine structure of the areolae help distinguish *Halamphora* from *Amphora*, the genus from which *Halamphora* species have been removed.

Remarks: like *Amphora* species, *Halamphora* taxa are found in slower-moving, benthic habitats in streams and rivers. Stepanek and Kociolek (2012, 2015) and Kociolek et al. (2014) have described new species from the western USA.

Order Naviculales, Family Cymbellaceae

Cymbella C.A. Agardh (Fig. 5.4m)

Cells growing at the ends of long, mucilaginous stalks, or free living. Valves asymmetrical about the apical axis, symmetrical about the transapical axis. Raphe with external distal ends deflected towards the dorsal margin. Stigmata placed around the central area towards the ventral margin. Apical pore fields present in most species, not bisected by the distal raphe ends.

Remarks: Biraphid species with asymmetry about the apical axis were mostly assigned to the genus *Cymbella* (e.g. Patrick and Reimer 1975; Krammer and Lange-Bertalot 1986), but a more narrow circumscription of the genus, with the orientation of features as described above, is how the genus is currently recognized. *Cymbella* is attached on a variety of substrates in flowing waters.

Cymbopleura (K. Krammer) K. Krammer (Fig. 5.4n)

Valves slightly asymmetrical to the apical axis, with apices protracted, variously formed. Striae nearly parallel to slightly radiate, distinctly punctate, sometimes appearing lineate. Raphe narrow, slightly to distinctly lateral, with external distal ends deflected dorsally and external proximal ends deflected ventrally. Stigmata and apical pore fields are both wanting.

Remarks: Members of this genus were previously assigned to *Cymbella* (e.g. Patrick and Reimer 1966; Krammer and Lange-Bertalot 1986; Krammer 2003). This unique character suite differentiates species assigned to this genus from other cymbelloid taxa. Members of *Cymbopleura* are encountered across a wide range of habitats.

Didymosphenia M. Schmidt (Fig. 5.4o)

Cells attached by long mucilaginous stalks secreted at the footpole. Valves asymmetrical to transapical axis, symmetrical to the apical axis. Wedge shaped in girdle view. Large valves with a distinct raphe, one to several stigmata present in the central area, a large spine may be present at the headpole, and a large, apical pore field not bisected by the distal raphe ends is present at the footpole.

Remarks: despite its symmetry features, which would suggest a close affinity to gomphonemoid diatoms, *Didymosphenia* was shown by Kociolek and Stoermer (1989) to be more closely allied with cymbelloid diatoms than gomphonemoid diatoms, a hypothesis supported recently with molecular data (Nakov et al. 2014). The most commonly reported species of the genus is *D. geminata*, which can form a great abundance of muco-polysaccharide as stalks and clog streams and dam faces. Known as "rock snot" this nuisance alga has been exported across the northern hemisphere, and from there to localities in the southern hemisphere (Whitton et al. 2009).

Encyonema F.T. Kützing (Fig. 5.4p)

Cells free living or living in mucilaginous tubes as a colony. Valves asymmetrical to the apical axis, symmetrical to the transapical axis. Biraphid valves with external distal raphe ends deflected towards the ventral margin. Stigmata wanting or positioned dorsally. Apical pore fields absent.

Remarks: the filamentous colonies of *Encyonema* can appear visible macroscopically, and are common in slow-flowing waters. An excellent treatment of the species of *Encyonema* is presented by Krammer (1997a, b).

Encyonopsis K. Krammer (Fig. 5.4q)

Valves weakly asymmetrical to the apical axis, or asymmetry not evident. Cells small to large, with external distal raphe ends deflected ventrally. Stigmata central or, more usually, absent. Apical pore fields absent.

Remarks: Krammer (1997b) described this genus for species that did not conform to the symmetry features, or had fine structural differences, found in other cymbelloid diatoms. The genus is widely reported. Graeff and Kociolek (2013) described new species from the western USA.

Reimeria J.P. Kociolek and E.F. Stoermer (Fig. 5.4r)

Valves asymmetrical to the apical axis, symmetrical to the transapical axis. Rectangular in girdle view, septa and pseudosepta absent. Small valves with filiform raphe, with external distal ends deflected to the ventral margin. Dorsal margin slightly convex, ventral margin slightly tumid around the center, where it is also unornamented. Stigma placed centrally. Small apical pore fields are bisected by the external distal raphe ends.

Remarks: a few species of this genus are widely reported from both lentic and lotic systems, including *R. sinuata* (Gregory) Kociolek and Stoermer (1987) and *R. uniseriata* (Sala et al. 1993). Levkov and Ector (2010) consider several of the species in this small genus.

Order Naviculales, Family Gomphonemataceae

Gomphoneis P.T.Cleve (Fig. 5.4s)

Cells terminate long mucilaginous stalks. Valves asymmetrical to the transapical axis, symmetrical to the apical axis. Frustules wedge-shaped in girdle view. Striae with double rows of areolae. Septa and pseudosepta present at both poles. In the current circumscription, there are two groups in this genus. The group that is most like the generitype, *G. elegans* (Grunow) Cleve (an extinct species from western North America, Kociolek and Stoermer 1988), valves with apical pore fields not distinct from the striae, with either no stigmata or four stigmoids. Longitudinal lines wanting in the extant species. The second group has longitudinal lines on either side of the axial area, a single stigma, and apical pore fields differentiated from the striae.

Remarks: this genus has been studied in several publications by Kociolek and Stoermer (e.g., 1988, 1989), with most species being found in streams in the western USA. It has also been reported as an introduced species in streams in Europe, New Zealand, and South America.

Gomphonema C.G.Ehrenberg (Fig. 5.4t)

Cells occur at the ends of mucilaginous stalks secreted from the narrow footpole attached to a variety of substrate types. Frustules occur in mucilage masses or singly. Valves asymmetrical to the transapical axis, symmetrical to the apical axis, and wedge shaped in girdle view. Striae usually uniseriate, with areolae typically "c" or "s" shaped, although other configurations have been described. Apical pore fields at the footpole differentiated from the striae. Usually one stigma is present in the central area (but some species lack stigmata and a few possess two stigmata). Septa and pseudosepta present at both poles.

Remarks: This is a large, highly variable genus (Fourtanier and Kociolek 2011), with about 800 described taxa (Kociolek and Kingston 1999; Reichardt 1999, 2005, 2007). Species occur worldwide.

Order Naviculales, Family Rhoicospheniaceae

Gomphosphenia H. Lange-Bertalot (Fig. 5.4u)

Valves small, asymmetrical to the transapical axis, symmetrical to the apical axis. Valve may be flexed about the transapical axis; usually wedge shaped in girdle view. Valves with short striae, usually formed by one or a few, elongate areolae. Raphe straight, filiform. Apical pore fields, septa, pseudosepta, and stigmata all lacking.

Remarks: members of the genus found in flowing waters are usually very small, flexed, and thus may be confused with small monoraphid diatoms. *Gomphosphenia* taxa are biraphid and differ in symmetry. Thomas et al. (2009) and Kociolek et al. (2014) described species of the genus found in flowing waters, and Camburn et al. (1978) reported species from South Carolina (taxa were referred to *Gomphonema* at that point). Usually found in fast-flowing sections of mountainous streams.

Rhoicosphenia A. Grunow (Fig. 5.4v)

Cells grow at the end of mucilaginous stalks secreted from the footpole. Frustules heterovalvate, with one valve having an elongate raphe, the other valve with short raphe branches. Valves asymmetrical to the transapical axis, symmetrical to the apical axis. Valves strongly flexed about the transapical axis, with one valve concave and bearing two full raphe branches, the valve with the convex margin has two reduced raphe branches. Septa and pseudosepta present. Stigmata absent. Apical pore fields undifferentiated from the striae.

Remarks: while there usually is one species commonly recorded from around the world, *R. abbreviata*, in fast-flowing waters, several new taxa have been described from Lake Ohrid (Levkov and Nakov 2008; Levkov et al. 2010), California, and India (Thomas et al. 2015), and several additional species from the Pacific Northwest of the USA (Thomas and Kociolek 2015). We expect other new species will be discovered as well.

Order Naviculales, Family Naviculaceae

Amphipleura F.T. Kützing (Fig. 5.4w)

Cell free living or enclosed in mucilaginous tubes, found in hard-water environments. Valves biraphid, symmetrical to both axes. Raphe branches short, confined to near the apices. The central sternum runs the length of the valve, but bifurcates to accommodate the raphe near the valve terminus. Raphe occurs within a rib, giving the impression of "needle eyes" near the valve terminus. Striae barely resolvable in the LM, organized perpendicular to the central sternum.

Remarks: *Amphipleura* is closely related to *Frustulia* and allies (Round et al. 1990). These taxa share the characteristic of the raphe bordered by ribs on either side.

Brachysira F.T. Kützing (Fig. 5.5a)

Cells free living or growing at the ends of mucilaginous stalks. Valves symmetrical to the apical and transapical axes (though some species and specimens may appear weakly asymmetrical to the transapical axis), rhomboid to lanceolate in shape, with

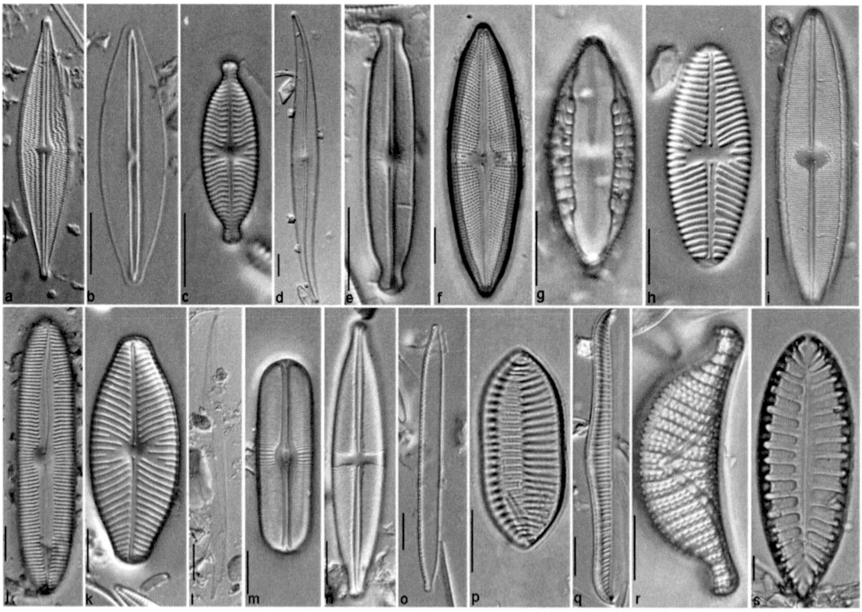

Fig. 5.5 (**a**) *Brachysira*. (**b**) *Frustulia*. (**c**) *Geissleria*. (**d**) *Gyrosigma*. (**e**) *Kobayasiella*. (**f**) *Luticola*. (**g**) *Mastogloia*. (**h**) *Navicula*. (**i**) *Neidium*. (**j**) *Pinnularia*. (**k**) *Placoneis*. (**l**) *Pleurosigma*. (**m**) *Sellaphora*. (**n**) *Stauroneis*. (**o**) *Nitzschia*. (**p**) *Tryblionella*. (**q**) *Rhopalodia*. (**r**) *Epithemia*. (**s**) *Surirella*. Scale bars = 10 μm

narrowly rounded to acute apices. Raphe filiform, straight, contained within a narrow axial area. Axial area bordered externally on both sides by a thin ridge. The central area is barely distinguished. Striae are distinctly punctate, and the areolae are not aligned between striae, giving the impression of longitudinal undulations to the striae.

Remarks: This genus is usually benthic in acid to circumneutral waters. A monograph of the genus has been published by Lange-Bertalot and Moser (1994).

Frustulia L.Rabenhorst (Fig. 5.5b)

Cells free-living or living within mucilaginous tubes. Valves symmetrical to both axes, raphe contained in a rib, connecting with helictoglossae at the valve apices. The distinctive shape of the joint rib/helictoglossa is termed a "porte-crayon" like the eighteenth century writing implement. Striae are perpendicular to the central sternum, such that they appear to form longitudinal and perpendicular rows.

Geissleria H. Lange-Bertalot and D. Metzeltin (Fig. 5.5c)

Valves relatively small, obtuse to elliptical-lanceolate, with protracted to rounded ends. Raphe straight, filiform, with external distal raphe ends hooked or deflected onto the valve mantle. Striae comprised of fine lineolate areolae. At the apices, one to four striae are interrupted to form an annulus, which is distinctive under the SEM. Isolated punctum present or absent in the central area.

Remarks: this genus was previously included in *Navicula*, within a group called "Annulatae."

Gyrosigma A.H. Hassall (Fig. 5.5d)

Valves weakly to strongly sigmoid, usually elongated but some species shorter and more robust in outline. Raphe filiform, indistinct, with external proximal and distal ends deflected. Axial area follows the contour of the valve. Central area small, usually orbicular. Striae distinctly punctate, forming two patterns of ornamentation; perpendicular and parallel to the axial area.

Remarks: this genus is usually found in slow flowing waters.

Kobayasiella H. Lange-Bertalot (Fig. 5.5e)

Valves linear to linear-lanceolate with protracted, capitate apices. Striae difficult to resolve or not resolvable in LM. Raphe filiform, straight, with dilated proximal ends and sharply deflected distal ends. In some species longitudinal lines between the raphe and valve margins are evident

Remarks: this genus occurs in acid habitats.

Luticola D.G. Mann in F.E. Round, R.M. Crawford, and D.G. Mann (Fig. 5.5f)

Valves symmetrical to apical and transapical axis, with rounded to protracted apices. Striae are distinctly punctate, distinct stigma generally present in the central area. External proximal raphe ends deflected in the same direction in the central area. External distal raphe ends deflected or hooked onto the valve mantle.

Remarks: the genus is often found in aerophilous habitats, in the spray zones or banks of cold, oxygen-rich waters, amongst bryophytes.

Mastogloia Thwaites ex W. Smith (Fig. 5.5g)

Cells occur singly. Valves symmetrical to the apical and transapical axis, lanceolate to elliptical-lanceolate with rounded to protracted ends. Axial area narrow, with a filiform to complex raphe. Striae distinctly punctate, the areolae may be in complex patterns (quincunx). Girdle band next to the valve (valvocopula) with several square to rectangular chambers or locules. Locules evident in valve and girdle view.

Navicula J.B.M. Bory de Saint-Vincent (Fig. 5.5h)

Valves symmetrical to both axes. Striae comprised of line-shaped (lineolate) areolae. Raphe filiform, proximal ends dilated, distal ends hooked. Internally, raphe on an elevated ridge. Central area may be expanded, but not forming a staurose.

Remarks: though many genera have been excised from *Navicula*, it is still a large, taxonomically difficult genus. Krammer and Lange-Bertalot (1986, 1991) and Hoffmann et al. (2013) are excellent resources for *Navicula* identification. Species of *Navicula* occur mostly in benthic habitats.

Neidium E. Pfitzer (Fig. 5.5i)

Valves with distinctly punctate striae. Raphe has generally with external proximal raphe ends deflected opposite one another in the central area, and distal ends that appear forked. One to several longitudinal canals run the length of the valve, positioned near the margin on either side of the valve face. Interruptions in the striae on one (the secondary) side of the axial area, termed Voigt faults, are usually distinct.

Remarks: the genus is widely distributed in lotic and lentic habitats, in both acid and circum-neutral conditions. New species of the genus have been described recently by Siver and colleagues (e.g., Siver and Hamilton 2005; Stachura-Suchoples et al. 2004, 2010; Thomas and Kociolek 2008; Kociolek et al. 2014; Liu et al. 2016).

Pinnularia C.G.Ehrenberg (Fig. 5.5j)

Cells single or in chains (e.g., *P. cardinalis* (Ehrenberg) W. Smith). Valves linear-elliptical, some very large (>250 μm). Striae appearing wide but without resolvable areolae ('costate'). The striae are actually multiseriate, but the areolar openings are below the resolving capability of LM. Axial area distinct. Raphe lateral to complex, with dilated external proximal ends and bayonet-shaped distal ends. The central area may contain thickenings evident in LM. Internally, a lamina or plate of silica may exist in some species, and the edge of this plate can give the impression of longitudinal lines.

Remarks: this large and complex (both taxonomically and morphologically) genus has been treated recently by Krammer (1992a, b). Recent phylogenetic analysis suggests *Caloneis* a genus once separated out from *Pinnularia* belongs in the genus *Pinnularia*. Round et al. (1990) hinted at this previously.

Remarks: the genus occurs across a wide spectrum of habitats.

Placoneis C.Mereschkowsky (Fig. 5.5k)

Cells single. Valves symmetrical to both the apical and transapical axis. Valves lanceolate, with protracted apices. Striae uniseriate, comprised of rounded areolae. Axial area narrow, straight, expanded to form a distinct central area. Isolated punctum may be present in the central area. Raphe with dilated external proximal ends and distal ends recurved in the same, or opposite directions.

Remarks: Cox (1987) has reviewed this genus, showing that the nature of the chloroplast helps to differentiate it from *Navicula*. The genus is now thought to be part of the Cymbellales, including *Cymbella* and its allies and *Gomphonema* and its allies. Phylogenetic analyses addressing these relationships include Kociolek and Stoermer (1989) and Nakov et al. (2014).

Pleurosigma W. Smith(Fig. 5.5l)

Valves weakly to strongly sigmoid, usually elongated. Raphe filiform, indistinct, with external proximal and distal ends appearing straight. Axial area follows the contour of the valve. Central area small, usually orbicular. Striae distinctly punctate, forming three patterns of ornamentation; perpendicular, parallel, and oblique to the axial area.

Remarks: This genus is usually found in waters of elevated conductance and slower flows.

Sellaphora C. Mereschkowsky(Fig. 5.5m)

Cells occur singly. Valves linear-lanceolate to capsule-shaped, with rounded to capitate apices. Central sternum straight, externally with a hooded structure (conopeum) extending laterally and covering a small portion of the striae near the axial area. Central area can be laterally expanded, but not comprised of a stauros. Raphe straight, with dilated external proximal ends. Striae uniseriate, comprised of rounded areolae. Internal thickenings expanding to the margin near the helictoglossae produce a "T" shape at the ends of the valve.

Remarks: members of this genus occur benthic habitats in stream and rivers. Commonly-reported species include *S. pupula* (Kützing) Mereschkowsky, *S. laevissima* (Kützing) D.G. Mann and *S. bacillum* (Ehrenberg) D.G. Mann among others. Mann and colleagues (e.g., Mann 1989; Mann et al. 2004) have studied many aspects of the biology of species in this genus.

Stauroneis C.G. Ehrenberg(Fig. 5.5n)

Cells solitary, though they may be stuck together after division to form small, weak chains. Valves linear to lanceolate, with axial area straight and expanded into a wide, internally thickened fascia (stauros). Raphe filiform or undulate, with external proximal ends dilated and external distal ends curved or hooked in the same direction. Striae distinct, uniseriate. Infolding of silica from the valve at the apices (pseudosepta) may be present in some species.

Remarks: the genus is found widely in freshwaters, and several species are commonly associated with aerophilous habitats.

Order Bacillariales, Family Bacillariaceae

Nitzschia A.H. Hassall(Fig. 5.5o)

Cells occurring singly, within mucilage tubes, or rarely forming spherical colonies with cells attached at one apex. Valves symmetrical to both axes, or sigmoid in outline. In most species the canal raphe is elevated in a keel, though in some species the raphe is flush against the valve. Raphe positioned on one valve margin. Raphe of one valve positioned on the margin on the opposite side of the other valve. Distinct fibulae mostly on the side of the raphe only, though in a few species, such as *N. amphibioides* Hustedt 1942 the fibulae decrease in thickness as they extend across the valve from margin to margin.

Remarks: *Nitzschia* is a very large genus, with hundreds of taxa described (Fourtanier and Kociolek 2011). It is also one of the most challenging groups taxonomically, since there have been no revisionary studies for nearly 150 years (the last revision was produced by Cleve and Grunow 1880); descriptions for many species have only been written in German. Many species of *Nitzschia* are facultative heterotrophs for some carbon sources (Cholnoky 1968; Tuchman et al. 2006). Many species are pollution indicators (Lowe 1974; Cholnoky 1968; Patrick 1977).

Tryblionella W. Smith(Fig. 5.5p)

Cells occurring singly, mostly epipelic. Valves symmetrical to both axes. Canal raphe is elevated in a keel. Raphe positioned on one valve margin. Raphe of one valve positioned on the margin on the opposite side of the other valve. Distinct fibulae mostly on the side of the raphe only. Striae usually distinct, extending across the valve face. A longitudinal fold occurs the length of the valve, appearing out of focus or unornamented relative to the rest of the valve. This feature distinguishes *Tryblionella* from *Nitzschia*.

Remarks: this genus occurs in waters of high nutrient content and elevated conductance.

Order Rhopalodiales, Family Rhopalodiaceae

Rhopalodia O.Müller(Fig. 5.5q)

Cells occur individually, in the benthos or on plants. Valves asymmetrical to the apical axis, with a slight elevation at the center of the valve and the distal ends deflected towards the ventral margin. Valves usually seen in girdle view. Raphe contained in a canal, located along the dorsal margin (not easy to see). Costae distinct, extend across the valve face. Striae distinctly punctate.

Remarks: *Rhopalodia* contains blue-green algal symbionts that appear to function like a nitrogen-fixing organelle (Prechtl et al. 2004; Trapp et al. 2012). The genus is found in hard-water environments, usually in nitrogen-poor conditions.

Epithemia F.T. Kützing (Fig. 5.5r)

Cells occur singly, usually epiphytic on plants. Valves asymmetrical to the apical axis, symmetrical to the transapcial axis. Valve with distinct areolae. Thickened internal costae extend across the valve face. Raphe contained in an evident canal, appearing biarcuate, extending from near the ventral margin near the apices, from slightly to most of the way towards the dorsal valve in the center of the valve. The canal opens into the valve interior through a series of holes or portules.

Remarks: Like *Rhopalodia*, *Epithemia* species may contain blue-green algal symbionts that fix elemental nitrogen (Floener and Bothe 1980). *Epithemia* species are common in alkaline, nitrogen-poor waters.

Order Surirellales, Family Surirellaceae

Surirella P.J.F. Turpin (Fig. 5.5s)

Cells often present in sandy or silty margins of streams; epipelic. Valves linear to elliptical in outline, not sigmoid, canal raphe in a keel elevated above the surface of the valve, extending around the periphery of the margin. Valves lack undulations along the center, and are not saddle shaped. Fibulae delicate or robust. Costae may be present on the valve surface. Striae in single or multiple rows, areolae usually very small. Valves sometimes covered in siliceous warts or small spines. In several species, a large subapical spine may be present on each valve.

Remarks: Ruck and Kociolek (2004) illustrate and describe the complex morphology of this genus and its close allies. Based on phylogenetic analysis of morphological (Ruck and Kociolek 2004) and molecular (Ruck and Theriot 2011) data, the genus *Surirella* is non-monophyletic. Taxonomic implications of these results await further work.

Acknowledgements We are grateful for the use of diatom micrographs in this chapter: Dennis Vander Meer (Rhithron Associates, Inc.) for figures 19, 20, 25, 30, 45, 48, and 71 and Laboratório de Ficologia, Departamento de Botânica, Universidade Federal do Paraná, Brazil, for figures 22, 26, 39, 53, 59, 65, and 75. We acknowledge the support of Rhithron Associates, Inc. for infrastructure.

References

Bahulikar RA, Kroth PG (2007) Localization of EPS components secreted by freshwater diatoms using differential staining with fluorophore-conjugated lectins and other fluorochromes. Eur J Phycol 42:99–208

Baldauf SL (2008) An overview of the phylogeny and diversity of eukaryotes. J Syst Evol 46:263–273

Biggs BJF (1996) Patterns in benthic algae of streams. In: Stevenson RJ, Bothwell ML, Lowe RL (eds) Benthic algae ecology: freshwater ecosystems. Academic, San Diego, pp 31–56

Bothwell ML, Taylor BW, Kilroy C (2014) The Didymo story: the role of low dissolved phosphorus in the formation of *Didymosphenia geminata* blooms. Diatom Res 29:229–236

Brant L (2003) A new species of *Meridion* (Bacillariophyceae) from western North Carolina. Southeastern Naturalist 2:409–418

Bukhtiyarova LN, Round FE (1996) Revision of the genus *Achnanthes sensu lato. Psammothidium*, a new genus based on *A. marginulatum*. Diatom Res 11:1–30

Camburn KE, Kingston JC (1986) The genus *Melosira* from soft-water lakes with special reference to northern Michigan, Wisconsin and Minnesota. In: Smol JP, Davis RW, Meriläinen J (eds) Diatoms and lake acidity. W. Junk Publishers, Dordrecht, pp 17–34

Camburn KE, Lowe RL, Stoneburner DL (1978) The haptobenthic diatom flora of Long Branch Creek, South Carolina. Nova Hedw 30:149–279

Carr JM, Hergenrader GL (1987) Occurrence of three *Nitzschia* (Bacillariophyceae) taxa within colonies of tube-forming diatoms. J Phycol 23:62–70

Cassie V (1989) A contribution to the study of New Zealand diatoms. Biblioth Diatomol 17:1–266

Cholnoky BJ (1968) The ecology of diatoms in Binnengewassern. J. Cramer, Academics

Cleve PT, Grunow A (1880) Contributions to the knowledge of Arctic diatoms. K Sven vetensk aka Handl 17: 1-121

Cox EJ (1981) Mucilage tube morphology of three tube-dwelling diatoms and its diagnostic value. J Phycol 17:72–80

Cox EJ (1987) *Placoneis* Mereschkowsky: the re-evaluation of a diatom genus originally characterized by its chloroplast type. Diatom Res 2:145–157

Reviers B (2003) Biology and phylogeny of algae, Volume 2. Belin, Paris

English J, Potapova M (2009) *Aulacoseira pardata sp. nov., A. nivalis comb nov., A. nivaloides comb. et stat. nov.,* and their occurrences in Western North America. Proc Acad Nat Sci Philadelphia 158:37–48

Falkowski P, Scholes RJ, Boyle E et al (2000) The global carbon cycle: a test of our knowledge of earth as a system. Science 290:291–296

Floener L, Bothe H (1980) Nitrogen fixation in *Rhopalodia gibba*, a diatom containing blue-greenish inclusions symbiotically. In: Schwemmler W, Shenk HEA (eds) Endocytobiology: endosymbiosis and cell biology. Walter de Gruyter and Company, Berlin, pp 541–552

Flower R, Jones V (1989) Taxonomic descriptions and occurrences of new *Achnanthes* taxa in acid lakes in the UK. Diatom Res 4:227–239

Fourtanier E, Kociolek JP (2011) Catalogue of diatom names. http://researcharchive.calacademy.org/research/diatoms/names/index.asp. Accessed 24 Feb 2015

Fry B, Wainright SC (1991) Diatom sources of 13C-rich carbon in marine food webs. Mar Ecol Prog Ser 76:149–157

Fungladda N, Kaczmarska I, Rushforth SR (1983) A contribution to the freshwater diatom flora of the Hawaiian Islands. Biblioth Diatomol 2:1–103

Furey PC, Lowe RL, Johansen JR (2011) *Eunotia* Ehrenberg (Bacillariophyta) of the Great Smoky Mountains National Park, USA. Biblioth Diatomol 56:1–134

Geitler L (1971) The unequal cell division in the formation of the inner Halen at *Meridion circulare* . Austrian Bot Z 119:442–446

Grady AE, Scanlon TM, Galloway JN (2007) Declines in dissolved silica concentrations in western Virginia streams (1988–2003): gypsy moth defoliation stimulates diatoms? J Geophys Res 112:1–11

Graeff C, Kociolek JP (2013) New or rare species of cymbelloid diatoms (Bacillariophyceae) from Colorado (USA). Nova Hedwigia 97:87–116

Håkansson H (2002) A compilation and evaluation of species in the genera *Stephanodiscus, Cyclostephanos* and *Cyclotella* with a new genus in the family Stephanodiscaceae. Diatom Res 17:1–139

Hill BH, Stevenson RJ, Pan YD et al (2001) Comparison of correlations between environmental characteristics and stream diatom assemblages characterized at genus and species levels. J North Am Benthol Soc 20:299–310

Hoagland KD, Rosowski JR, Gretz MR et al (1993) Diatom extracellular polymeric substances: function, fine structure, chemistry, and physiology. J Phycol 29:537–566

Hoffmann G, Werum M, Lange-Bertalot H (2013) diatoms in freshwater. Benthos of Central Europe. Koeltz Scientific Books, Konigstein

Hustedt F (1931) Diatoms from the Feforvatn in Norway. Arch Hydrobiol 22:537–545

Julius ML (2007) Why sweat the small stuff: the importance of microalgae in Hawaiian stream ecosystems. Bishop Mus Bull Cult Environ Stud 3:183–194

Karthick B, Kociolek JP (2011) Four new centric diatoms (Bacillariophyceae) from the Western Ghats, South India. Phytotaxa 22:25–40

Karthick B, Taylor JC, Mahesh MK, Ramachandra TV (2010) Protocols for collection, preservation and enumeration of diatoms from aquatic habitats for water quality monitoring in India. J Soil Water Sci 3:25–60

Kelly JR, Davis CS, Cibik SJ (1998a) Conceptual food web model for Cape Cod Bay, with associated environmental interactions; Report ENQUAD 98-04. Massachusetts Water Resources Authority, Boston

Kelly MG, Cazaubon AP, Coring E et al (1998b) Recommendations for the routine sampling of diatoms for water quality assessments in Europe. J Phycol 10:215–224

Khursevich GK, Kociolek JP (2012) A preliminary, worldwide inventory of the extinct, freshwater fossil diatoms from the Orders Thalassiosirales, Stephanodiscales, Paraliales, Aulacoseirales, Melosirales, Coscindiscales, and Biddulphiales. Nova Hedwigia 141:315–364

Kiss KT, Klee R, Ector L, Ács É (2012) Centric diatoms of large rivers and tributaries in Hungary: morphology and biogeographic distribution. Acta Bot Croat 71:311–363

Knudson BM (1952) The diatom genus *Tabellaria*. I. Taxonomy and morphology. Ann Bot 16:421–440
Knudson BM (1953) The diatom genus *Tabellaria*. II. Taxonomy and morphology of the plankton varieties. Ann Bot 17:131–155
Kobayasi H (1997) Comparative studies among four linear–lanceolate *Achnanthidium* species (Bacillariophyceae) with curved terminal raphe endings. Nova Hedwigia 65:147–163
Kociolek JP (2012) Diatoms of the Southern California Bight. http://dbmuseblade.colorado.edu/DiatomTwo/dscb_site/index.php. Accessed 2 Mar 2015
Kociolek JP, Kingston JC (1999) Taxonomy, ultrastructure and distribution of gomphonemoid diatoms (Bacillariophyceae: Gomphonemataceae) from rivers of the United States. Can J Bot 77:686–705
Kociolek JP, Lowe RL (1983) Scanning electron microscopic observations on the frustular morphology and filamentous growth habit of *Diatoma hiemale* var. *mesodon*. Trans Am Microsc Soc 102:281–287
Kociolek JP, Stoermer EF (1987) Ultrastructure of *Cymbella sinuata* (Bacillariophyceae) and its allies, and their transfer to *Reimeria*, gen. nov. Syst Bot 12:451–459
Kociolek JP, Stoermer EF (1988) Taxonomy, ultrastructure, and distribution of *Gomphoneis herculeana, G. eriense* and closely related species. Proc Acad Nat Sci Philadelphia 140:24–97
Kociolek JP, Stoermer EF (1989) Phylogenetic relationships and evolutionary history of the diatom genus *Gomphoneis*. Phycologia 28:438–454
Kociolek JP, Lamb MA, Lowe RL (1983) Notes on the growth and ultrastructure of *Biddulphia laevis* Ehr. (Bacillariophyceae) in the Maumee River, Ohio. Ohio J Sci 8:125–130
Kociolek JP, Lyon D, Spaulding S (2001) Revision of the South American species of *Actinella*. In: Jahn R, Kociolek JP, Witkowski A, Compere P (eds) Lange-Bertalot Festschrift. Studies on diatoms. Gantner Verlag KG, Ruggell, pp 131–166
Kociolek JP, Liu Y, Wang X (2011) Internal valves in populations of *Meridion circulare* (Greville) C.A. Agardh from the A'er Mountain region of northeastern China: implications for taxonomy and systematics. J Syst Evol 49:486–494
Kociolek JP, Laslandes B, Bennet D, Thomas E, Brady, M, Graeff C (2014) Diatoms of the United States. I. Taxonomy, ultrastructure and descriptions of fifty new species and other rarely reported taxa from lake sediments in the western USA. Biblioth Diatomol 61:1–188
Kociolek JP, Theriot EC, Williams DM, Julius M, Stoermer EF, Kingston JC (2015) Centric and Araphid Diatoms. In: Wehr J, Sheath R, Kociolek JP (eds) Freshwater Algae of North America. Academic Press, New York, pp 653–708
Koppen JD (1975) A morphological and taxonomic consideration of *Tabellaria* (Bacillariophyceae) from the northcentral United States. J Phycol 11:236–244
Koppen JD (1978) Distribution and aspects of the ecology of the genus *Tabellaria* Ehr. (Bacillariophyceae) in the northcentral United States. Am Midl Nat 99:383–397
Krammer K (1992a) Pinnularia. Die Gattung Pinnularia in Bayern. Hoppea 52:1–308
Krammer K (1992b) *Pinnularia.* A monograph of the European taxa. Biblioth Diatomol 26:1–353
Krammer K (1997a) The cymbelloiden diatoms. A monograph of the world's known taxa. Part 1. General and *Encyonema* Part. Biblioth Diatomol 36:1–382
Krammer K (1997b) The cymbelloiden diatoms. A monograph of the world's known taxa. Part 2. *Encyonema* Part, *Encyonopsis* and *Cymbellopsis*. Biblioth Diatomol 37:1–469
Krammer K (2003) Cymbopleura, Delicata, Navicymbula, Gomphocymbellopsis, Afrocymbella. In: Lange-Bertalot H (ed) Diatoms of Europe, diatoms of the European inland waters and comparable habitats. ARG Gantner Verlag KG, Rugell, pp 1–530
Krammer K, Lange-Bertalot H (1986) Bacillariophyceae, 1 Teil, Naviculaceae. In: Ettl H, Gerloff J, Heynig H, Mollenhauer D (eds) Süsswasserflora von Mitteleuropa, Band 2/1. Gustav Fischer, Stuttgart, pp 1–876
Krammer K, Lange-Bertalot H (1991) Bacillariophyceae, 3 Teil, Centrales, Fragilariaceae, Eunotiaceae. In: Ettl H, Gerloff J, Heynig H, Mollenhauer D (eds) Süsswasserflora von Mitteleuropa, Band 2/3. Gustav Fischer, Stuttgart, pp 1–576

Kulikovskiy M, Lange-Bertalot H, Witkowski A (2013) *Gliwiczia gen. nov.* a new monoraphid diatom genus from Lake Baikal with description of four species new for science. Phytotaxa 109:1–16

Lampert W, Sommer U (2007) Limnoecology: the ecology of lakes and streams, 2nd edn. Oxford University Press, Oxford

Lange-Bertalot H, G Moser (1994) *Brachysira*. Monograph of the genus. Biblioth Diatomol 29: 1-212

Lange-Bertalot H, Bak M, Witkowski A et al (2011) *Eunotia* and some related genera. In: Lange-Bertalot H (ed) Diatoms of Europe Vol 6: Diatoms of European inland waters and comparable habitats. ARG Gantner Verlag KG, Rugell, pp 1–747

Levkov Z (2009) Amphora sensu lato. In: Lange-Bertalot H (ed) Diatoms of Europe Vol 5: Diatoms of European inland waters and comparable habitats. ARG Gantner Verlag KG, Rugell, pp 1–916

Levkov Z, Ector L (2010) A comparative study of *Reimeria* species (Bacillariophyceae). Nova Hedwigia 90:469–489

Levkov Z, Nakov T (2008) *Rhoicosphenia tenuis*, a new diatom species from Lake Ohrid. Diatom Res 23:377–388

Levkov Z, Mihalic KC, Ector L (2010) A taxonomical study of *Rhoicosphenia* Grunow (Bacillariophyceae) with a key for identification of selected taxa. Fottea 10:45–200

Liu Q, Kociolek JP, Li B et al (2016) The diatom genus *Neidium* (Bacillariophyceae) from Zoige Wetland, China. Phytotaxa in press

Lowe RL (1974) Environmental requirements and pollution tolerance of freshwater diatoms. National Environmental Research Center, Office of Research and Development, U.S. Environmental Protection Agency, Cincinnati, OH

Lowe RL (1975) Comparative ultrastructure of the valves of some *Cyclotella* species (Bacillariophyceae). J Phycol 11:415–424

Ludwig TAV, Flores TL (1995) Diatomoflórula dos rios da região a ser inundada para a construção da usina hidrelétrica de Segredo, PR. I Coscinodiscophyceae, Bacillariophyceae (Achnanthales e Eunotiales) e Fragilariophyceae (Meridion e Asterionella). Arq Biol Tecn 38(2):31–65

Luttenton MR, Pfiester LA, Timpano P (1986) Morphology and growth habit of *Terpsinoe musica* Ehr. (Bacillariophyceae). Castanea 51:172–182

Mann DG (1989) The diatom genus *Sellaphora*: separation from *Navicula*. Br Phycol J 24:1–20

Mann DG, Droop SJM (1996) Biodiversity, biogeography and conservation of diatoms. Hydrobiologia 336:19–32

Mann DG, McDonald SM, Bayer MM et al (2004) The *Sellaphora pupula* species complex (Bacillariophyceae): morphometric analysis, ultrastructure and mating data provide evidence for five new species. Phycologia 43:459–482

Metzeltin D, Lange-Bertalot H (2007) Tropical diatoms of South America II. Special remarks on biogeography disjunction. In: Lange-Bertalot H (ed) Diversity-taxonomy-biogeography, iconographia diatomologica, annotated diatom micrographs, vol 18. ARG Gantner Verlag KG, Rugell, pp 1–877

Montoya-Moreno Y, Room S, Vouilloud A, Aguirre N et al (2013) List of diatoms of continental environments in Colombia. Biota Colomb 14:13–78

Morales E (2001) Morphological studies in selected fragilarioid diatoms (Bacillariophyceae) from Connecticut waters (USA). Proc Acad Nat Sci Philadelphia 151:105–120

Morales EA, Ector L, Fernández E et al (2011) The genus *Achnanthidium* Kütz. (Achnanthales, Bacillariophyceae) in Bolivian streams: a report of taxa found in recent investigations. Algol Stud 136(137):89–130

Müller O (1895) About axles, orientation and symmetry planes in Bacillariophyceae. Ber German Bot Ges 13:222–234

Nakov T, Ruck EC, Galachyants Y, Spaulding SA, Theriot EC (2014) Molecular phylogeny of the Cymbellales (Bacillariophyceae, Heterokontophyta) with a comparison of models for accommodating rate variation across sites. Phycologia 53:359–373

Oliveira LPH, Krau L (1970) General Hydrobiology, particularly applied to backers of schistosomes-hipereutrofia, poorly modern waters. Mem Inst Oswaldo Cruz 68:89–118

Patrick R (1977) Ecology of freshwater diatoms and diatom communities. In: Werner D (ed) The biology of diatoms. D. Blackwell Scientific Publications, Oxford, pp 284–322

Patrick R, Reimer CW (1966) The diatoms of the United States, exclusive of Alaska and Hawaii, Volume 1: Fragilariaceae, Eunotiaceae, Achnanthaceae, Naviculaceae. Academy of Natural Sciences, Philadelphia

Patrick R, Reimer CW (1975) The diatoms of the United States, exclusive of Alaska and Hawaii, Volume 2, Part 1—Entomoneidaceae, Cymbellaceae, Gomphonemaceae, Epithemaceae. Academy of Natural Sciences, Philadelphia

Péres F, Barthes A, Ponton E, Coste M, Ten-Hage L, Le-Cohu R (2012) *Achnanthidium delmontii sp. nov.*, a new species from French rivers. Fottea 12:189–198

Ponader KC, Potapova MG (2007) Diatoms of the genus *Achnanthidium* in flowing waters of the Appalachian Mountains (North America): ecology, distribution and taxonomic notes. Limnologica 37:227–241

Potapova M (2011) *Tabellaria vetteri*, a new diatom (Bacillariophyceae: Tabellariaceae) from Pennsylvania, USA. Proc Acad Nat Sci Philadelphia 161:35–41

Potapova M (2012) New species and combinations in monoraphid diatoms (Family Achnanthidiaceae) from North America. Diatom Res 27:29–42

Potapova M, Ponader KC (2004) Two common North American diatoms, *Achnanthidium rivulare sp. nov.* and *A. deflexum* (Reimer) Kingston: morphology, ecology and comparison with related specie. Diatom Res 19:33–57

Prechtl J, Kneip G, Lockhart PJ, Wenderoth K, Maier UG (2004) Intracellular spheroid bodies of *Rhopalodia gibba* have nitrogen-fixing apparatus of cyanobacterial origin. Mol Biol Evol 21:1477–1481

Qi YZ, Reimer CW, Mahoney RK (1984) Taxonomic studies of the genus *Hydrosera*. I. Comparative morphology of *H. triquetra* Wallich and *H. whampoensis* (Schwartz) Deby, with ecological remarks. Proc Int Diatom Symp 7:213–224

Reichardt E (1999) The revision of the genus *Gomphonema*. The species to *G. affine / insigne*, *G. angus tatum / micropus*, *G. acuminatum* and gomphonemoide diatoms from the Oberoligozän in Bohemia. Iconogr Diatomol 8:1–203

Reichardt E (2005) The identity of *Gomphonema entolejum* Østrup (Bacillariophyceae) and as well as revision änlicher species with further axial area. Nova Hedwigia 81:115–144

Reichardt E (2007) New and little known *Gomphonema* typologies (Bacillariophyeae) with areoles in double rows. Nova Hedwigia 85:103–137

Rimet F (2012) Recent views on river pollution and diatoms. Hydrobiologia 683:1–24

Rimet F, Couté A, Piuz A, Berthon V, Druart JC (2010) *Achnanthidium druartii sp. nov.* (Achnanthales, Bacillariophyta): a new species invading European rivers. Vie Milieu 60:185–195

Round FE, Crawford RM, Mann DG (1990) The diatoms. Cambridge University Press, Avon

Ruck E, Kociolek JP (2004) A preliminary phylogeny of the family Surirellaceae. Biblioth Diatomol 50:1–236

Ruck EC, Theriot EC (2011) Origin and evolution of the canal raphe in diatoms. Protist 162: 723–737

Sala SE, Guerrero JM, Ferrario ME (1993) Redefinition of *Reimeria sinuata* (Gregory) Kociolek and Stoermer and recognition of *Reimeria uniseriata* nov. sp. Diatom Res 8:439–446

Santos MS Tremarin IP Ludwig TAV (2011) Diatoms Periphytic in *Potamogeton polygonus* Cham. and Schltdl .: pioneer citations for the state of Paraná. Biota Neotrop 11:303–315

Schneck F, Torgan LC, Schwarzbold A (2007) Epilithic diatom community in high altitude stream impacted by fish farming in southern Brazil. Acta Limnol Bras 19:341–355

Simonsen R (1965) Ecological observations on the tropical diatom *Hydro Sera triquetra* Wallich and aerophilicity of diatoms. Int Rev complete Hydrobiol 50:49–56

Simonsen R (1979) The diatom system: ideas on phylogeny. Bacillaria 2:9–71

Siver P, Hamilton PB (2005) Observations on new and rare species of freshwater diatoms from Cape Cod, Massachusetts, USA. Can J Bot 83:362–378
Siver PA, Hamilton PB (2011) Diatoms of North America. The freshwater flora of waterbodies on the Atlantic Coastal Plain. Iconogr Diatomol 22:1–923
Siver PA, Hamilton PB, Stachura-Suchoples K, Kociolek JP (2005) Freshwater diatom floras of North America: Cape Cod, Massachusetts, USA. Iconogr Diatomol 14:1–512
Snoeijs PJM (1992) Studies in the *Tabularia fasciculata* complex. Diatom Res 7:313–344
Spaulding SA, Elwell L (2007) Increase in nuisance blooms and geographic expansion of the freshwater diatom *Didymopshenia geminata*. Recommendations for response. US Geolog Survey Open-File Report 2007–1425.
Stachura-Suchoples K, Kociolek JP, Siver PA (2004) *Neidium pseudodensistriatum* sp. nov., a new epiphytic diatom from Florida (USA) and comparison with *N. densestriatum* (Østrup) Krammer. Proc Int Diatom Symp 17:359–370
Stachura-Suchoples K, Kociolek JP, Siver PA (2010) *Neidium palpebrum* sp. nov., a new diatom from Florida (USA): comparison with *N. sacoense* Reimer and some remarks on biogeography. Diatom Res 25:385–395
Stepanek JG, Kociolek JP (2012) Several new species from the genera *Amphora* Ehrenberg *ex* Kützing and *Halamphora* (Cleve) Levkov from the western United States. Diatom Res 28:61–76
Stepanek JG, Kociolek JP (2015) Three new species of the diatom genus *Halamphora* (Bacillariophyta) from the prairie pothole lakes region of North Dakota, USA. Phytotaxa 197:27–36
Stevenson RJ (1997) Scale-dependent causal frameworks and the consequences of benthic algal heterogeneity. J N Am Benthol Soc 16:248–262
Stevenson RJ, Bahls L (1999) Periphyton protocols. In: Barbour MT, Gerritsen J, Snyder BD, Stribling JB (eds) Rapid bioassessment protocols for use in streams and wadeable rivers: periphyton, benthic macroinvertebrates and fish. EPA 841-B-99-002, 2nd edn. U.S. Environmental Protection Agency, Office of Water, Washington, DC, pp 97–118
Stevenson RJ, Pan Y, Van Dam H (2010) Assessing environmental conditions in lakes and streams with diatoms. In: Stoermer EF, Smol JP (eds) The diatoms: application for environmental and earth sciences. Cambridge University Press, Cambridge, pp 57–85
Tanaka H (2007) Taxonomic studies of the genera *Cyclotella* (Kützing) Brébisson, *Discostella* Houk *et* Klee, and *Puncticulata* Håkanson in the family Stephanodiscaceae Glezer et Makarova (Bacilariophyta) in Japan. Biblioth Diatomol 53:1–205
Taylor JC, Harding WR, Archibald CGM (2007) Method manual for the collection, preparation and analysis of diatom samples. Report to the Water Research Commission, Pretoria
Thomas EW, Kociolek JP (2008) Taxonomy and ultrastructure of two new *Neidium* species (Bacillariophyceae) from lakes in the Sierra Nevada mountains of northern California (USA). Diatom Res 23:471–482
Thomas EW, Kociolek JP (2015) Taxonomy and ultrastructure of several new *Rhoicosphenia* (Bacillariophyta) from California, USA. Phytotaxa 204:001–021
Thomas EW, Kociolek JP, Lowe RL, Johansen JR (2009) Taxonomy, ultrastructure and distribution of gomphonemoid diatoms (Bacillariophyceae) from Great Smoky Mountains National Park (U.S.A.). Nova Hedwigia 135:201–238
Thomas EW, Kociolek JP, Karthick B (2015) Four new *Rhoicosphenia* species from fossil deposits in India and North America. Diatom Res 30:35–54
Trapp EM, Adler S, Zauner S, Maier UG (2012) *Rhopalodia gibba* and its endosymbionts as a model for early steps in a cyanobacterial primary endosymbiosis. Endocytobiosis Cell Res 23:21–24
Tuchman N, Schollett M, Rier S et al (2006) Differential heterotrophic utilization of organic compounds by diatoms and bacteria under light and dark conditions. Hydrobiologia 561:167–177
Wang Q, Hamilton PB, Kang F (2014) Observations on attachment strategies of periphytic diatoms in changing lotic systems (Ottawa, Canada). Nova Hedwigia 99:239–253

Whitton BA, Ellwood NTW, Kawecka B (2009) Biology of the freshwater diatom *Didymosphenia*: a review. Hydrobiologia 630:1–37
Williams DM (1985) Morphology, taxonomy and inter-relationships of the ribbed araphid diatoms from the genera *Diatoma* and *Meridion* (Diatomaceae: Bacillariophyta). Biblioth Diatomol 8:1–228
Williams DM, Round FE (1987) Revision of the genus *Fragilaria*. Diatom Res 2:267–288
Wojtal AZ, Ector L, Van De Vijver B, Morales EA, Blanco S, Piatek J, Smieja A (2011) The *Achnanthidium minutissimum* complex in southern Poland. Algol Stud 136(137):211–238

Chapter 6
Brown Algae (Phaeophyceae) in Rivers

John D. Wehr

Abstract Freshwater brown algae can be abundant in streams, but represent just seven species in the Phaeophyceae, a class of ~2000 species, most from marine environments. Freshwater species do not form parenchyma, but are based on one of the three filamentous growth forms: (1) uniseriate, creeping filaments infrequently or frequently branched; (2) complex branched forms producing basal and vertical series of filaments forming a crust; and (3) multiseriate, frequently branched forms. Their evolutionary history and relation to marine taxa is unclear, with molecular data suggesting some closely related to existing marine taxa, while others forming a separate clade of freshwater forms. Most species are known from few locations worldwide, with all but a few reports from sites in the northern hemisphere. All species are benthic and most colonize epilithic substrata, but a few are epiphytes on macrophytes or other algae. A key to all known species is provided.

Keywords Algae • Benthic • Biodiversity • Brown algae • Genus, Phaeophyceae • River

Introduction

The Phaeophyceae, or brown algae, represent a large and heterogeneous class within the Heterokontophyta (van den Hoek et al. 1995) or Stramenopiles, which also includes diatoms and chrysophytes (Patterson 1989). Currently more than 2000 species have been described within ~285 genera, at least 50 families, and 19 orders (de Reviers et al. 2007). The forms of algae that comprise the class include complex, macroscopic thalli with parenchyma and meristematic regions, many of which are large conspicuous seaweeds, and simpler, mostly microscopic, filamentous

J.D. Wehr (✉)
Louis Calder Center—Biological Field Station, Fordham University, Armonk, NY 10504, USA
e-mail: wehr@fordham.edu

O. Necchi Jr. (ed.), *River Algae*, DOI 10.1007/978-3-319-31984-1_6

forms (van den Hoek et al. 1995; Wehr 2015). Of the many and diverse taxa, nearly all are marine in habit, and only 6 or 7 are currently recognized from freshwater environments. All freshwater forms are filamentous, but may develop colonies or other forms recognizable in the field. The macroscopic appearances of freshwater taxa result from different branching patterns and type of cellular organization. However, no members of the Phaeophyceae are unicellular in the vegetative phase—the predominant morphology in other golden-brown groups. Brown algae have cell walls composed of alginates, sulfated fucans, and cellulose, in average proportions of ~ 3:1:1 (Kloareg and Mabeau 1987; Michel et al. 2010). Walls may also be supplemented with phlorotannins (halogenated/sulfated phenolic compounds; Vreeland et al. 1998), and lesser quantities of proteins (Quatrano and Stevens 1976).

The brownish color of phaeophytes is the result of the carotenoid pigment, fucoxanthin, and in some species, various phaeophycean tannins. These tannins accumulate in cytoplasmic inclusions known as physodes, and are characteristic of the group. They have been suggested to function as an inducible defense against herbivory or as protection against UV exposure (Schoenwaelder 2002; Lüder and Clayton 2004). In addition to chlorophylls a, c_1, and c_2, chloroplasts also contain various other xanthophylls including β-carotene, diadinoxanthin, neoxanthin and violoxanthin. Cells contain one to several chloroplasts, which may be discoid or ribbon-like and may or may not contain pyrenoids (Pueschel and Stein 1983; Silberfeld et al. 2011). The ultrastructure of plastids is consistent with other members of the heterokont algae, having thylakoids in stacks of three enclosed by a girdle lamella. There is also a peripheral endoplasmic reticulum, which is contiguous with the nuclear envelope. The primary storage product is laminarin, a water-soluble β-1,3-glucan that accumulates in the cytoplasm. Cells may also produce and store mannitol, sucrose, glycerol, or oil droplets (Graham et al. 2009).

The first description of freshwater brown algae was published more than 100 years ago (*Pleurocladia* was described by Braun in 1855 and *Heribaudiella* by Gomont in 1896). Despite this relatively long history of their study, their diversity, biogeography, and ecology are still incompletely known. Most records of freshwater phaeophytes have come from streams and rivers, but several reports describe populations from littoral zone of lakes (Waern 1952; Schloesser and Blum 1980; Kann 1993; Wehr 2015). It is surprising that so few records exist, as some species, such as *Heribaudiella fluviatilis*, can at times cover large areas of stones in small and medium-sized streams and rivers (Holmes and Whitton 1977a; Kann 1978a; Wehr and Stein 1985). It has been mentioned that field researchers may mistake the brown crusts as cyanobacteria, diatoms, or even lichens (Holmes and Whitton 1975). Newer data suggest that the worldwide distribution may be wider than was previously thought (Eloranta et al. 2011), although this relatively common alga has yet to be described from the southern hemisphere.

Classification and Diversity

Because of their relatively simple morphology, thallus organization, and reproductive structures, traditional classification schemes place nearly all of the freshwater members of the Phaeophyceae in a simple, ancestral lineage, with most as members of the order Ectocarpales (five genera) or Sphacelariales (one genus, two species) (Van den Hoek et al. 1995; Graham et al. 2009). From a morphological standpoint, this made sense, as freshwater members exhibit far less morphological complexity than do marine taxa. However, *rbc*L sequences from most of the recognized freshwater taxa indicate a polyphyletic origin, with some members only distantly related to *Ectocarpus* (McCauley and Wehr 2007). Further, these data also suggest there may be greater genetic diversity within some clades. Recent molecular studies on marine taxa also suggest that traditional classifications in which simple forms (like most freshwater species) are thought to be ancestral to complex lineages may not be true (de Reviers et al. 2007). While DNA sequence information is presently incomplete, published molecular data (McCauley and Wehr 2007) tentatively suggest five lineages of freshwater brown algae, as follows: (a) Ectocarpales, Ectocarpaceae: *Ectocarpus subulatus* Kützing, *Pleurocladia lacustris* A. Braun; (b) Sphacelariales, Sphacelariaceae: *Sphacelaria lacustris* Schloesser and Blum, *S. fluviatilis* C.-C.Jao; (c) Heribaudiellales, Heribaudiellaceae (provisional): *Heribaudiella fluviatilis* (Areschoug) Svedelius, *Bodanella lauterborni* Zimmermann; (d) Ralfsiales, Ralfsiaceae: *Porterinema fluviatile* (H.C. Porter) Waern.

The nomenclature has varied over the years and synonymies are numerous in the older literature, but these have been largely resolved, based on current data (McCauley and Wehr 2007; Eloranta et al. 2011; Wehr 2015; see also entries in Guiry and Guiry 2015). However, much of the phylogeny is unsettled, so in this chapter only morphological traits will be used to differentiate taxa. However, because of uncertainty regarding the reliability of certain morphological features, such as branching patterns, colony shape, or presence of hairs as taxonomic attributes, taxa described in this account may become synonymous or be split into several species with further study. The scheme used here is artificial until further phylogenetic relationships are resolved.

Even if further molecular studies determine reveal greater cryptic diversity than previously thought, the number of freshwater brown algal taxa will remain few compared with marine members of the class, or with other freshwater groups. Their diversity within a freshwater habitat is similarly low. It is uncommon to observe more than one freshwater phaeophyte in a single habitat, although Kann (1993) observed filamentous *P. lacustris* and crust-forming *H. fluviatilis* occurring together on stones in the littoral zone of Lake Erken, Sweden, and Kusel-Fetzmann (1996) noted that the *Pleurocladia* may grow as an epiphyte on *Heribaudiella* in certain Austrian streams. A survey of 121 stream-reaches in western Canada and USA identified 25 locations with *H. fluviatilis* present, but none contained other species of brown algae (Wehr and Stein 1985). However, upper and middle reaches of the

Green River in Wyoming have extensive populations of *Heribaudiella*, while ~ 200 km downstream (below Flaming Gorge Dam), stones in the river were colonized by *Pleurocladia* (Wehr unpublished data). Israelsson (1938) has suggested that different species of freshwater brown algae rarely co-occur because of strongly different ecological requirements. Given the few species of freshwater brown algae overall, it is not surprising that their local diversity within specific habitats or along water courses is low.

Morphology and Reproduction

The growth patterns and microscopic morphologies of freshwater brown algae occur in one of three types: (1) uniseriate (single axis), alternately or dichotomously branched filaments (Fig. 6.1a, b) (e.g., *Bodanella, Ectocarpus, Pleurocladia, Porterinema*); (2) a two-phase system, with prostrate, irregularly branched filaments (Fig. 6.1c), which produce a series of densely packed, vertical filaments (Fig. 6.1d) that are usually dichotomously branched, and forming a crustose morphology (e.g., *Heribaudiella*); and (3) uni- and (progressively) multiseriate filaments (Fig. 6.1e) with alternate or opposite branching patterns (e.g., *Sphacelaria*).

Life cycles are incompletely known for most freshwater species, but observations of field populations and culture isolates suggest that all have an isomorphic alternation of generations. Not all stages have been observed in freshwater species, but the basic life cycle pattern known for *Ectocarpus* is thought to be a fundamental model for the group (Coelho et al. 2012). On diploid filaments (sporophytes), terminal unilocular (single compartment) sporangia release motile, haploid, biflagellate zoospores (meiospores). Most appear as enlarged spherical, ovate, or club-shaped structures (Fig. 6.1f). In freshwater species that have been studied, released zoospores can disperse and eventually settle on various substrata, form germination tubes, and develop into new filaments (Kumano and Hirose 1959; Müller and Geller 1978; Yoshizaki et al. 1984).

Gametophyte filaments produce plurilocular (multiple compartment) sporangia, which release motile zoospores (gametes) that fuse to form zygotes and produce new sporophytes. The biflagellate zoospores produced by *Heribaudiella* and *Porterinema* are pear-shaped with two laterally inserted flagella, and possess a single parietal chloroplast and an apical stigma (Kumano and Hirose 1959; Dop 1979). Both types of sporangia (and presumably ploidy levels) have been observed in *Ectocarpus* and *Heribaudiella* (Svedelius 1930; West and Kraft 1996; Charrier et al. 2008; Wehr 2015). In *Pleurocladia* and *Porterinema*, only unilocular sporangia have been observed (Kusel-Fetzmann 1996; Dop 1979). Reproduction, while documented in marine *Sphacelaria* species, is unknown in freshwater forms. All known freshwater species likely have a diplohaplontic life history with diploid and haploid vegetative phases, and which are isomorphic (both diploid and haploid stages identical or very similar).

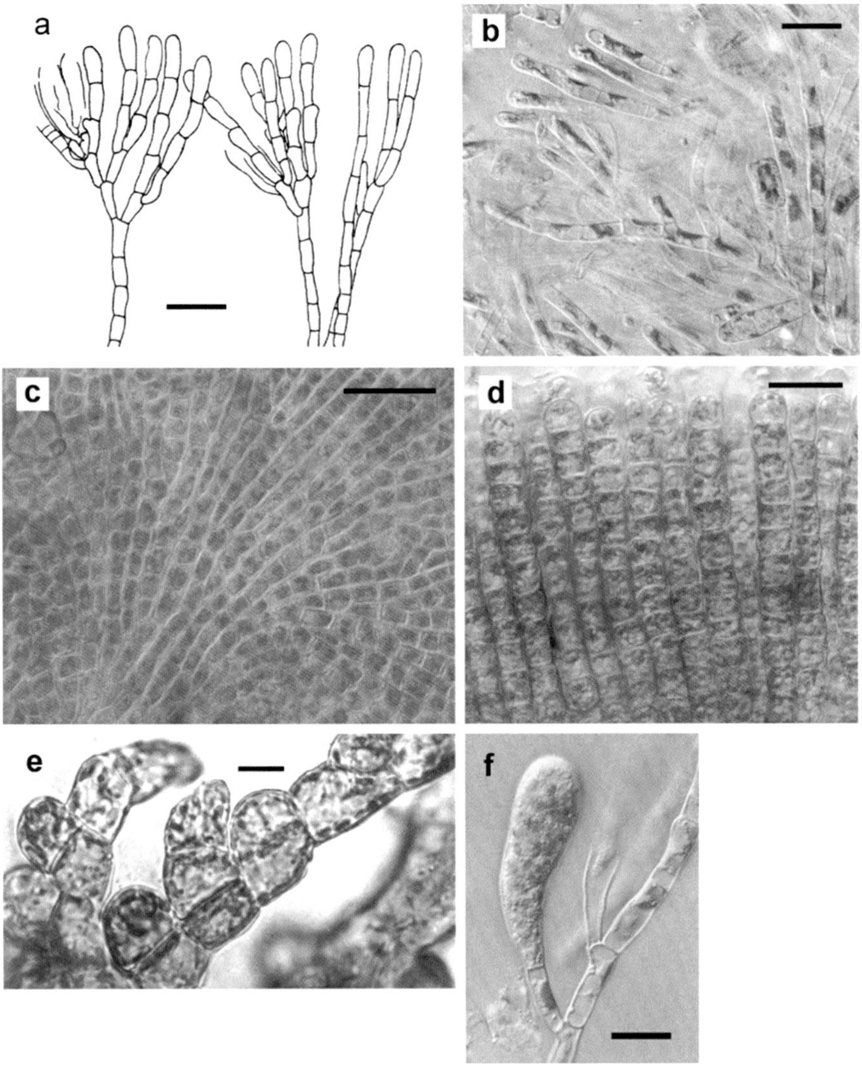

Fig. 6.1 (**a**–**f**) Three basic filamentous growth forms of freshwater phaeophytes. (**a**–**b**) *Pleurocladia*: simple, uniseriate filaments, with alternate or irregular branching. (**c**–**d**) *Heribaudiella*: (**c**) two-phase system, with prostrate, irregularly branched filaments; (**d**) series of densely packed, vertical filaments. (**e**) *Sphacelaria*: uni- and multiseriate filaments with alternate or opposite branching patterns; note cell divisions in two planes. (**f**) *Pleurocladia*: terminal unilocular (single compartment) sporangium. Scale bars: **a**–**d** = 50 μm; **e** = 20 μm; f = 10 μm

Colonies or thalli based on each of these developmental patterns can form loose macroscopic aggregations, hemispherical tufts, or crusts on solid substrata, which can be recognized with the naked eye. Most often, freshwater phaeophytes colonize rocks and boulders in rivers, although *Pleurocladia* can also grow epiphytically on

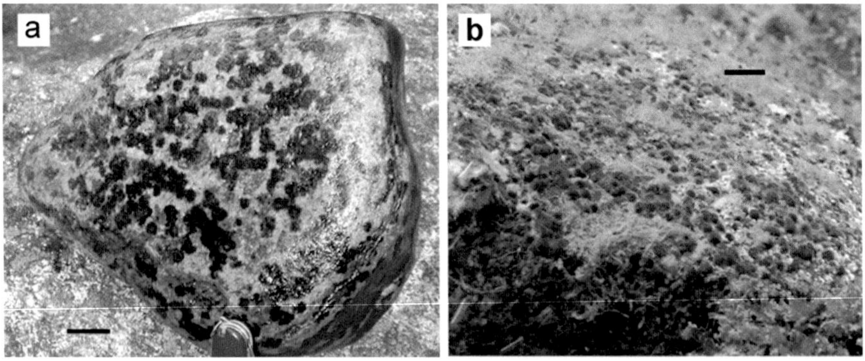

Fig. 6.2 (**a–b**) Macroscopic appearance of two freshwater phaeophytes. (**a**): *Heribaudiella*: crustose form. (**b**) *Pleurocladia*: macroscopic cushions or tufts. Scale bars: **a** = 2 cm; **b** = 1 cm

aquatic plants (Waern 1952), and its spores can germinate and develop on artificial substrata, including several types of plastic (Kann and Tschamler 1976). From small filaments, each of the species expands into larger networks to form hemispherical cushions or distinct crusts; none form true parenchymatous (tissue-like) thalli, as seen in many seaweeds. Crustose forms, which can be recognized with the naked eye (Fig. 6.2a), range from a few vertically arranged cells up to roughly 30 cells tall (1–2 mm), and create colonies of perhaps a few mm to 50 cm^2 in area (Wehr and Stein 1985; Kusel-Fetzmann 1996). Filamentous forms can aggregate to form macroscopic tufts 1–10 mm in size (Fig. 6.2b).

Most species are capable of forming hyaline, multicellular filaments, or hairs (Waern 1952; Wilce 1966; Dop 1979; Schloesser and Blum 1980; Yoshizaki et al. 1984; Wujek et al. 1996; Eloranta et al. 2011). Given that hairs appear to be environmentally induced, such as in response to low inorganic phosphorus (Wehr 2015) or long photoperiods (Dop 1979), they should not be used as diagnostic features. Some species, such as *Pleurocladia*, occur in more calcareous environments, and become encrusted with $CaCO_3$, which may cause the thallus to appear a pale brown macroscopically; microscopically carbonates may even cover filaments in a carbonate tube (Kirkby et al. 1972; Kusel-Fetzmann 1996; Wehr et al. 2013).

Ecology

Substrata

All freshwater species have a benthic habit, most colonizing epilithic substrata in streams and lakes. And of those species that occur in running waters, most are reported in rocky, clear-water streams (Fritsch 1929; Allorge and Manguin 1941; Jao 1941; Holmes and Whitton 1975; Kann 1978a, b; Wehr and Stein 1985; Yoshizaki and Iura 1991; Kusel-Fetzmann 1996; Wehr et al. 2013). *H. fluviatilis* forms dark brown crusts on rocks and grows rather slowly, so must in streams colonize cobbles or boulders that are large enough to be resistant to disturbance during flood events. In the turbulent Rhine River, *Heribaudiella* colonized rocks at depths down to 2 m, presumably to avoid effects of scouring and high discharge periods (Backhaus 2006). Some studies further suggest this alga also has a preference for geologically resistant rocks, such as basalt, quartz, schist, and gneiss (Allorge and Manguin 1941; Wehr and Stein 1985), although Kusel-Fetzmann (1996) also reported this alga colonizing bricks in one Austrian stream. In contrast, *P. lacustris* seems to be indifferent to type of surface. For example, *P. lacustris* grows on rocks and boulders in mountain streams in Austria, California, and Utah (Ekenstam et al. 1996; Kusel-Fetzmann 1996; Wehr et al. 2013), on rocks and pebbles (Kahlert et al. 2002) and on glass slides (Kann 1993) in Lake Erken, Sweden. This species is also found as an epiphyte on the filamentous alga *Cladophora* in Lake Michigan (Young et al. 2010), is epizooic on zebra mussel shells (Carter and Lowe 2001), epiphytic on reeds and other emergent macrophytes (Waern 1952; Kirkby et al. 1972; Szymanska and Zakrys 1990), and even grows as inside the tissues of aquatic plants and macroalgae (Waern 1952). *P. fluviatile,* a species less often reported in fresh waters, also occurs in running waters on reed stems and is able to colonize submerged glass slides in slowly flowing ditches and eutrophic lakes (Dop and Vroman 1976). But in general, most populations of freshwater phaeophytes, especially *H. fluviatilis, P. lacustris,* and *S. fluviatilis*, develop substantial populations on colonize epilithic substrata in flowing waters (Jao 1943; Kann 1978a; Wehr and Stein 1985; Kusel-Fetzmann 1996; Wehr et al. 2013).

Salinity

As the vast majority (>99 %) of the members of the Phaeophyceae occur in marine environments, it may be predicted that some freshwater species can occupy habitats across a range of salinities. This is clearly true for the euryhaline species of *Ectocarpus* spp. and *P. fluviatile*, taxa that are in fact more often reported from marine or brackish environments (Dop 1979; West and Kraft 1996). The first discovery of *E. subulatus* (as *E. siliculosus*) in a freshwater river in Australia was from a waterfall just 40 m above the high tide level, with a fairly high specific conductivity of 3.0 mS cm^{-1}, but co-occurring with freshwater species of green algae (West and Kraft 1996). In culture the alga tolerates a wide range of salinities to nearly full seawater (West and Kraft 1996; Wehr unpublished data). Recent experiments with this freshwater strain have shown this that alga differed from a marine strain in having specific reversible, morphological, physiological, and transcriptomic changes when exposed to different salinities (Dittami et al. 2012). A species of *Ectocarpus* (recorded as *E. confervoides*) also occurs in the River Werra (Germany), in sites polluted potassium mine waste, along with several common freshwater species (e.g., *Anabaena, Cyclotella, Scenedesmus*) (Geissler 1983).

But of the other species, only *Pleurocladia lacustris* has been collected from slightly brackish waters (0–8 %). It is a common epiphyte on reeds in near-fresh sites in wetlands connected to the Baltic Sea (Waern 1952), and colonizes boulders just above the high tide line in Arctic Canada and Greenland (Wilce 1966). The majority of sites in which *P. lacustris* has been recorded are calcareous freshwater streams and rivers (Israelsson 1938; Ekenstam et al. 1996; Wehr et al. 2013). Indeed some of these sites have been described as travertine streams, in which colonies of *Pleurocladia* form calcareous nodules on rocks (Wehr et al. 2013). The lentic populations are also strongly calcareous, which may suggest a demand for high cation levels rather than strictly Na concentrations. The most widely observed freshwater phaeophyte, *H. fluviatilis*, has no know marine or brackish populations (Wehr 2015).

Interestingly, there are no reports of any members of the Phaeophyceae from inland saline lakes, habitats that occur on most continents. However, *P. fluviatile* has been collected from many brackish and freshwater sites in Europe (Waern 1952). Wilce et al. (1970) described a North American population from a freshwater site adjacent to a salt marsh in Massachusetts.

Water Quality and Nutrients

Ecological data for most freshwater phaeophytes are few, general consensus (citations below) indicates few if any populations of freshwater phaeophytes occur in acid (< pH 6.0), or in strongly nutrient-enriched environments. There have been several studies that provide more specific information for *H. fluviatilis*. Israelsson (1938) characterized this species as broadly tolerant of oligotrophic to eutrophic conditions. More specific data suggest this species prefers neutral to moderately alkaline water (≥pH 7.0), but fairly broad Ca, P, and N levels (Wehr 2015). An assessment of benthic algae in more than 100 streams and rivers in Finland, Norway, and Sweden characterized *H. fluviatilis* as an indicator of relatively low total P (periphyton index of trophic status indicator value = 4.98; Schneider and Lindstrom 2011), while the mean total dissolved phosphorus (TDP) level of *H. fluviatilis* sites in British Columbia was ~14 μg/L (Wehr and Stein 1985). One population from an Oregon river had elevated soluble reactive phosphorus (62.5 μg L^{-1} P) but low inorganic N (NO_3^-: 5.9, NH_4^+ 64.9 μg L^{-1} N; Wehr and Perrone 2003). Early reports in Europe also suggested that *H. fluviatilis* was most common in low-calcium rivers (Kann 1966, 1978a, b), although it frequently co-occurred with the encrusting red alga *Hildenbrandia rivularis* a species more common in Ca-rich streams (e.g. Fritsch 1929; Holmes and Whitton 1975, 1977a; Kusel-Fetzmann 1996).

As mentioned earlier, *P. lacustris* appears to prefer alkaline, hard-water systems (pH ≥ 8.0; Ca > 30 mg/L), often with $CaCO_3$ precipitate intermixed with or even enclosing the filamentous colonies (Kusel-Fetzmann 1996). Nutrient data for this species are fewer, although some reports in Europe suggest it is most common in moderately nutrient-rich, alkaline, environments (Israelsson 1938; Waern 1952; Kusel-Fetzmann 1996). Populations in California occur in calcareous mountain streams with relatively low TDP (9.5 μg/L) and TDN (63.6 μg/L; Wehr et al. 2013). The species may be adapted to low inorganic P concentrations, as most populations (including those from California) produce extensive hyaline hairs that are likely sites of alkaline phosphatase activity (Wehr 2015). Interestingly, sites in the UK where *P. lacustris* has been reported in past decades but are no longer observed have experienced nutrient pollution (Kirkby et al. 1972; Brodie et al. 2009). Elevated nutrients may have caused its local or regional extinction (Wehr 2011). No rigorous studies have yet been conducted on the nutrient requirements of this species in streams, but *P. lacustris* comprised a major percentage of a periphyton assemblage (5 – 20 % of biovolume) with high N:P ratios and elevated alkaline phosphatase activity in an epilithic lake habitat (Kahlert et al. 2002). None of the other freshwater phaeophytes have been sampled from enough river locations to draw even general conclusions about possible nutrient requirements or limits.

Light

The light requirements of freshwater phaeophytes have not been studied experimentally, although a number of field studies suggest that *H. fluviatilis* preferentially colonizes shaded reaches of streams (Kusel-Fetzmann 1996; Eloranta et al. 2011). Its presence in deep water (to 2 m) in the River Rhine also suggests a tolerance of low light levels (Backhaus 2006). But these reports contrast with results from river surveys in North America and the UK, which have reported this species in both shaded and open habitats (broad streams or areas devoid of riparian vegetation (Holmes and Whitton 1977a, b; Wehr and Stein 1985). In culture, both *H. fluviatilis* and *P. lacustris* grow well at irradiances of ~100 μmol photons m^{-2} s^{-1} (Wehr unpublished data)

However, there are reports of freshwater phaeophytes from very deep locations in lakes, including *B. lauterborni* (>15 m; Geitler 1928; Müller and Geller 1978; Kann 1982), *S. lacustris* (5–15 m; Schloesser and Blum 1980). Culture studies with this alga found that optimum growth and reproduction occurs under reduced light levels (1000 lux) and short-day (8 L:16D) conditions (Schloesser 1977; Schloesser and Blum 1980). *Sphacelaria lacustris* contrasts with *S. fluviatilis,* the latter been reported as a dominant member of the benthic algal assemblage in both open and shaded sections of the Kialing River in China (Jao 1943, 1944).

Distribution

Presently, perhaps a few hundred populations of freshwater brown algae have been reported worldwide, most from streams and rivers. The large majority of these reports are for just two species: *H. fluviatilis* and *P. lacustris*. The lack of reports for other species may be due to their relatively small size and under-reporting, but others may have a very limited distribution. Additional confusion has arisen due to synonymies and reports of new species based on limited morphological information (Wehr 2015).

The first described species of freshwater phaeophyte was *Pleurocladia lacustris* A. Braun (1855), a species that despite more than 150 years of study, has been since reported from less than 30 freshwater (or intermittently fresh) locations globally

(Wehr et al. 2013). Most are in western Europe, but there are also populations in North America and one in Australia. Distances between these reported populations can be hundreds or thousands of miles, assuming that no other unknown populations occur nearby. Many but not all freshwater populations have been reported from streams or lakes distant from the ocean. In northern Europe, distribution patterns of *Pleurocladia* indicate very few if any nearby marine populations from which freshwater populations may have invaded (Israelsson 1938; Waern 1952; Szymanska and Zakrys 1990; Kusel-Fetzmann 1996). It has been suggested that the freshwater history of this species may be quite old, perhaps pre-glacial (Waern 1952). But given the occurrence of several marine or brackish populations elsewhere, all of which appear morphologically identical, the means of dispersal and the possibility of adaptation to low salinity are topics worthy of careful study (Wilce 1966; Wehr et al. 2013). Data such as these have led some to suggest this and other freshwater phaeophytes may have isolated or disjunct distributions (Wehr 2015), but future discoveries may alter that perspective.

Heribaudiella fluviatilis appears to be the most widely reported species brown algae from freshwater habitats. None have been reported from brackish or marine environments. Populations are known from three continents, with locations in Europe, western North America (one from eastern North America), Russia, Japan, and China (Wehr and Stein 1985; Kusel-Fetzmann 1996; Wehr 2015; Johansen et al. 2007). These include temperate, boreal, subarctic, and Mediterranean biomes, from ~63° N latitude (Sweden, Northwest Territories; Israelsson 1938; Sheath and Cole 1992), south to ~29° N latitude (China; Jao 1941), and from western North America, across Europe to Japan. None are yet known from subtropical or tropical climates, or any sites in the southern hemisphere. Its North American distribution exhibits an apparent anti-coastal distribution, with greatest abundances in mountainous interior regions (Wehr and Stein 1985). Similar patterns have been seen in the British Isles (Holmes and Whitton 1977a, b; Wehr 2011).

Beyond these two species, the distribution of the other freshwater members of the Phaeophyceae is poorly known. Some may indeed be rare. For example, *S. lacustris* is thus far known only from one location in western Lake Michigan (Schloesser and Blum 1980), and has not been collected or observed in the field since the late 1970s. A presumably related species, *S. fluviatilis*, is presently reported from only two known locations, a stream in south-central China and a small lake in Michigan (Jao 1943; Thompson 1975; Wujek et al. 1996). *Bodanella lauterborni* is apparently known from only three locations in western Europe. Each of these has been described only as a morphospecies, and thus may represent fewer (or more) taxa than has been thus far recognized. No molecular data has yet emerged to determine their phylogenetic relationships. The two reports of freshwater brown algae from the southern hemisphere (both in Australia) include a single population of *E.subulatus* (as *E. siliculosus*; Dittami et al. 2012) from a river near the ocean (West and Kraft 1996), and a population of *P. lacustris* from a limestone spring (S. Skinner pers. comm.).

The wider question of the origin of freshwater phaeophytes, presumably from marine ancestors, has not been adequately studied. And because more than 99 % of all known species within the Phaeophyceae occupy marine habitats, questions concerning their dispersal, adaptation and ultimate evolution, to species that occupy freshwater rivers, need to be addressed. Some taxa apparently are not closely related to marine species (McCauley and Wehr 2007), with the exception of *E.* and *P. fluviatile* (Dop 1979; West and Kraft 1996). Recently, stepwise adaptation and genetic changes have been demonstrated in the freshwater population of *E. subulatus* (Dittami et al. 2012). Other taxa, such as *P. fluviatile*, may have a similar history.

Key to Genera of Brown Algae in Rivers (Based on Morphospecies Concepts Only)

1a	Macroscopic thalli small hemispherical tufts or broadly spreading filaments	2
1b	Thalli not in tufts or spreading filaments; low creeping form or in crusts	3
2a	Filaments uni- and multiseriate; with numerous disc-shaped chloroplasts	*Sphacelaria*
2b	Filaments uniseriate, usually branching to one side; single (rarely two) parietal chloroplasts; basal filaments curved; unilocular sporangium ovoid or irregular	*Pleurocladia*
3a	Thalli crustose, forming dark brown patches on stones; branched basal filaments with short, densely packed upright filaments, many discoid chloroplasts, unilocular sporangia club-shaped, plurilocular (rare) elongate	*Heribaudiella*
3b	Thalli not crustose, spreading or simple, variously branched	4
4a	Filaments sparingly branched, vegetative cells narrow, cylindrical, chloroplasts several, ribbon-like (occasionally spiral); plurilocular sporangia narrow-elongate	*Ectocarpus*
4b	Filaments frequently or irregularly branched, cells inflated or quadrate, chloroplasts few to several, plurilocular sporangia unknown or (if present) broad or inflated in shape	5
5a	Branched filaments with prostrate and erect forms	6
5b	Branched filaments prostrate only, creeping along substrata; may have rhizoid-like branches, chloroplasts many	*Bodanella*
6a	Basal filaments often curved or arching, erect filaments spreading; cells with single (rarely two) chloroplast, unilocular sporangia ovoid or irregular	*Pleurocladia*
6b	Basal and erect filaments irregularly arranged, with two parietal, lobed chloroplasts; plurilocular sporangia in crown-shaped clusters of four or more	*Porterinema*

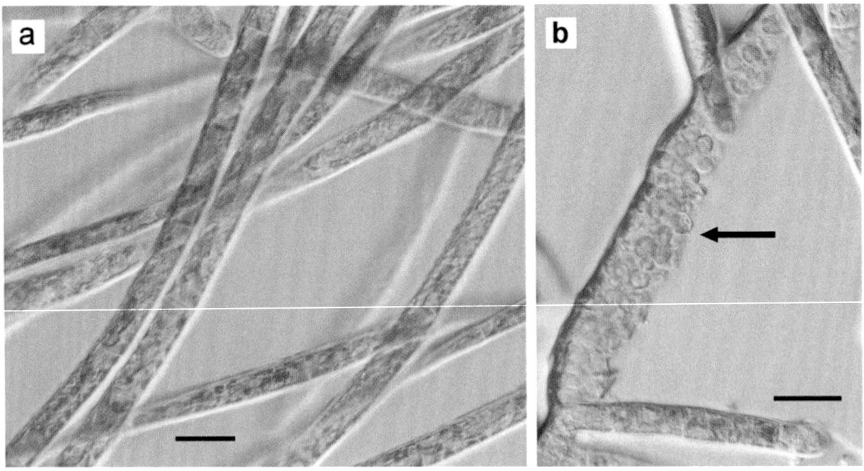

Fig. 6.3 (**a**–**b**) *Ectocarpus subulatus* (cultured material). (**a**) cylindrical (isodiametric) vegetative with multiple ribbon-like chloroplasts. (**b**) older filaments with terminal plurilocular sporangium (*arrow*) and numerous zooids being released. Scale bars = 20 μm

Description of Taxa

Ectocarpales, Ectocarpaceae: Genus Ectocarpus Lyngbye. *One Species Described from Freshwater*

Ectocarpus subulatus Kützing (originally described as *Ectocarpus siliculosus* (Dillwyn) Lyngbye) (Fig. 6.3a–b)

Uniseriate, sparingly or irregularly branched; thalli mostly erect filaments (up to 20 cm in length) consisting of cylindrical (isodiametric) cells 15–40 μm diameter, up to four times as long as broad (Fig. 6.3a). Chloroplasts several (2–4), ribbon-like (occasionally spirally arranged), *Ectocarpus subulatus* Kützing lobed and parietal, with pyrenoids, arrangement variable (e.g., net-like, ribbons, spiral). Plurilocular sporangia terminal (Fig. 6.3b), narrowly ellipsoid, conical or linear, with numerous divisions; length 70–200 (up to 500) μm × 15–35 μm diameter. Unilocular sporangia unknown in fresh water; common in marine populations.

Remarks: An ecologically and geographically widespread marine and estuarine alga (Müller 1979) was reported (originally as *E. siliculosus* [Dillwyn] Lyngbye) from a freshwater waterfall of the Hopkins River (Australia), 25 km from the Hopkins River mouth and 40 m above sea level (West and Kraft 1996). The site included a mix of freshwater green (e.g., *Mougeotia, Cladophora*), and marine red (*Caloglossa leprieurii*) algal taxa (West and Kraft 1996). Another species, *E. confervoides* (Roth) Kjellman, has been observed in the River Werra (Germany) in sites polluted by potassium mine waste (Geissler 1983), occurring along with several freshwater taxa (e.g., *Anabaena, Cyclotella, Scenedesmus*). The taxonomy of the freshwater population is being revised and moved to *E. subulatus* Kützing based on ITS, Rubisco spacer, and cox1 and cox2 genes (Stache-Crain et al. 1997; A. Peters, pers. comm.). A unialgal culture of the Australian population is maintained at the Culture Collection of Algae and Protozoa (CCAP 1310/196), and the River Werra strain has been deposited in the SAG culture collection in Gottingen (Kusel-Fetzmann and Schagerl 1992; http://sagdb.uni-goettingen.de/). Several important molecular analyses have recently been conducted on the freshwater *E. subulatus* strain (see Dittami et al. 2012).

Ectocarpales, Ectocarpaceae: Genus Pleurocladia A. Braun

Pleurocladia lacustris A. Braun

Thalli small, brown to pale brown or tan (depending on calcification) hemispherical tufts or cushions up to 3 mm diameter x 100–300 μm high) attached to rocks, mussel shells, or plants (angiosperms, mosses, macroalgae). Occasionally an endophyte within the tissues of aquatic plants. When calcified appear macroscopically pale tan or gray; in less calcareous regions, colonies may be gelatinous. At low magnification (100–200X) colonies appear as a dense cluster of radiating, branched filaments, often with crystals of $CaCO_3$. Two filament systems are usually evident: (a) a creeping or basal system, with (infrequently) branched filaments usually consisting of narrow or inflated cells 8–16 μm diameter (occasionally elongate), often exhibiting centrifugal or arched growth patterns, which give rise to (b) upright, irregularly (alternate or opposite) branched filaments, which are usually narrower (6–12 μm), cells elongate (12–35 μm long), and more nearly isodiametric. Vegetative cells contain one (rarely two) large golden-brown parietal chloroplast (with pyrenoids); darker granules (physodes?) and lipid bodies may be common. Unilocular sporangia common, single; clavate or globose (15–30 μm in diameter × 25–60 [–80] μm long); borne laterally or terminally. Plurilocular sporangia uncommon (Waern 1952); linear-elongate, narrow. Long (100–300 μm) multicellular hairs (5–7 μm diameter), which are induced by low inorganic P concentrations (Wehr 2015), are common in field populations. Hairs arise from upright filaments, giving the colony a fuzzy or whitish macroscopic appearance.

Remarks: occurs in rivers, streams and lakes. *Pleurocladia lacustris* is currently known from many sites in Europe, several streams and lakes in North America (including the Arctic), and one location in Australia (Wehr et al. 2013). The alga also colonizes brackish ponds and upper intertidal habitats, especially if intermittently fresh and saline (Waern 1952; Wilce 1966). A few recent field collections have been made, with historical and contemporary herbarium specimens placed in major herbaria (e.g., NY, USA), and a unialgal culture from a German population has been deposited in the SAG culture collection in Gottingen (Kusel-Fetzmann and Schagerl 1992; http://sagdb.uni-goettingen.de/). Molecular data based on 18S rDNA and *rbc*L sequences place this alga within the Ectocarpales (McCauley and Wehr 2007) (Fig. 6.4).

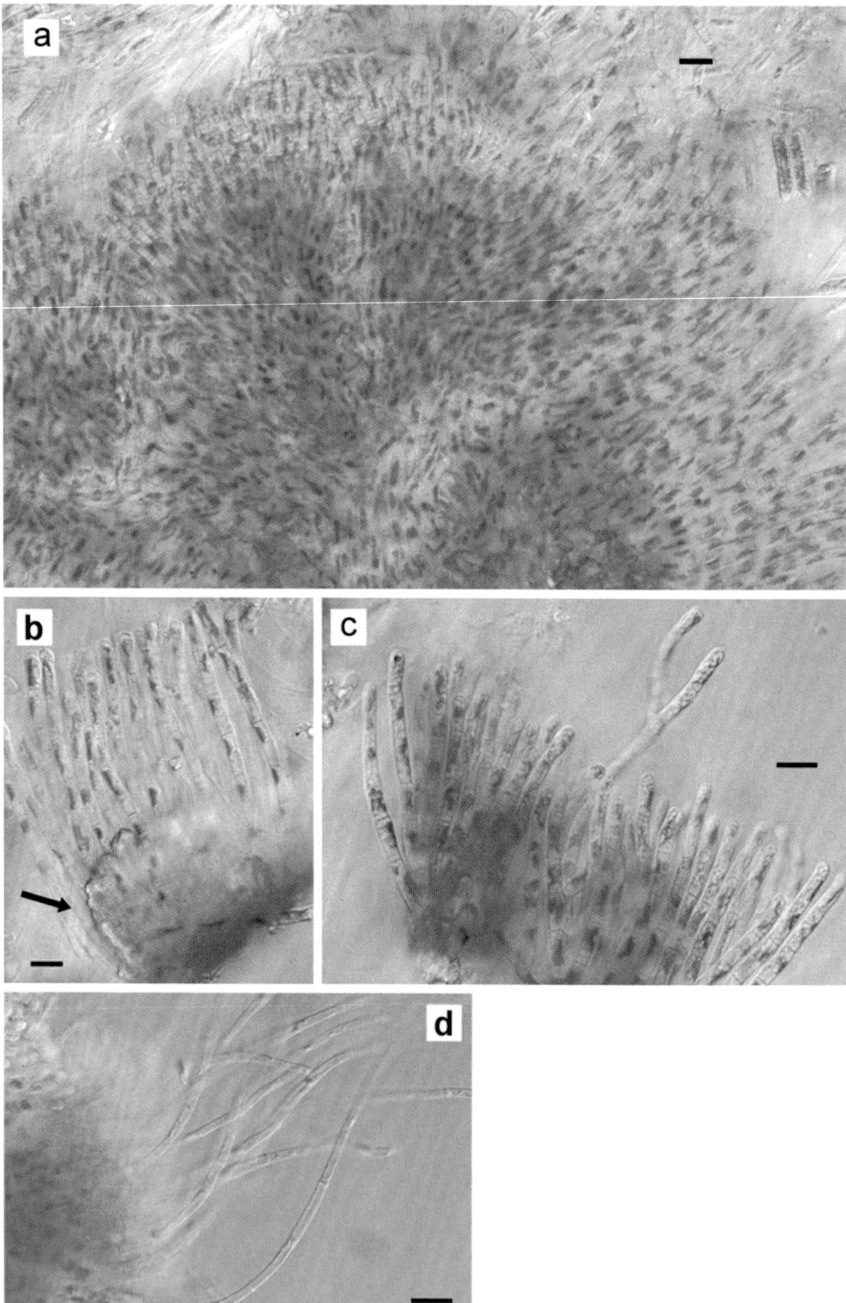

Fig. 6.4 (**a**–**b**) *Pleurocladia lacustris*. (**a**) view of cluster of filaments that form hemispherical cushion-like colonies. (**b**) Series of erect filaments with calcified (+$CaCO_3$) basal cells (*arrow*). (**c**) Spreading form of filaments with branches forming to one side of the main axes. (**d**) Terminal hyaline hairs arising from vegetative filaments. Scale bars = 20 μm

Sphacelariales, Sphacelariaceae: Genus Sphacelaria Lyngbye

Thalli small, up to ~3 mm in size, although multiple colonies may become confluent. Filaments form brown tufts or cushions on rocks in streams or lakes, which become calcified in hard water environments. Growth occurs from creeping (basal) and erect filaments; rhizoidal cells form where creeping filaments contact the substrata. Composed of uniseriate (younger) and multiseriate (older, usually basal) filaments; branches arise from upper position, resulting in an apical growth pattern. Cells in the main axis may be rectangular or inflated (12–25 μm diameter); broader prior to lateral cell division; cells nearly cylindrical on erect filaments. Cells contain numerous (10–20), small (3–8 μm), peripheral, disc-shaped chloroplasts and physodes; pyrenoids unknown. Multicellular hairs (>500 μm length) may develop from basal and erect filaments in some plants (function or factors unknown). Sporangia uncommon and differ with species; plurilocular sporangia are unknown in for either freshwater species.

The two described freshwater species have been separated on the basis of branching pattern.

Branching pattern opposite	*Sphacelaria fluviatilis*
Branching pattern irregular or alternate	*Sphacelaria lacustris*

Sphacelaria fluviatilis C.-C. Jao (Figs 6.1e and 6.5a–d)

Filament branching opposite (rarely alternate). Lateral cell divisions (from which arise multiseriate axes) are frequent; multiseriate axes common; some cells apically inflated. Unilocular and plurilocular sporangia are unknown, but main axes produce dense clusters of (sessile?) vegetative propagules (16–32). Hairs have not been observed.

Remarks: two populations have been described, from rapidly flowing water (occasionally subaerial) in the Kialing River China (Jao 1943, 1944), and in the shallow (≤1 m) littoral zone of Gull Lake, Michigan (Thompson 1975; Wujek et al. 1996). No recent field collections, herbarium specimens, or molecular data are known.

Sphacelaria lacustris Schloesser and Blum (Figs. 6.1e and 6.5a–d)

Filaments branching is alternate or irregular. Lateral cell divisions (and hence multiseriate axes) are infrequent, alternate, or irregular. Hairs have not been observed. Unilocular sporangia present, club shaped (Schloesser and Blum 1980). Plurilocular sporangia are unknown. Small clusters of gemmae-like propagules common; sessile or borne on short (1–2 celled) branches.

Remarks: The only known population of *S. lacustris* was reported from rocks in western Lake Michigan at depths of 5–15 m (Schloesser and Blum 1980). Herbarium specimens from cultured material were deposited in the National Herbarium (Smithsonian Institution). No recent field collections, herbarium specimens, or molecular data are known.

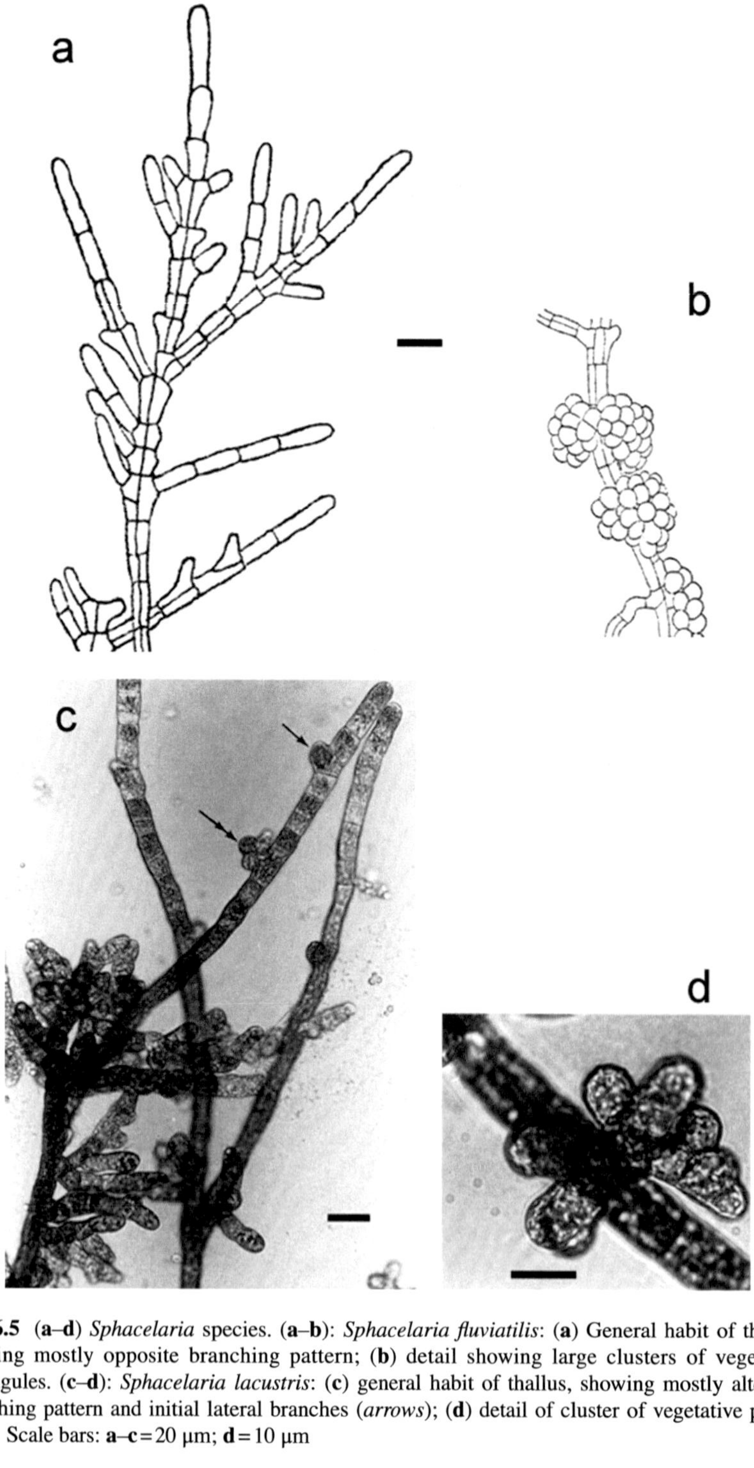

Fig. 6.5 (**a**–**d**) *Sphacelaria* species. (**a**–**b**): *Sphacelaria fluviatilis*: (**a**) General habit of thallus, showing mostly opposite branching pattern; (**b**) detail showing large clusters of vegetative propagules. (**c**–**d**): *Sphacelaria lacustris*: (**c**) general habit of thallus, showing mostly alternate branching pattern and initial lateral branches (*arrows*); (**d**) detail of cluster of vegetative propagules. Scale bars: **a**–**c** = 20 μm; **d** = 10 μm

Heribaudiellales, Heribaudiellaceae: Heribaudiella Gomont

Heribaudiella fluviatilis (Areschoug) Svedelius (Figs. 6.1c–d, 6.2a, 6.6a–c)

Thalli olive-brown to dark brown crusts on rocks in streams and lakes; individual colonies 1–10 cm in diameter with rounded or irregular outline, but with distinct margins; multiple colonies may coalesce to cover entire rocks. Thalli formed from dense networks of multiply branched, creeping filaments, which give rise to tightly packed vertical filaments. Filaments in the vertical series are infrequently (usually dichotomously) branched. Vertical series of filaments do not easily separate under pressure and may be missed during microscopic preparation unless prepared in thin sections or if fairly strong pressure is applied to the cover slip. Cells rectangular, 8–15 µm diameter, 5–15 (–20) cells long. Many oval or discoid chloroplasts (4–10 per cell; without pyrenoids), oil droplets, physodes present. Multicellular hyaline hairs may be present (up to 1 mm long). Unilocular sporangia terminal on vertical filaments, ovoid or clavate; 10–25 µm wide × 15–35 µm long. Biflagellate zoospores pyriform or irregular shape (≈6-8 µm). Plurilocular sporangia uncommon, also produced terminally in narrow-celled columns four (rarely 8) cells tall; immature plurilocular sporangia difficult to distinguish from smaller vegetative filaments.

Remarks: many species and synonyms resembling this species have been described based on morphological features, but all were united under *H. fluviatilis* (Areschoug) Svedelius (see Svedelius 1930; Wehr 2015; Eloranta et al. 2011). *Heribaudiella fluviatilis* is the most widely observed freshwater brown alga, with perhaps several hundred known populations worlwide Wehr and Stein 1985; Wehr 2015). Frequently co-occurs with the crustose red alga *Hildenbrandia rivularis* throughout Europe, but this association has not been seen in any populations from North America. Many recent field collections have been made, with herbarium specimens placed in major herbaria (e.g., NY), and an unialgal culture isolated from a stream in Austria has been deposited in the SAG culture collection in Gottingen (Kusel-Fetzmann and Schagerl 1992; http://sagdb.uni-goettingen.de/). Molecular data based on 18S rDNA and *rbc*L sequences place this alga in a group distant from the Ectocarpales, and sister to the Sphacelariales (McCauley and Wehr 2007).

Heribaudiellales, Heribaudiellaceae (Provisional Name): Genus Bodanella Zimmermann.

Bodanella lauterborni Zimmermann (Fig. 6.6d)

Thalli basal or creeping filaments, forming pseudoparenchymatous patches or disks on rocky substrata. No erect filaments. Filaments uniseriate, irregularly branched, composed of irregularly shaped cells that may be inflated, quadrate, angular, or ovoid; cells 10–16 µm wide, 10–25 µm long. Filaments look superficially like the creeping phase of *Heribaudiella*, although creeping filaments do not form pseudoparenchymatous crusts (McCauley and Wehr 2007). Filaments occasionally

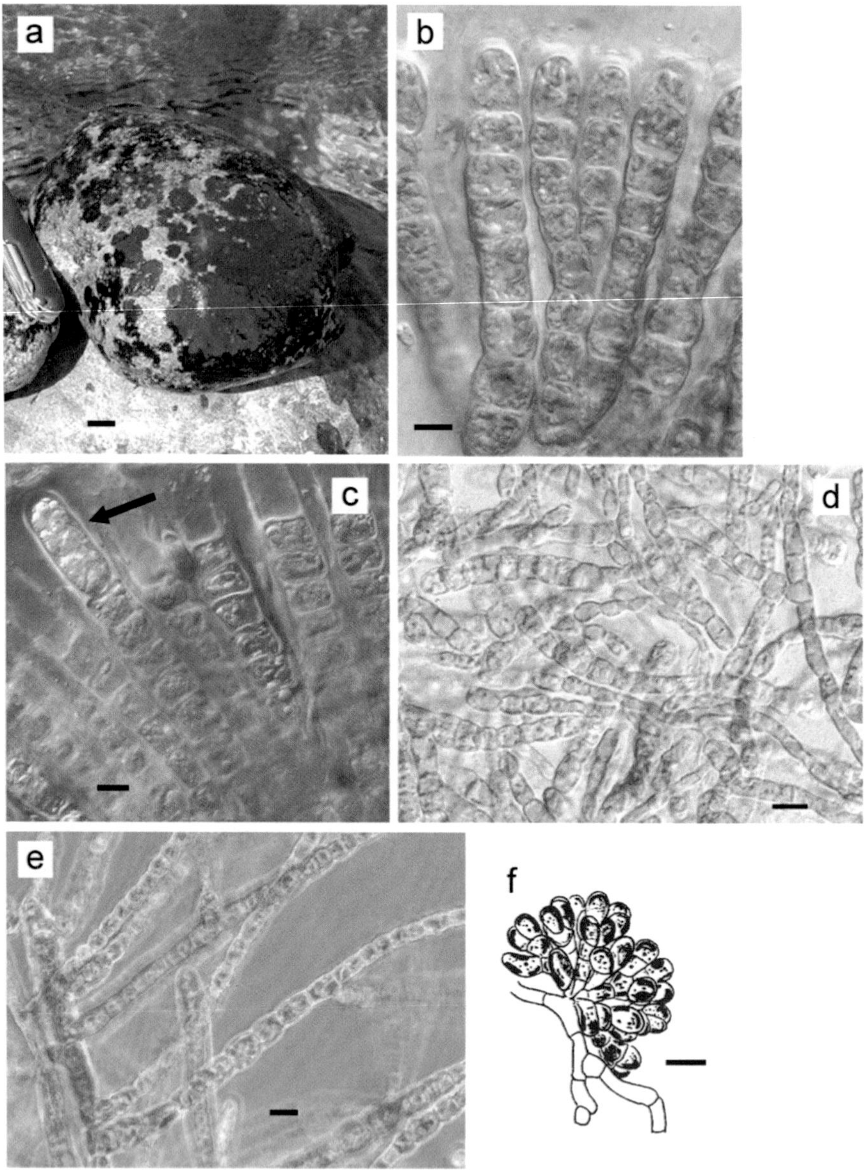

Fig. 6.6 (**a**–**c**) *Heribaudiella fluviatilis*: (**a**) appearance of multiple colonies coalescing to cover a large portion of a cobble; (**b**) detail of vertical filaments showing dichotomous branching, many discoid chloroplasts (*golden brown*) and refractive oil droplets; (**c**) terminal unilocular sporangium (*arrow*) and several empty sporangia from which zoospores were released. (**d**) *Bodanella lauterborni* (cultured material): creeping uniseriate filaments with inflated cells and irregular branching (cultured material). (**e**–**f**) *Porterinema fluviatile*: (**e**) creeping uniseriate filaments with inflated cells and irregular branches. (**f**) Cluster or "crown" of plurilocular sporangia on a pedicel. Scale bars: **a** = 1 cm; **b**–**f** = 10 μm

produce terminal, short, narrow hairs (6–10 μm diameter), or basal rhizoid-like filaments. Parietal chloroplasts small, numerous (10–15 per cell; without pyrenoids) and discoid. Unilocular sporangia ovoid or globose; 15–20 μm wide × 25–30 μm long. Zoospores pyriform (10–12 μm × 5–6 μm), with laterally inserted flagella. Plurilocular sporangia have not been observed.

Remarks: *Bodanella lauterborni* was named for its original location, Bodensee (Lake Constance, Germany-Switzerland-Austria), where it colonizes deep (15–35 m) limestone substrata (with *Hildenbrandia* and *Cladophora*). Not known from North America. Thus far, the worldwide distribution consists of three European lacustrine populations, in Lake Constance (Zimmermann 1928; Müller and Geller 1978), Lunzer Untersee (Austria: Geitler 1928), and Traunsee (Austria: Kann 1982). No recent field collections or herbarium specimens are known. A strain of *B. lauterborni* in culture is maintained at both the SAG and UTEX culture collections. Molecular data based on 18S rDNA and *rbc*L sequences place this alga nested within the *Heribaudiella* clade described above, and is closely related to *H. fluviatilis* (McCauley and Wehr (2007).

Ralfsiales, Ralfsiaceae: Genus Porterinema Waern

Porterinema fluviatile (H.C. Porter) Waern (Figs. 6.6e–f)

Thalli forming a single layer, brown disc-shaped plates of loosely arranged filaments. An epiphyte or endophyte on or in other algae (e.g., *Rhizoclonium, Enteromorpha*); also colonize stones and artificial substrata (e.g., glass slides). Thalli creeping, of irregularly branched filaments, with basal cells inflated on proximal ends. Occasionally producing short erect filaments (2–5 elongate cells; 6–8 μm diameter × 8–15 μm long) (Fig. 6.6e). Vegetative cells with 1 (occasionally up 3) plate-like or lobed, parietal chloroplasts (pyrenoids unknown). Terminal, multicellular hairs (3–8 μm diameter × 100–200 μm long) are common, and may be sheathed at their base. Plurilocular sporangia common, intercalary (occasionally terminal); typically in four-celled clusters or "crowns" on pedicels (short filaments) or sessile (Fig. 6.6f); sometimes produced in clusters of up to 32 sporangia. Unilocular sporangia rare.

Remarks: One species, *P. fluviatile*, is distributed mainly among brackish sites and slowly-flowing streams and ditches in Europe and North America. A few truly freshwater sites are known in Europe (Waern 1952; Dop 1979), and one site in North America: a stream draining into a salt marsh near Ipswich, Massachusetts (Wilce et al. 1970). Several taxa have been regarded as synonymies (see Wehr 2015; Eloranta et al. 2011). Molecular data based on 18S rDNA and *rbc*L sequences did not clearly resolve the position of *P. fluviatile* within Phaeophyceae (McCauley and Wehr 2007). Multiple estuarine populations have been collected, but there are no known freshwater specimens in major herbaria. Unialgal cultures are available through the SAG culture collection (Kusel-Fetzmann and Schagerl 1992; http://sagdb.uni-goettingen.de/).

Acknowledgements My thanks go to Alissa Perrone, Kam Truhn, Robert Sheath, Rosalina Stancheva, and Xian Wang for assistance in field sampling over the years. Also thanks go to Tim Entwisle, Elsa Kusel-Fetzmann, Dieter Müller, Jan Simons, Steve Skinner, John West, and Robert Wilce for advice on phaeophyte populations, and general information on the biology of the brown algae.

References

Allorge P, Manguin E (1941) Algues d'eau douce des Pyrénées basques. Bull Soc Bot France 88:159–191

Backhaus D (2006) Litorale Aufwuchsalgen im Hoch- und Oberrhein. Carolinea 64:5–68

Braun A (1855) Decade XLV+XLVI. In: Rabenhorst L (Ed.) (1848–1860) Die Algen Sachsens, Respective Mittel-Europas, Dresden

Brodie J, Andersen RA, Kawachi M et al (2009) Endangered algal species and how to protect them. Phycologia 48:423–438

Carter RL, Lowe RL (2001) Distribution and abundance of a previously unreported brown alga, Pleurocladia lacustris, in the littoral zone of northeastern Lake Michigan [abstract]. Proc North Am Benthol Soc, La Crosse, WI, USA

Charrier B, Coelho SM, Le Bail A et al (2008) Development and physiology of the brown alga *Ectocarpus siliculosus*: two centuries of research. New Phytol 177:319–332

Coelho SM, Scornet D, Rousvoal S et al (2012) Ectocarpus: a model organism for the brown algae. Cold Spring Harb Protoc 2011:193–198

de Reviers B, Rousseau F, Draisma SGA (2007) Classification of the Phaeophyceae from the past to the present and current challenges. In: Brodie J, Lewis J (eds) Unravelling the Algae: the past, present, and future of algal systematics. CRC Press (Taylor and Francis Group), Boca Raton, pp 267–284

Dittami SM, Gravot A, Goulitquer S et al (2012) Towards deciphering dynamic changes and evolutionary mechanisms involved in the adaptation to low salinities in *Ectocarpus* (brown algae). Plant J 71:366–377

Dop AJ (1979) *Porterinema fluviatile* (Porter) Waern (Phaeophyceae) in the Netherlands. Acta Bot Neerl 28:449–458

Dop AJ, Vroman M (1976) Observations on some interesting freshwater algae from the Netherlands. Acta Bot Neerl 25:321–328

Ekenstam D, Bozniak EG, Sommerfeld MR (1996) Freshwater *Pleurocladia* (Phaeophyta) in North America. J Phycol Suppl 32:15

Eloranta P, Kwandrans J, Kusel-Fetzmann E (2011) Rhodophyta and Phaeophyceae, Süßwasserflora von Mitteleuropa, vol 7. Spektrum Akademischer Verlag, Heidelberg

Fritsch FE (1929) The encrusting algal communities of certain fast-flowing streams. New Phytol 28:165–96

Geissler U (1983) Die salzbelastete Flußstrecke der Werra—ein Binnenlandstandort für *Ectocarpus confervoides* (Roth) Kjellman. Nova Hedwigia 37:193–217

Geitler L (1928) Über die Tiefenflora an Felsen im Lunzer Untersee. Arch Protistenkd 62:96–104

Gomont M (1896) Contribution à la flore algologique de la Haut-Auvergne. Bull Soc Bot France 43:373–393

Graham LE, Graham JM, Wilcox LW (2009) Algae, 2nd edn. Benjamin Cummings, San Francisco

Guiry MD, Guiry GM (2015) AlgaeBase. World-wide electronic publication, National University of Ireland, Galway. http://www.algaebase.org. Accessed 02 December 2015

Holmes NTH, Whitton BA (1975) Notes on some macroscopic algae new or seldom recorded for Britain: *Nostoc parmelioides, Heribaudiella fluviatilis, Cladophora aegagropila, Monostroma bullosum, Rhodoplax schinzii*. Vasculum 60:47–55

Holmes NTH, Whitton BA (1977a) The macrophytic vegetation of the River Tees in 1975: observed and predicted changes. Freshwater Biol 7:43–60
Holmes NTH, Whitton BA (1977b) The macrophytic vegetation of the River Swale, Yorkshire. Freshwater Biol 7:545–558
Israelsson G (1938) Über die Süsswasserphaeophycéen Schwedens. Bot Not 1938:113–128
Jao C-C (1941) Studies on the freshwater algae of China. VII. *Lithoderma zonatum,* a new freshwater member of the Phaeophyceae. Sinensia 12:239–244
Jao C-C (1943) Studies on the freshwater algae of China. XI. *Sphacelaria fluviatilis,* a new freshwater brown alga. Sinensia 14:151–154
Jao C-C (1944) Studies on the freshwater algae of China. XII The attached algal communities of the Kialing River. Sinensia 15:61–73
Johansen JR, Lowe RL, Carty S et al (2007) New algal species records for Great Smoky Mountains National Park, with an annotated checklist of all reported algal taxa for the park. Southeastern Nat 6:99–134
Kahlert M, Hasselrot AT, Hillebrand H et al (2002) Spatial and temporal variation in the biomass and nutrient status of epilithic algae in Lake Erken, Sweden. Freshwater Biol 47:1191–1215
Kann E (1966) Der Algenaufwuchs in einigen Bächen Österreichs. Verh Int Ver Theor Angew Limnol 16:646–54
Kann E (1978a) Systematik und Ökologie der Algen österreichischer Bergbäche. Arch Hydrobiol Suppl 53:405–643
Kann E (1978b) Typification of Austrian streams concerning algae. Verh Int Ver Theor Angew Limnol 20:1523–1526
Kann E (1982) Qualitative Veränderungen der litoralen Algenbiocönose österreichischer Seen (Lunzer Untersee, Traunsee, Attersee) im Laufe der letzen Jahrzehnte. Arch Hydrobiol Suppl 62:440–490
Kann E (1993) Der litorale Algenaufwuchs im See Erken und in seinem Abfluβ (Uppland, Schweden). Algol Stud 69:91–112
Kann E, Tschamler H (1976) Algenaufwuchs unter natürlichen Bedingungen auf Kunststoffen. Chem Kunststoff Akt 1976:63–71
Kirkby SM, Hibberd DJ, Whitton BA (1972) *Pleurocladia lacustris* A. Braun (Phaeophyta)—a new British Record. Vasculum 57:51–56
Kloareg B, Mabeau S (1987) Isolation and analysis of the cell walls of brown algae: *Fucus spiralis, F. ceranoides, F. vesiculosus, F. serratus, Bifurcaria bifurcata* and *Laminaria digitata.* J Exp Bot 38:1573–1580
Kumano S, Hirose H (1959) On the swarmers and reproductive organs of a phaeophyceous freshwater alga of Japan, *Heribaudiella fluviatilis* (Areschoug) Svedelius. Bull Jpn Soc Phycol 7:45–51 (In Japanese)
Kusel-Fetzmann EL (1996) New records of freshwater Phaeophyceae from lower Austria. Nova Hedwigia 62:79–89
Kusel-Fetzmann E, Schagerl M (1992) Verzeichnis der Sammlung von Algen-Kulturen an der Abteilung für Hydrobotanik am Institut für Pflanzen-physiologie der Universität Wien. Phyton 32:209–234
Lüder UH, Clayton MN (2004) Induction of phlorotannins in the brown macroalga *Ecklonia radiata* (Laminariales, Phaeophyta) in response to simulated herbivory—the first microscopic study. Planta 218:928–937
McCauley LR, Wehr JD (2007) Taxonomic reappraisal of the freshwater brown algae *Bodanella, Ectocarpus, Heribaudiella, and Pleurocladia* (Phaeophyceae) on the basis of *rbc*L sequences and morphological characters. Phycologia 46:429–439
Michel G, Tonon T, Scornet D et al (2010) The cell wall polysaccharide metabolism of the brown alga *Ectocarpus siliculosus.* Insights into the evolution of extracellular matrix polysaccharides in Eukaryotes. New Phytol 188:82–89
Müller DG (1979) Genetic affinity of *Ectocarpus siliculosus* (Dillw.) Lyngb. from the Mediterranean, North Atlantic and Australia. Phycologia 18:312–318

Müller DG, Geller W (1978) Einige Beobachtungen an Kulturen der Süsswasser-Braunalge *Bodanella lauterborni* Zimmermann. Nova Hedwigia 29:735–41
Patterson DJ (1989) Stramenopiles: chromophytes from a protisan perspective. In: Leadbeater, BSC, Diver WL (es) The chromophyte algae: problems and perspectives; Systematics Association Special Volume No. 38. Clarendon Press, Oxford, pp 357–379
Pueschel CM, Stein JR (1983) Ultrastructure of a freshwater brown alga from western Canada. J Phycol 19:209–215
Quatrano RS, Stevens PT (1976) Cell wall assembly in *Fucus* zygotes: I. Characterization of the polysaccharide components. Plant Physiol 58:224–231
Schloesser R (1977) The identification of a new freshwater brown alga from the Lake Michigan sublittoral zone. M.Sc. Thesis, University of Wisconsin, Milwaukee
Schloesser RE, Blum JL (1980) *Sphacelaria lacustris* sp. nov., a freshwater brown alga from Lake Michigan. J Phycol 16:201–207
Schneider SC, Lindstrom E-A (2011) The periphyton index of trophic status PIT: a new eutrophication metric based on non-diatomaceous benthic algae in Nordic rivers. Hydrobiologia 665:143–155
Schoenwaelder MEA (2002) The occurrence and cellular significance of physodes in brown algae. Phycologia 41:125–139
Sheath RG, Cole KM (1992) Biogeography of stream macroalgae in North America. J Phycol 28:448–60
Silberfeld T, Racault M-FLP, Fletcher RL et al (2011) Systematics and evolutionary history of pyrenoid-bearing taxa in brown algae (Phaeophyceae). Eur J Phycol 46:361–377
Stache-Crain B, Müller DG, Goff LJ. (1997) Molecular systematics of *Ectocarpus* and *Kuckuckia* (Ectocarpales, Phaeophyceae) inferred from phylogenetic analysis of nuclear and plastid-encoded DNA sequences. J Phycol 33:152–168.
Svedelius N (1930) Über die sogenanntem Süsswasser-Lithodermen. Z Bot 23:892–918
Szymanska H, Zakrys B (1990) New phycological records from Poland. Arch Hydrobiol Suppl 87:25–32
Thompson RH (1975) The freshwater brown alga *Sphacelaria fluviatilis*. J Phycol Suppl 11:5
Van den Hoek C, Mann DG, Jahns HM (1995) Algae: an introduction to phycology. Cambridge University Press, Cambridge
Vreeland V, Waite JH, Epstein L (1998) Polyphenols and oxidases in substratum adhesion by marine algae and mussels. J Phycol 34:1–8
Waern M (1952) Rocky-shore algae in the Öregund Archipelago. Acta Phytogeogr Suecica 30:1–298
Wehr JD (2011) Phylum Phaeophyta (Brown Algae). In: John DM, Whitton BA, Brook AJ (eds) The freshwater algal flora of the British Isles. Cambridge University Press, Cambridge, pp 354–357
Wehr JD (2015) Brown Algae. In: Wehr JD, Sheath RG, Kociolek JP (eds) Freshwater algae of North America, ecology and classification. Academic, San Diego, pp 851–871
Wehr JD, Perrone AA (2003) A new record of *Heribaudiella fluviatilis*, a freshwater brown alga (Phaeophyceae) from Oregon. West N Am Naturalist 63:517–523
Wehr JD, Stein JR (1985) Studies on the biogeography and ecology of the freshwater phaeophycean alga *Heribaudiella fluviatilis*. J Phycol 21:81–93
Wehr JD, Stancheva R, Sheath RG, Truhn K (2013) Discovery of new populations of the rare freshwater brown alga *Pleurocladia lacustris* A. Braun in California streams. West N Am Naturalist 73:148–157
West JA, Kraft GT (1996) *Ectocarpus siliculosus* (Dillwyn) Lyngb. from Hopkins River Falls, Victoria—the first record of a freshwater brown alga in Australia. Muelleria 9:29–33
Wilce RT (1966) *Pleurocladia lacustris* in Arctic America. J Phycol 2:57–66
Wilce RT, Webber EE, Sears JR (1970) *Petroderma* and *Porterinema* in the New World. Mar Biol 5:119–135

Wujek DE, Thompson RH, Timpano P (1996) The occurrence of the freshwater brown alga *Sphacelaria fluviatilis* Jao from Michigan. Mich Bot 35:111–114

Yoshizaki M, Iura K (1991) Notes on *Heribaudiella fluviatilis* from Chiba Prefecture and Ibaraki Prefecture. Chiba Seibutu-si 40:37–39 (In Japanese)

Yoshizaki M, Miyaji K, Kasaki H (1984) A morphological study of *Heribaudiella fluviatilis* (Areschoug) Svedelius (Phaeophyceae) from Central Japan. Nankiseibutu 26:19–23 (In Japanese)

Young EB, Tucker RC, Pansch LA (2010) Alkaline phosphatase in freshwater *Cladophora*-epiphyte assemblages: regulation in response to phosphorus supply and localization. J Phycol 46:93–101

Zimmermann,W (1928) Über Algenbestände aus der Tiefenzone des Bodensees. Zur Ökologie und Soziologie der Tiefseepflanzen. Z Bot 20:1–28 + 2 pl

Chapter 7
Heterokonts (Xanthophyceae and Chrysophyceae) in Rivers

Orlando Necchi Jr.

Abstract The heterokont classes Xanthophyceae and Chrysophyceae are introduced with its key characteristics and typical benthic river genera. Two relatively widespread genera are the xanthophyte siphonous *Vaucheria* and the filamentous *Tribonema*. The heterokont classs (Chrysophyceae) genus *Hydrurus* is also listed, which consists of a branched colony and can be widespread in cold mountain rivers with turbulent waters.

Keywords Algae • Benthic • Biodiversity • Chrysophyceae • Heterokontophyta • Xanthophyceae • Yellow-green algae

Introduction

The purpose of this chapter is to introduce the algal groups common and occasionally abundant in stream habitats but not diverse enough in terms of representatives to justify a specific chapter. One major group treated here is the class Xanthophyceae (yellow-green algae). The genus *Hydrurus* of the class Chrysophyceae (golden algae) can also be widespread and abundant in some drainage basins and is briefly listed as well.

The Xanthophyceae (=Tribophyceae), or yellow-green algae, do not have fucoxanthin masking the chlorophylls *a* and *c* and so they have a greenish color (Graham et al. 2009). The xanthophytes have the storage polymer chrysolaminarin but also produce cytoplasmic lipid droplets. The cell walls are composed primarily of cellulose, with silica sometimes present. Members of the class occur primarily in freshwater and they reproduce by various means but some produce thick-walled cysts to persist over the non-growing season.

Members of this xanthophytes are generally not as common as the other stream algal groups described in this book but some taxa can be periodically widespread

O. Necchi Jr. (✉)
Department of Zoology and Botany, São Paulo State University,
Rua Cristóvão Colombo, 2265, São José do Rio Preto, SP 15054-000, Brazil
e-mail: orlando@ibilce.unesp.br

O. Necchi Jr. (ed.), *River Algae*, DOI 10.1007/978-3-319-31984-1_7

and abundant. The coenocytic genus *Vaucheria*, for example, is distributed in rivers and streams from most biomes in North America (Sheath and Cole 1992) and south-eastern Brazil (Necchi et al. 2000) and it is composed of many species. Another genus of this class (*Tribonema*) can also be found in streams, but it is not so species rich and widespread as *Vaucheria*.

The Chrysophyceae, or golden algae, have large amounts of the accessory pigment fucoxanthin in their chloroplasts, masking chlorophylls a, c_1, and c_2, a storage polymer chrysolaminarin in vacuoles, and a variety of cell coverings (Graham et al. 2009). The majority of these algae occur in freshwater habitats and they employ a silica-walled resting stage, the stomatocyst, to persist through the non-growing season. Benthic, lotic members of the golden algae are not very diverse and widespread as members of other algal groups but one genus (*Hydrurus*) can be quite widespread and it is particularly distributed in cold, mountain streams where the colonial thalli are firmly attached to hard rock or large stones in turbulent water (Wehr and Sheath 2015).

Phylogenetic Relationships of Heterokont Algae

Within the Phylum Heterokontophyta, two classes are treated here: Chrysophyceae and Xanthophyceae (=Tribophyceae). Both were shown to be monophyletic within the phylum by Riisberg et al. (2009). For the Xanthophyceae I followed the taxonomic scheme by Maistro et al. (2009), who recognized the two orders treated here (Tribonematales and Vaucheriales) as monophyletic, whereas for the Chrysophyceae, Kawai and Nakayama (2015) was adopted.

Sample Collection and Preservation

Informative sources describing in details procedures, equipment, and tools for collection and preservation of these algal groups are essentially the same as described for other groups, particularly green and red algae (Chaps. 3 and 4). Thus, no descriptions are presented here and readers should search for more detailed information in those chapters.

Taxonomic Key to the Genera of Xanthophyceae and Chrysophyceae in Rivers

1a	Thalli golden colored (Chrysophyceae)	*Hydrurus*
1b	Thalli yellow-green or green (Xanthophyceae)	2
2a	Thalli coenocytic, consisting of siphons lacking cross walls	*Vaucheria*
2b	Thalli filamentous, consisting of chain of cells	3
3a	Cell walls with one section, not forming H-shaped pieces	*Xanthonema*

3b	Cell walls with two sections, forming H-shaped pieces	4
4a	Filaments short, breaking into fragments, with short cells (length/diameter ratio ≤1.5)	*Bumilleria*
4b	Filaments long, not breaking into fragments, with elongate cells (length/diameter ratio ≥2)	*Tribonema*

Descriptions of Heterokontophyta genera in rivers

Phylum Heterokontophyta

Class Xanthophyceae (= Tribophyceae): Order Tribonematales

Bumilleria Borzi (Figs 7.1a–b)

Filaments unbranched, straight, often constricted at cross walls, short; filaments break apart into fragments in some species. Cells cylindrical to cubic, short (length/diameter ratio ≤1.5), thin-walled, with one to several parietal, disc-shaped chloroplasts; cell walls in two sections, with H-shaped intercalary segments of the cell wall every 2–4 cells, usually evident between cells but sometimes only at broken ends; each cell with one to several discoid parietal chloroplasts with pyrenoids visible only on staining; oil globules often present. Asexual reproduction by biflagellate zoospores released by cell wall disruption.

Remarks: *Bumilleria* is probably a cosmopolitan genus, with only five species known (Guiry and Guiry 2015), which are usually found associated with other filamentous algae in streams, rivers, and ponds, mostly reported in North America and Europe (Johnson 2002; Ott et al. 2015).

Tribonema Derbès et Solier (Figs 7.1c–d)

Filaments unbranched, straight, non-constricted, long. Cells cylindrical or less often barrel-shaped, elongate (length/diameter ratio ≥2.0), usually thin-walled, with one to many parietal, disc-shaped chloroplasts, lacking pyrenoids; cell walls in two sections, with H-shaped pieces usually evident at the end of broken ends. Asexual reproduction by zoospores, aplanospores, and cysts; sexual reproduction isogamous.

Remarks: *Tribonema* is a cosmopolitan and diverse genus with 28 species currently accepted (Guiry and Guiry 2015), which is relatively well represented in streams and rivers. It occurs as free-floating masses or entangled to other filamentous algae, bryophytes, and macrophytes or less frequently it is found attached, especially in young stages. The genus can be misidentified as *Microspora*, which also has H-shaped pieces, but the latter has chloroplasts with starch; a simple test to distinguish these two genera is to apply Lugol's solution, which colours the starch in *Microspora* dark purple.

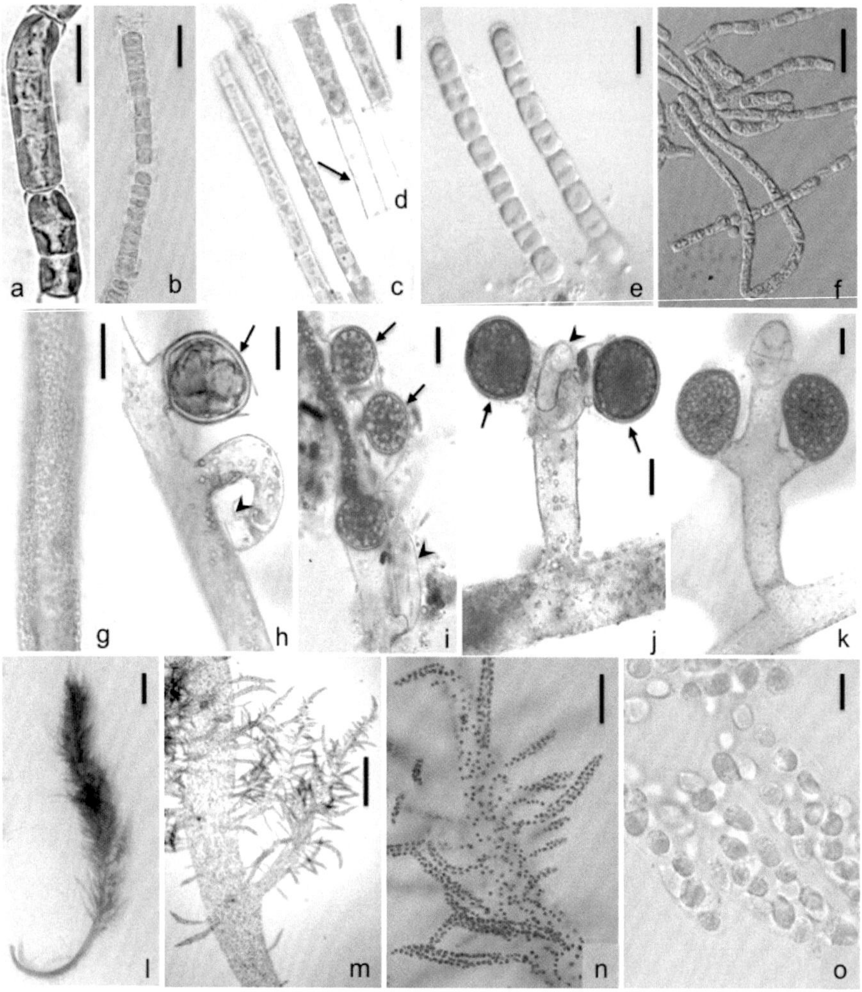

Fig. 7.1 (**a–b**) *Bumilleria*. (**c–d**) *Tribonema*: (**c**) filaments with vegetative cells; (**d**) empty cell with H-shaped cell wall (*arrow*). (**e–f**) *Xanthonema*. (**g–k**) *Vaucheria*: (**g**) detail of a siphon; (**h–k**) oogonia (*arrows*) and antheridia (*arrowheads*). (**l–o**) *Hydrurus*. Scale bars: fig. **l** = 5 mm; fig. **m** = 500 μm; fig. **n** = 250 μm; fig. **k** = 50 μm; figs **f–j**, o = 25 μm; figs **a–e** = 10 μm. Image authors: (**a**) Coimbra Collection of Algae (ACOI); (**b**) Photobucket; (**c–d**, **h–k**, **o**) C. Carter; (**e**) Y. Tsukii; (**f**) Culture Collection of Autotrophic Organisms (CCALA; (**l–n**) I. Bárbara

Xanthonema P.C. Silva (= *Heterothrix* Pascher) (Figs 7.1e–f)

Filaments unbranched, straight, slightly curved or undulated, often constricted at cross walls, short; filaments fragile and breaking apart into fragments or single cells in some species. Cells cylindrical, with one or two parietal, disc-shaped chloroplasts without pyrenoids. Asexual reproduction by zoospores and aplanospores. Akinetes also reported.

Remarks: *Xanthonema* is a cosmopolitan genus, with 14 species presently recognized (Guiry and Guiry 2015), which occurs in a variety of aquatic habitats and also as subaerial; few species are found in stream habitats, mostly in distrophic or oligotrophic water bodies.

Class Xanthophyceae (= Tribophyceae): Order Vaucheriales

Vaucheria De Candolle (Figs 7.1g–k)

Thalli coenocytic, multinucleate, cylindrical, branched, consisting of interwoven, grass- to dark-green, sparingly branched siphons, forming macroscopic felt-like or cushion-like patches; siphons attached by colorless rhizoids; chloroplasts numerous, parietal, disc-shaped to ellipsoidal; with or without pyrenoids; usual storage products oil or fat. Asexual reproduction by zoospores, aplanospores or akinetes. Sexual reproduction oogamous, female gametangia (oogonia) bearing single large eggs and male gametangia (antheridia) producing several heterokont sperms. Gametangia isolated by septa, arranged in either regular bisexual groups or loose arrangements of one or more oogonia and/or antheridia on monoecious or dioecious plants. Antheridia usually tubular, curved or straight, sessile or stalked. Oogonia spherical, ovoid or kidney-shaped, sessile or stalked; mature oogonia have distinctive beak with a pore; fertilization usually through pore in oogonial wall. Sexual reproductive structures is required to identify species of this genus.

Remarks: *Vaucheria* is a cosmopolitan and the most species-rich genus among the Xanthophyceae with 79 species currently accepted (Guiry and Guiry 2015). The genus is very well represented in stream habitats. Some species occurs in marine coastal habitats, particularly in salt marshes. The most important taxonomic characters to distinguish species in the genus are the size of siphons, as well as arrangement, shape, and size of gametangia (Johnson 2002; Ott et al. 2015).

Class Xanthophyceae: Other Potential River Inhabitants

A number of microscopic forms could be overlooked in environmental samples due to their small size and presence in low quantities. Thus, more xanthophytes could be potentially found in streams and general identification keys for Xanthophyceae should be applied (e.g., Ott et al. 2015). The branched filamentous genus *Heterococcus* can be found epillithic in streams (Rybalka et al. 2013) or as a lichen photobiont of the aquatic members of the fungal family Verrucariaceae (Thüs et al. 2011).

Class Chrysophyceae: Order Hydrurales

Hydrurus C. Agardh (Figs 7.1l–o)

Thalli macroscopic, consisting of branched mucilaginous colonies up to 30 cm in length. Cells distributed in the colonial matrix, oval with a two-lobed chloroplast containing a pyrenoid.

Zoospores tetrahedral, with one long and one short flagellum. Stomatocysts lenticular, with an equatorial wing.

Remarks: *Hydrurus* is common in cold, clear, fast-flowing mountain streams attached to firm substrata (Guiry and Guiry 2015). It has a peculiar and unpleasant smell described as foetid. During warm weather, the macroscopic form degrades and cysts are formed.

Acknowledgements I am thankful to the following people for kindly sharing their images to be used in this chapter: Ignacio Bárbara, Chris Carter, and Yuuji Tsukii. The help in image editing by Cauê Necchi is greatly appreciated.

References

Graham LE, Graham JM, Wilcox LW (2009) Algae, 2nd edn. Benjamin Cummings, San Francisco

Guiry MD, Guiry GM (2015) AlgaeBase. World-wide electronic publication, National University of Ireland, Galway. http://www.algaebase.org; Accessed 01 December 2015

Johnson LR (2002) Phylum Xanthophyta (yellow-green algae). In: John DM, Whitton BA, Brook AJ (eds) The freshwater algal flora of the British Isles: an identification guide to freshwater and terrestrial algae. Cambridge University Press, Cambridge, pp 243–270

Kawai H, Nakayama T (2015) Class Chrysophyceae Pascher. In: Frey W (ed) Syllabus of plant families, Photoautotrophic Eukaryotic Algae. Borntraeger Science Publishers, Stuttgart, pp 120–127

Maistro S, Broady PA, Andreoli C et al (2009) Phylogeny and taxonomy of Xanthophyceae (Stramenopiles, Chromalveolata). Protists 160:412–426

Necchi O Jr., Branco CCZ, Branco LHZ (2000) Distribution of stream macroalgae in São Paulo State, southeastern Brazil. Algol Stud 97:43–57

Ott DW, Oldham-Ott CK, Rybalka N et al (2015) Xanthophyte, Eustigmatophyte and Raphidophyte algae. In: Wehr JD, Sheath RG, Kocyolek JP (eds) Freshwater algae of North America, Ecology and Classification. Academic, San Diego, pp 485–536

Riisberg I, Orr RJS, Klubeg R et al (2009) Seven gene phylogeny of heterokonts. Protists 160:191–204

Rybalka N, Wolf M, Andersen RA et al (2013) Congruence of chloroplast- and nuclear-encoded DNA sequence variations used to assess species boundaries in the soil microalga *Heterococcus* (Stramenopiles, Xanthophyceae). BMC Evol Biol 13:1471–2148

Sheath RG, Cole KM (1992) Biogeography of stream macroalgae in North America. J Phycol 28:448–460

Thüs H, Muggia L, Pérez-Ortega S et al (2011) Revisiting photobiont diversity in the lichen family Verrucariaceae (Ascomycota). Eur J Phycol 46:399–415

Wehr JD, Sheath RG (2015) Habitats of freshwater algae. In: Wehr JD, Sheath RG, Kociolek JP (eds) Freshwater algae of North America, ecology and classification. Academic, San Diego, pp 13–74

Chapter 8
The Spatio-Temporal Development of Macroalgae in Rivers

Eugen Rott and John D. Wehr

Abstract In shallow clear rivers macroalgae are a diverse component of biota and especially easily recognizable and easily accessible organisms, facilitating studies of spatial and temporal variation across environmental gradients. Their macroscopic form and appearance makes field studies of their limits and requirements in space and time possible. Some taxa occupy spatially restricted microhabitats, particularly in headwaters, while others are abundant along the length of a river. Spatial variation can occur within a site, between different stream types, or across wide regions. Temporal variation in growth or colonization by a particular species at a particular site can be driven by seasonal changes in discharge and nutrients, which may lead to regular or irregular temporal sequences. More stable temporal dynamics are observed in taxa with high adaptive capacities to resist physical disturbances or to re-establish quickly after floods. We examined the most common soft-bodied macroscopic algae (SBM) from two geographically extended datasets of temperate streams, from alpine to lowland regions in Austria and a shallower altitudinal gradient of southeastern New York State. Morphological and functional characters, combined with key environmental variables (based on median and multivariate statistics), are used to analyze general trends and causalities for species-specific spatial and temporal niches. These results provide strong arguments in favor of using a combination of on-site studies of growth form and phenology, with ecophysiological and molecular studies in the lab to improve our understanding of the factors regulating stream macroalgae occurrence in space and time in the future.

Keywords Austria • Functional species-groups • Growth form • Lotic algae: near-natural condition • New York State • Niche descriptors • Size and spatial scales

E. Rott (✉)
Biology Faculty, Institute of Botany, University of Innsbruck,
Sternwartestraße 15, 6020 Innsbruck, Austria
e-mail: Eugen.Rott@uibk.ac.at

J.D. Wehr
Louis Calder Center—Biological Field Station, Fordham University,
53 Whippoorwill Rd., Box 887, Armonk, NY 10504, USA

O. Necchi Jr. (ed.), *River Algae*, DOI 10.1007/978-3-319-31984-1_8

Introduction

Algae are often a neglected or under-estimated component of biota in flowing water, owing in part to many ecosystem studies that suggest their quantitative importance in organic matter flux is minor compared to allochthonous organic matter inputs (Tank et al. 2010). In addition, it is challenging to study benthic algal assemblages that often consist of a large number of small-sized individuals of various taxonomic affiliations (microalgae), even though certain single key taxa can dominate > 90 % of periphyton biomass within a stream habitat (Rott et al. 2006a). Especially in clear rivers, SBM (= soft-bodied macroscopic algae excluding diatoms, but including cyanobacteria) are common and may be dominated by several clearly recognizable taxa (Parker et al. 1973; Holmes and Whitton 1981; Sheath and Cole 1992). Although algal species and their assemblages, similar to many other microbial organisms, are often assumed to be spatially ubiquitous (Fenchel and Finlay 2004), we hypothesize that most SBM have species-specific defined habitats, leading to recurring growth periods in many locations. In specific ecological niches SBM can play a significant role in ecosystem function and may act to buffer impacts of changing environmental conditions in aquatic systems (Caron and Countway 2009). Thus it is important that the ecological spectra of key SBM be characterized within streams segments, among different river types, along the river continua, and within larger ecoregions (Sheath and Vis 2013).

Biggs et al. (1998) applied several aspects of Grime's (1977) theory originally developed for higher plants to define CRS strategies of benthic algae (competitor, ruderal, stress-tolerant) along disturbance and productivity (= eutrophication) gradients, although in this model SBM only partly explained the overall relationship between habitat duration and productivity. Recognizing the challenge of understanding the importance of temporal vs. spatial drivers, we suggest that better knowledge of SBM species niches across environmental gradients, with special attention to scales of space and time, could allow a basis for a broader concept. In particular, theories pertaining to temporal patterns of SBM differ greatly. One theory to explain temporal/seasonal variability in SBM is based on differences in niche requirements among species, leading to competitive exclusion of taxa over time, whereas the opposite theory assumes that on the long run resilience strategies of the key SBM would be more prevalent under near-natural conditions (Lindstrøm et al. 2004). In addition, life cycle differences and other adaptations, as well as external biotic factors (e.g. grazing, Pringle 1996; McNeely and Power 2007), are additional potential factors, collectively resulting in what appears to be unpredictable seasonal patterns of dominant taxa over longer time periods in rivers (Whitford and Schumacher 1963; Biggs 1996).

Niche differences of SBM among river types and to some extent along the river continuum are expected to be most prominent where strong gradients occur, such as within alpine areas. These have partly been recorded in case studies of periphyton in alpine stream types in Switzerland (Hieber et al. 2001; Uehlinger et al. 2010), Poland (Kawecka 1980, 1981), and Colorado, USA (Ward 1986; Vavilova and

Lewis 1999), and bioregion-specific taxa for near natural reference sites, including the Alps were proposed for Austria (AT) by Pfister and Pipp (2013). Even if we could confirm the specific river type/bioregional niche differentiation in some taxa (Grimm 1995), SBM temporal variation would still be unclear if seasonal occurrences of non-persisting taxa (ephemerals) would be of similar importance in river assemblages, and had their own responses to effects of local climate (temperature, light, runoff pattern) in temperate rivers.

Guides and floras illustrating morphological characters of river macroalgae are mainly available for Australia (Entwisle et al. 1997), Europe (Germany: Gutowski and Förster 2009; Poland: Eloranta and Kwandrans 2012; UK: John et al. 2011) and North America (Sheath and Cole 1992; Wehr and Sheath 2015; Wehr et al. 2015), but with few attempts to study underlying principles (e.g. for Germany: Förster et al. 2004). In this chapter we focus on morphologically identical and specific SBM ("morphospecies"; several documented with colour illustrations) and their environment across the major taxonomic groups, including cyanobacteria, from approximately 400 streams (irregularly sampled) from across AT and 60 streams (systematically sampled) in southeastern New York (NY) State in the USA (~25 % smaller total area than AT) with the aim to clarify the broader features of SBM niches. Form-functional patterns and phenology from these and other studies will be used to generate broad insights into features of habitat niches for macroalgal species in relation to space and time.

Spatial Aspects

Although niches of stream organisms vary both in space and time and often simultaneously, spatial and temporal aspects in SBM in this chapter are first considered separately and along a scaled-up approach, in order to better visualize the major structural and temporal environmental variables involved.

Microhabitat Scale

Macroalgae have features allowing to some extent macroscopic recognition in the field (Sheath and Cole 1992). The concept of a freshwater macroalga (SBM) in this chapter refers to species that may either form (1) macroscopic thalli or plant-like structures as individuals (a single macro-specimen), or (2) species whose individuals are microscopic, but produce macroscopic growths.

Size scales. The area or space covered by algal species that can be regarded as macroalgae in streams is determined by the average size ranges (min/max) of individuals of a species within a habitat (Tables 8.1 and 8.2), as well as the capacity to form visible aggregates (e.g. networks, patches, tufts, turfs) in such environments.

Table 8.1 Morphological characteristics and general habitat preferences for SBM species occurring in Austrian rivers (approx. 400 rivers/1300 sites at 140–2080 m a.s.l., stream orders 1–9, sampled 1984–2001), supplemented by key taxa from springs and spring streams (indicated by[a]). For abbreviations see below

Taxa	MSI	MCA	GFT	SF	H	FA	RZ	BR	SO	Seas	Alt	Colour
Cyanobacteria / Cyanophytes*												
Ammatoidea simplex	µm	mm^2	co	-	Y	-	-	KH	<7	Sp+A	>300	bs
Calothrix braunii agg.	µm	mm^2	co	nf	Y		-	KH	<7	A-Sp	any	br
Chamaesiphon fuscus	µm	mm^2	ca	sp	-	RB	-	UZA	<7	W+A	any	dd
Chamaesiphon geitleri	µm	dm^2	ca	sp	-	RB	-	KH	<7	W+A	>300	dd
Chamaesiphon polonicus	µm	cm^2	cr	-	-	RP	SZ	UZA	any	W+A	any	rr
Chamaesiphon polymorphus	µm	mm^2	ca	-	-	RP	-	GG	any	all	any	bg
Chamaesiphon starmachii	µm	dm^2	ca	pp	-	RB	-	UZA	<7	A-Sp	any	dd
Clastidium rivulare	µm	mm^2	ca	sp	-	RP	-	any	<7	A+W	>300	dd
Clastidium setigerum	µm	mm^2	ca	sp	-	RP	-	VZA	<6	A-Sp	<300	dd
Chlorogloea microcystoides	µm	mm^2	gl	-	-	-	-	any	n.d.	all	n.d.	bl
Dichothrix gypsophila	µm	cm^2	co	cr, nf	Y	-	-	KH	<6	Su+A	any	br
Gloeocapsa alpina agg.	µm	cm^2	co	-	-	-	SP	VZA	<6	W	any	re
Homoeothrix crustacea	µm	cm^2	pu	hc	Y	RP	-	KV	<7	all	<1300	lb
Homoeothrix gracilis	µm	mm^2	co	-	Y	RP	-	VZA	<7	A+W	any	br
Homoeothrix janthina	µm	mm^2	co	-	Y	RB	RF	UZA	<7	W+A	any	bs
Homoeothrix varians	µm	mm^2	co	-	Y	RB	RF	KV	any	W+A	any	bs
Hydrococcus cesatii	µm	mm^2	co	-	-	-	-	AV	<7	W+A	any	bl
Hydrococcus rivularis	µm	mm^2	co	-	-	RP	-	VZA	any	W	any	bl
Hydrocoleum homoeotrichum	mm	cm^2	tu	-	-	RP	-	VZA	<7	all	any	db
Leptolyngbya foveolarum	µm	mm^2	co	-	-	-	-	FH	any	W+A	any	db
Oscillatoria limosa	mm	cm^2	ca	sf	-	-	-	GG	any	Sp	any	ib
Phormidium autumnale	µm	cm^2	ca	sm	-	RB	-	UZA	<7	W+A	any	db
Phormidium incrustatum	µm	cm^2	co	hc	-	RP	-	any	<7	all	<1300	gg
Phormidium retzii	µm	cm^2	ca	-	-	RP	-	any	any	Su+Sp	any	bl
Phormidium setchellianum	µm	cm^2	co	sk	-	RP	-	n.d.	<6	Sp	<1300	rv
Phormidium subfuscum	µm	cm^2	co	sk	-	RP	-	any	any	A-Sp	any	vb
Pleurocapsa aurantiaca	µm	cm^2	co	-	-	-	-	KV	<7	W+A	any	or
Pleurocapsa minor	µm	cm^2	co	-	-	-	SH	any	any	W+A	any	db
Rivularia periodica agg.*	µm	cm^2	pu	ca, nf	Y	RP	-	n.d.	<3	all	<1300	db
Schizothrix fasciculata	µm	cm^2	pu	-	-	-	-	KH	<6	A+Sp	>300	ge
Schizothrix semiglobosa	µm	cm^2	pu	wc	-	RB	-	KH	<5	A+Sp	>300	bs
Schizothrix tinctoria	µm	cm^2	co	-	-	-	-	VZA	<6	W+A	>300	db
Siphonema polonicum	µm	m^2	cr	-	-	RP	-	KH	<7	W+A	any	or
Tolypothrix distorta agg.	mm	cm^2	tu	nf	-	-	-	KH	<5	A+Sp	any	bs
Xenotholos kerneri	µm	mm^2	co	-	-	-	-	any	<6	Sp+A	any	bg
Chlorophyta s.l.												
Chaetophora elegans	mm	mm^2	tu	ge	-	-	-	GG	any	Sp	any	gr
Cladophora glomerata	cm	$>m^2$	sh	rf	-	-	-	FL	any	all	<1300	gr
Chara spp.	cm	$>m^2$	sh	ec	-	-	-	n.d.	<6	all	any	cg
Draparnaldia acuta	mm	cm^2	tu	ge	Y	-	-	n.d.	<3	Su	any	gr
Gongrosira debaryana	µm	cm^2	co	-	-	-	-	GG	<7	W+A	<1300	gr
Gongrosira incrustans	µm	cm^2	cr	ca	-	-	-	KH	any	A	any	gr
Klebsormidium rivulare	mm	cm^2	co	-	-	RP	-	UZA	<7	A	any	gr
Microspora amoena	mm	mm^2	co	-	-	-	-	VZA	<7	W+A	any	gr
Oocardium stratum *	µm	$>m^2$	pu	ca	-	RP	-	KH	<3	all	<1300	gp
Protoderma viride	µm	cm^2	ca	-	-	-	-	FL	any	W+A	<800	yg
Sphaerobotrys fluviatilis	µm	cm^2	ca	ge	-	RB	-	FH	<7	W+Sp	any	bs
Stigeoclonium tenue	µm	cm^2	co	-	Y	-	-	FH	any	all	<1300	gr

(continued)

Table 8.1 (continued)

Tetraspora gelatinosa agg.	mm	cm^2	tu	ge	-	-	-	any	<7	Su	<800	gr
Trentepohlia aurea	mm	cm^2	so	ve	-	-	SP	VZA	<7	A	>800	ro
Ulothrix zonata	mm	cm^2	tu	gf	-	RP	SZ	VZA	any	W	any	gr
Zygnema spp.	mm	cm^2	tu	ge	-	-	-	n.d.	<5	A+Sp	>300	gr
Rhodophyta												
Audouinella hermannii	mm	cm^2	tu	-	Y	-	SH	GG	<7	W+A	any	vg
Audouinella pygmaea	µm	mm^2	tu	-	Y	-	SH	FH	<7	all	<800	vg
Bangia atropurpurea	cm	cm^2	tu	fi	-	RP	SZ	KH	<7	W+A	<800	re
Batrachospermum gelatinosum	cm	cm^2	sh	sm	Y	RP	SH	any	<7	W	<1300	vi
Hildenbrandia rivularis	cm	dm^2	cr	rs	-	RP	SH	GG	any	all	<1300	re
Lemanea fluviatilis	cm	dm^2	tu	tt	-	RB	RF	VZA	<7	W+A	>300	vg
Chrysophyceae												
Hydrurus foetidus	cm	$>m^2$	tu	ge	-	RB	RF	UZA	<7	A-Sp	any	gd
Phaeodermatium rivulare	µm	mm^2	ca	ge	-	RB	-	UZA	<7	W	any	gb
Phaeophyta												
Heribaudiella fluviatilis	cm	cm^2	ca	rs	-	RP?	SH	AV	any	W	<800	br
Xanthophyta												
Vaucheria spp.	cm	dm^2	ch	ve	-	-	-	any	<7	Sp+A	<1300	og

Abbreviations: *MSI* maximum size of individual specimens; *MCA* maximum area covered by aggregates; *GFT* growth-form type: *ca* coating, *ch* cushions, *co* cover, *cr* crust, *gl* globules, *pu* pustules, *sh* shrub-like, *so* sods, *tu* tuft; *SF* specific features: *ca* calcified, *cr* crust, *fi* filamentous, *ge* gelatinous, *gf* gelatinous filaments, *hc* hard calcified, *nf* nitrogen fixing, *pp* spots or patches, *rf* rough filaments, *rs* radial spreading, *sf* soft filaments; *sm* soft mucilage, *sp* spots, *tt* thick threads, *ve* velvet, *wc* weakly calcified; *H* hair-forming species, *Y* yes; *FA* Flow adaptation: *RB* rheobiontic, *RP* rheophilous; *RZ* preferred river zone: *RF* riffle, *SH* shade, *SP* spray zone, *SZ* splash zone; *BR* Preferred bioregion, abbreviations and location of bioregions see Fig. 8.6, n.d. insufficient data; *SO* preferred stream order acc. to Horton and Strahler in Allen (1995); *Seas* preferred season: *A* autumn; *Sp* spring, *Su* summer; *W* winter; most preferred season named first; *Alt* preferred altitude; Colour: *bg* bluegreen, *bl* blue, *br* brown, *bs* brownish/clear brown, *cg* clear green, *db* dark blue, *dd* dark brown, *gb* golden brown, *gd* golden to dark brown, *ge* grey brown, *gg* grey blue, *gr* green, *ib* intense blue, *lb* light blue green, *og* olive green, *or* orange red, *re* red, *ro* red or orange, *rr* rusty red, *rv* red violet, *vb* violet brown, *vg* violet grey, *vi* violet, *yg* yellow green

[a]Within Cyanobacteria identification, we followed the traditional species concepts of Komárek and Anagnostidis (1998, 2005) and Komárek (2013), and did not refer to other recently proposed changes based on molecular studies

The sizes of SBM individuals in streams are highly variable, covering a range of 5–6 orders of magnitude. They range from small microalgae (<10 µm) capable of forming visible aggregates (<1 mm^2) to individuals whose upper size limits are comparable to higher aquatic plants (several cm or dm to >1 m length). Several SBM can eventually cover an entire microhabitat (defined by Frissell et al. 1986, as an area of 1 m^2 or more).

These extremes have been recorded for *Cladophora* networks in eutrophic rivers worldwide (e.g. Winter and Duthie 2000 for North America). Macroalgae start their development covering very small elements or patches of microhabitats and become visible at a scale of ~1 mm^2. In small headwaters (first and second order), they may expand to cover the entire area of a stream. In contrast, in larger, fast-flowing streams (>third order) several microhabitats may coexist beside each other in

Table 8.2 Morphological characteristics and general habitat preferences of SBM species from streams and rivers in southeastern New York State (USA) (60 wadeable river sites × 3 seasons = 180 time-date combinations between 2005 and 2007)

Taxa	MSI	MCA	GFT	SF	H	FA	RZ	SO	Seas	Alt	Colour
Cyanobacteria / Cyanophytes*											
Capsosira brebissonii	µm	cm^2	cr	nf	-	RP	RF	< 3	Su-A	< 500	dg
Chamaesiphon amethystinus	µm	mm^2	ca		-	RP	-	< 5	all	< 500	dd
Chamaesiphon curvatus	µm	mm^2			-	RP	-	< 5	all	< 600	dd
Chamaesiphon polonicus	µm	mm^2	cr	sp	-	RP	SZ	any	all	any	dd
Chamaesiphon polymorphus	µm	mm^2	ca		-	RP	SZ	any	all	> 200	dd
Chlorogloea rivularis	µm	mm^2	co		-	RB	-	< 4	A-W	< 400	bg
Coleodesmium wrangelii	mm	cm^2	tu	nf, rf	Y	RP	SZ	< 4	all	> 200	bg
Homoeothrix varians	µm	mm^2	co		Y	RB	-	any	Su-A	any	bs
Nostochopsis lobata	mm	cm^2	gl	ge, nf	-	-	-	> 3	Su	< 300	bg
Leptolyngbya foveolarum	µm	mm^2	co		-	-	-	< 5	all	any	db
Phormidium autumnale	µm	cm^2	co	sm	-	RP	-	any	A+Su	any	db
Phormidium retzii	µm	cm^2	ca		-	RP	-	any	Sp-A	any	bl
Stigonema mamillosum	mm	cm^2	sh	fi, nf	-	-	-	< 4	Su	< 400	dd
Chlorophyta s.l.											
Cladophora glomerata	cm	>m^2	so	rf	-	-	-	any	all	< 300	gr
Draparnaldia acuta	mm	cm^2	tu	ge	Y	-	-	< 4	all	< 400	gr
Gongrosira fluminensis	µm	cm^2	cr		-	RP	RF	any	Su-A	any	dg
Gongrosira incrustans	µm	cm^2	cr	ca	-	-	-	any	Su	any	gr
Klebsormidium rivulare	mm	cm^2	co	fi	-	RP	-	< 5	A + Sp	any	gr
Klebsormidium sp.	mm	cm^2	co	fi	-	RP	-	< 5	n.d.	any	gr
Microspora amoena	mm	cm^2	co	fi	-	-	-	< 5	Sp+Su	> 200	gr
Mougeotia spp. agg.	mm	cm^2	co	fi	-	-	-	any	Su-A	> 100	gr
Oedogonium spp. agg.	mm	cm^2	co	fi	-	-	-	any	Su-A	any	gr
Rhizoclonium hieroglyphicum	mm	dm^2	so	rf	-	-	-	< 4	Su-A	> 200	gr
Sirogonium sp.	mm	cm^2	co	fi	-	-	-	any	Su-A	> 200	gr
Spirogyra spp.	mm	cm^2	co	fi	-	-	-	any	Su-A	> 200	gr
Stigeoclonium tenue agg.	µm	cm^2	co	fi	Y	-	-	any	A-Sp	any	gr
Tetraspora gelatinosa	mm	dm^2	tu	ge	-	-	-	< 5	Sp-A	any	gr
Tetraspora lubrica	mm	cm^2	tu	ge	-	-	-	< 5	all	> 200	cg
Ulothrix aequalis	mm	cm^2	co	fi	-	-	-	< 5	A	any	gr
Ulothrix variabilis	mm	cm^2	co	fi	-	-	-	any	W-Sp	< 500	gr
Ulothrix zonata	mm	cm^2	co	fi	-	RP	-	any	W-Sp	any	gr
Zygnema sp.	mm	cm^2	co	fi	-	-	-	any	Su	any	gr
Rhodophyta											
Audouinella hermannii	mm	cm^2	tu		Y	RP	-	< 5	all	any	re
Batrachospermum gelatinosum	cm	cm^2	sh	sm	Y	-	SH	< 3	Sp+W	< 500	og
Batrachospermum helminthosum	cm	cm^2	sh	sm	Y	RP	SH	< 4	Sp+W	< 500	og
Lemanea fluviatilis	cm	dm^2	tu	tt	-	RB	RF	< 5	Sp+Su	any	rr
*Sheathia americana***	µm	mm^2	sh	sm	Y	RP	SH	< 4	all	< 400	og
Chrysophyceae											
Chrysocapsa maxima	mm	cm	co	ge	-	RB	RF	< 5	A-W	any	gd
Hydrurus foetidus	cm	dm^2	tu	ge	-	RB	RF	< 5	Sp+W	> 300	gd
Xanthophyta											
Tribonema affine	cm	dm^2	ch	fi	-	-	-	< 3	A+Sp	> 200	cg
Tribonema vulgare	cm	dm^2	ch	fi	-	-	-	< 5	Sp+A	> 300	cg
Vaucheria sp.	cm	dm^2	ch	ve	-	-	-	< 5	Sp+Su	any	gr

Streams ranging from headwater streams to mid-sized rivers (1st to 5th order), sites located at 5–544 m a.s.l. Abbreviations given in Table 8.1

[a]Within Cyanobacteria identification we followed the traditional species concepts of Komárek and Anagnostidis (1998, 2005) and Komárek (2013), and did not refer to other recently proposed changes based on molecular studies

[b]Formerly as Batrachospermum anatinum

reaches that have a prevalence of stony substrata. Even in these cases the primary colonization within the mosaic of habitats can temporally be modified in a later stage by rapidly growing assemblages of different algae, and eventually cover the entire streambed. Weathered boulders and larger rocks offer cavities facilitating persistence of vegetative and/or resting stages, which enables recovery after extreme flood events. Under harsh conditions single colonized microhabitats may remain quite small, such that several species of SBM may share the same stone (Fig. 8.1a).

Growth strategies. SBM can follow different growth strategies leading to convergent growth forms, including the following:

- *Extended filamentous floating mats* formed by (1) single non-branched extended filaments with regular cells and cell division (*Klebsormidium*, *Zygnema*) or (2) firmly attached branched filaments with apical growth determination and multinucleate cells (siphonocladal *Cladophora*) (Whitton 1970; Dodds and Gudder 1992), or (3) based on a network of infinitely-growing, weakly branching, multinucleate siphon-like trichomes (*Vaucheria*)
- *Tufts* formed by large consistent single threads (>> 1 cm long, > 1 mm wide) formed e.g. by (1) pseudo-parenchymatous single tubes originating from a central axis (*Lemanea*) or (2) corticated filaments with nodular shrub-like branching (*Batrachospermum*) (Sheath and Hambrook 1990; Eloranta and Kwandrans 2012), or (3) tree- or chandelier like nodular branched specimens with apical cell differentiation (e.g. *Chara*) (Becker and Marin 2009)
- *Appressed patches* (from > 1 mm^2 to extensive covers) formed by (1) large single specimen clonal growth based tissue-like patches densely packed by radial spreading and branching, tightly adherent to the rock (*Hildenbrandia rivularis* (Liebmann) J. Agardh) (Sheath and Hambrook 1990), or by (2) colonial clonal growth of minute polar attached cells forming patches on the substratum (from >1 mm^2 to larger patches) by adherence of several layers of exocytes within multi-layered parent walls (most epilithic *Chamaesiphon*- species except, *C. polonicus* Rostafinski (Hansgirg)) (Komárek and Anagnostidis 1998).
- *Clonal spheroid calcifications* (from 1 mm diameter to extended covered areas) formed by (1) laterally branching single cells connected by mucilage tubes and re-calcified as single crystals per clone (*Oocardium stratum* Nägeli; Rott et al. 2010a, 2012), or (2) irregular globular, soft calcified colonies (2–10 mm) based on growth and aggregation of non- polar trichomes (*Schizothrix semiglobosa* Komárek and Anagnostidis 2005), or (3) solid calcified (crystallized) pustules based on polar differentiated, false-branched filaments with subapical meristematic zones forming seasonal intensified calcification rings (calcified *Rivularia* spp. potentially different clones growing together; Berrendero et al. 2008).
- *Clonal extended mucilage masses* (from 1 to 15 cm) formed by (1) anchored mucilage tubes with internal cell differentiation comprising three types of cells, smaller lateral, intermediate-sized central, and larger main stem cells (*Hydrurus foetidus* (Villars) Travesan; Klaveness and Lindstrøm 2011); or (2) mucilage without internal cell differentiation (*Tetraspora* spp.)

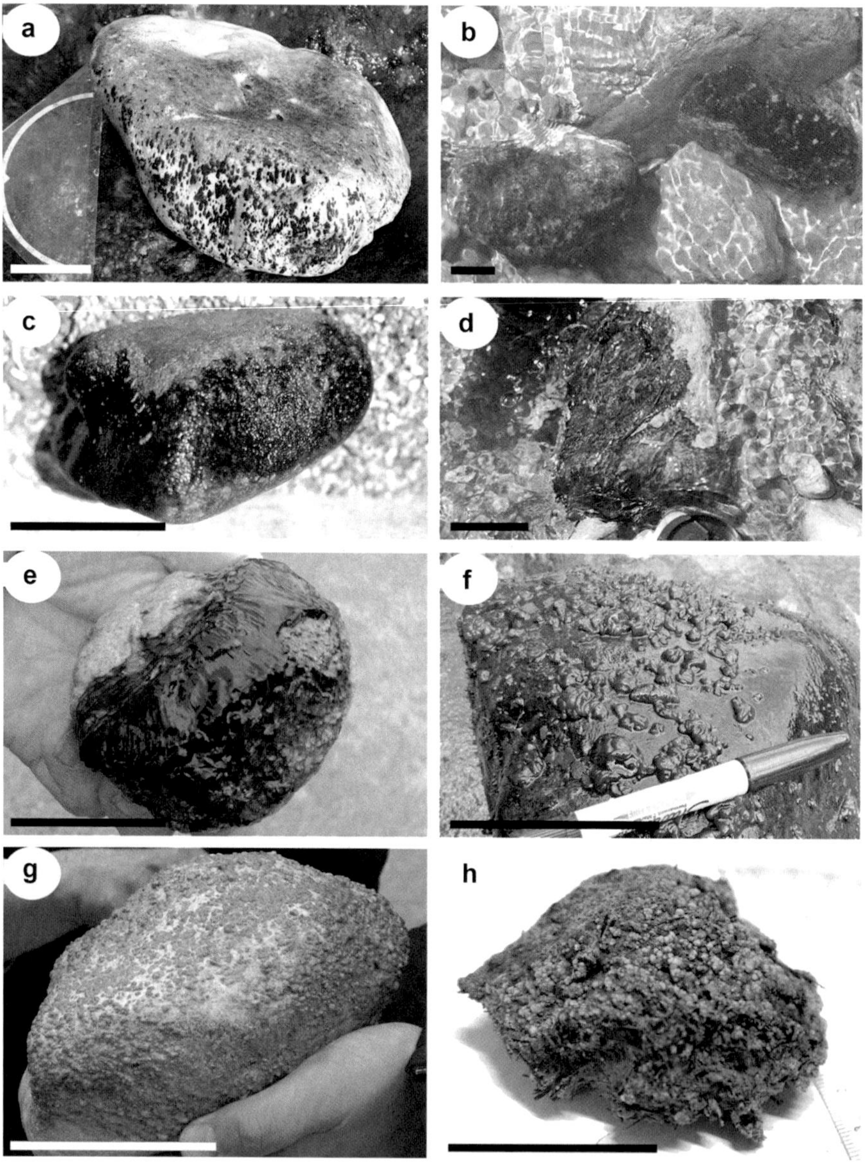

Fig. 8.1 (**a**–**h**) Growth form types of SBM, part 1: (**a**) *Chamaesiphon geitleri* (brown spots) and *Chamaesiphon polonicus* (*red brown crust*). (**b**). *Siphononema polonicum.* (**c**) *Hildenbrandia rivularis.* (**d**) *Phormidium autumnale.* (**e**) *Phormidium setchellianum.* (**f**) *Nostochopsis lobata.* (**g**) *Schizothrix semiglobosa.* (**h**) *Oocardium stratum.* (**f**) from USA, all others from Austria. (**d**, **h**) In headwater streams and springs; (**a**, **b**, **e**–**g**) in small rivers; (**c**) in large rivers. Scale bars = 5 cm

Growth form types (Figs. 8.1 and 8.2). SBM are largely characterized by a specific geometric arrangement to flow, resulting into variable flow resistance or resilience of which the major recorded types (see also Steinman et al. 1992) observed were:

- *Thin coatings* (<5 mm) (Fig. 8.1a: *Chamaesiphon geitleri* H. Luther dark spots) and crusts (Fig. 8.1b: *Siphononema polonicum* (Raciborski) Geitler and Fig. 8.1c: *Hildenbrandia rivularis* (Liebmann) J. Agardh) within the boundary layer.
- *Soft gelatinous covers/coatings* (<5 mm thick) (Fig. 8.1d: *Phormidium autumnale* Gomont; Fig. 8.1e: *Phormidium setchellianum* Gomont (originally not differentiated from *P. autumnale*) forming a flexible ground layer.
- *Subspherical covers* comprising: (1) Lobate, mucilagineous globules (1–10 mm) (Fig. 8.1f: *Nostochopsis lobata* Wood ex Bornet and Flahault); (2) Slightly calcified (Fig. 8.1g: *Schizothrix semiglobosa*); or (3) Solid calcified coenobia (Fig. 8.1h: *Oocardium stratum*) with somewhat enhanced flow resistance.
- *Gelatinous covers* (0.5 cm to ~10 cm thick) with variable resistance to flow from (1) low resistance type with soft mucilage close to the substratum (e.g. in Fig. 8.2a: *Phaeodermatium rivulare* Hansgirg brownish covers mixed with young, soft stages of *Hydrurus foetidus*; Fig. 8.2c: *Draparnaldia acuta* (C. Agardh) Kützing) to (2) high resistance-type mucilage extended into flow (in Fig. 8.2c: *Hydrurus foetidus*).
- *Tuft-like stands* (<1 cm) (e.g. Fig. 8.2d: *Audouinella hermannii* (Roth) Duby) with high extension resistance.
- *Networks* extended into flow consisting of single filaments with enhanced anchoring strength by firmly attached rhizoid cells (Fig. 8.2e background: *Cladophora glomerata* (Linnaeus) Kützing).
- *Loosely attached velvet cushions* (Fig. 8.2e center: *Vaucheria* sp., see Ott and Brown 1974) with low flow resistance.
- *Shrub-like tufts* of branched gelatinous masses of complex filaments (Fig. 8.2f: *Batrachospermum gelatinosum* (Linnaeus) De Candolle; so-called frog-spawn alga).
- *Cartilaginous tufts* of highly flow-resistant tubes (Fig. 8.2g: *Lemanea fluviatilis* (Linnaeus) C. Agardh).
- *Shrub-like, usually calcified stands* (Fig. 8.2h: *Chara* sp.) with somewhat enhanced flow resistance.

Colour and texture. The colour scale of SBM (see examples in Fig. 8.2a–j) ranges from dark blue/black to green, yellow, brown, orange and dark red. Indeed, the range in macroalgae covers the entire visible spectrum (Tables 8.1 and 8.2). Colour may also vary over time with aging of cells (Fig. 8.2g, h), exposure to desiccation (Algarte et al. 2013), in response to the spectral composition of available light (e.g. chromatic adaptation: Bautista and Necchi 2007; Sobczyk et al. 1993; Tonetto et al. 2012), or UV exposure (DeNicola and Hoagland 1996). The true colours are related to the specific intracellular pigmentation in each taxonomic group (major chlorophylls—green (Fig. 8.3a–i) plus accessory pigments (e.g. xanthophylls and carotenoids for

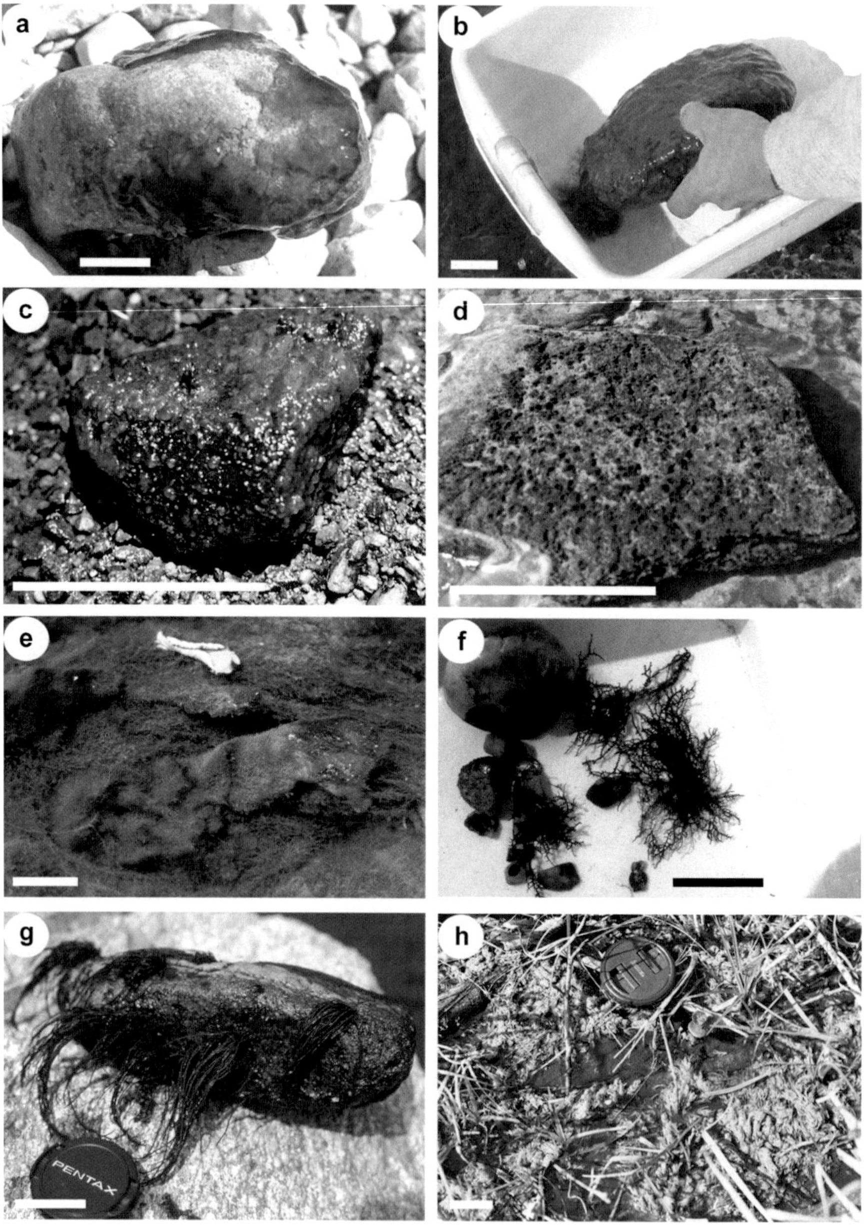

Fig. 8.2 (**a**–**h**) Growth form types of SBM, part 2: (**a**) *Phaeodermatium rivulare* (*thin yellow cover*) and *Hydrurus foetidus* (thick mucilage). (**b**) *Hydrurus foetidus*. (**c**) *Draparnaldia acuta*. (**d**) *Audouinella hermannii*. (**e**) *Vaucheria sp.* velvet aspect (center). *Cladophora glomerata* (netlike on the back). (**f**) *Batrachospermum gelatinosum*. (**g**) *Lemanea fluviatilis*. (**h**) *Chara* sp. Fig. (**e**) from USA, all others from Austria. Scale bars = 5 cm

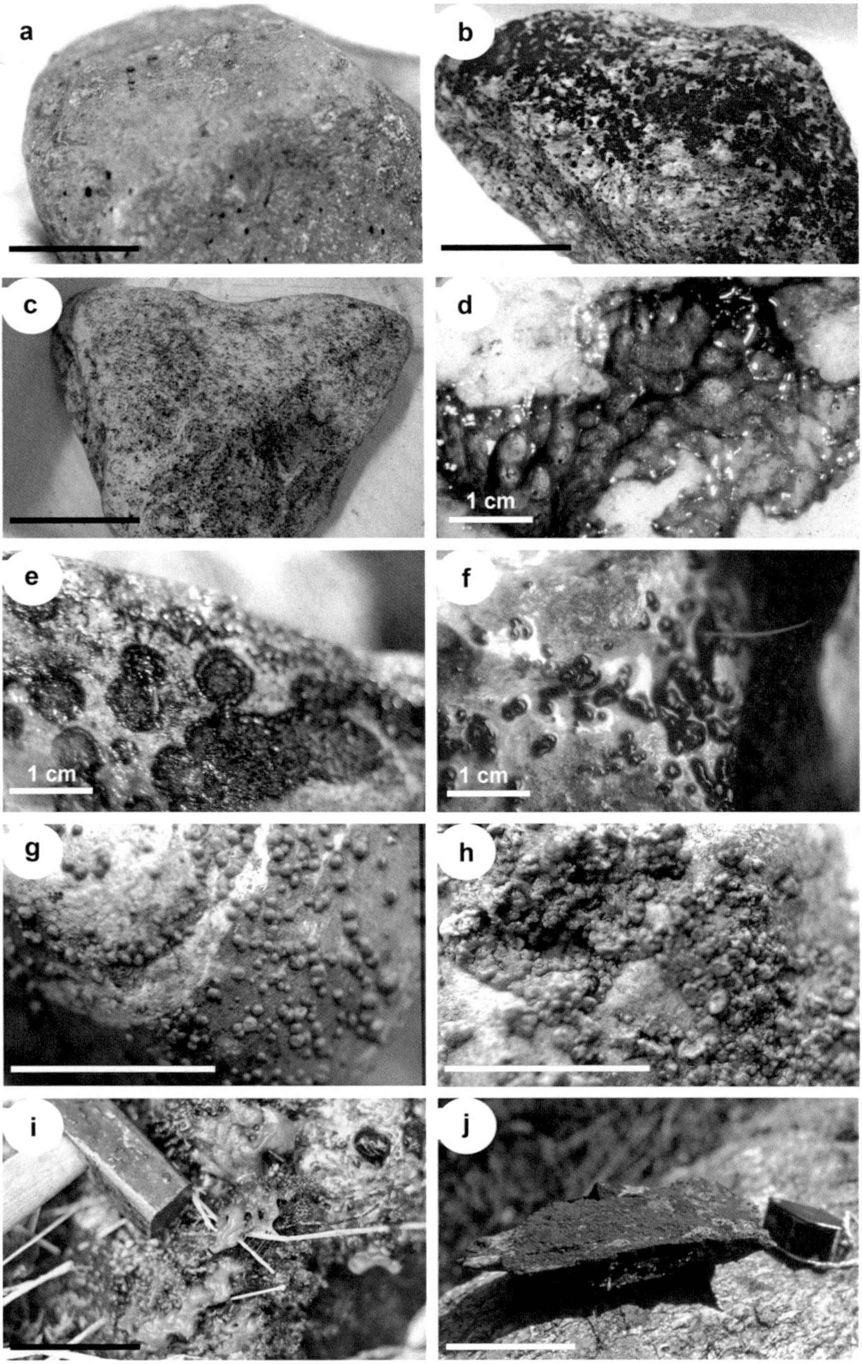

Fig. 8.3 (**a**–**j**) Colours and textures of key SBM: (**a**) encrusted *Gongrosira incrustans*. (**b**) black spots *Chamaesiphon starmachii*. (**c**) *red brown crusts* of *Chamaesiphon polonicus*. (**d**) close up of mucilagineous *Phormidium autumnale*. (**e**) circular clonal growth of *Hildenbrandia rivularis*. (**f**) *Rivularia periodica* in Obenlüneschloß. (**g**, **h**) *Schizothrix semiglobosa*. (**i**) Green aggregations of *Zygnema sp*. (*Rivularia* in the right upper corner). (**j**) Spray zone with *Trentepohlia aurea*. Substrata: limestone in (**a**, **c**, **d**, **f**–**j**), granite in (**b**, **e**, **j**). Scale bars = 5 cm, when not otherwise indicated

green algae; see Becker and Marin 2009). For example, carotenoids create the rusty red colour in *Trentepohlia aurea* (Linnaeus) C.F.P. Martius (Fig. 8.3j), phycocyanin creates the blue colour in Cyanobacteria (Fig. 8.3d), and phycoerythrin creates the red appearance in Rhodophyta (Fig. 8.3e). Two UV-shielding pigment types scytonemins and microsporine-like amino acids (MAA) are synthesized by many cyanobacteria (Garcia-Pichel and Castenholz 1991; Castenholz and Garcia-Pichel 2012) but only partly known for *Chamaesiphon* species (Fig. 8.3b, c) and *Rivularia* spp. (Fig. 8.3f) yet. Whereas MAAs as intercellular pigments have most likely no effect on colour, scytonemin is transported into the sheath layers and thus can contribute to the yellow or brownish colour of the cell wall.

Specific morphological characters related to uter cell wall features (e.g. calcite crystals on outer cell wall in Fig. 8.3a, excreted mucilage in Fig. 8.3d–i) lead to algal growth with specific surface textures and feel. These range from silky smooth (e.g. filamentous *Zygnema* spp. (Fig. 8.3i), to skinny gelatinous (Fig. 8.3d: *Phormidium autumnale*), rough and slimy (*Cladophora*), spongy (Fig. 8.3e: *Vaucheria*), smooth and slightly calcified (Fig. 8.3g, h: *Schizothrix semiglobosa*), to hard and encrusted with a variable smooth slimy surface (Fig. 8.3f: *Rivularia* spp. and right upper corner in Fig. 8.3i). For calcareous streams, very rough forms are typical, such as (Fig. 8.3a: *Gongrosira incrustans*) and hard, calcified crusts (Fig. 8.1h: *Oocardium stratum*). Texture is only one easily recognizable feature describing the complex relationship between the microhabitat niche and surface near-flow mechanics (e.g. Sheath and Hambrook 1988, 1990).

Reach Scale

Zonation. Spatial characteristics at reach scale conditions of fast-flowing mountain s result in alternating habitat differentiation preferred by specialized SBM taxa (see specific preferences in Table 8.1). The number of co-occurring SBM may increase from lower order headwater streams and lower numbers in forested catchments, to a maximum richness in small, open canopy rivers or braided open canopy sections of mountain rivers, given stable substrata. This pattern is in accordance with the River Continuum Concept (Vannote et al. 1980). According to findings for high-altitude Alpine streams (Rott et al. 2006b), mid-order streams and rivers (3–6) show the maximum habitat richness comprising the following habitats:

- Spray zone, water-level fluctuation zone or permanent wetter perimeter habitats
- Within the wetted perimeter niches
  - Along stable lateral banks on gliding/depositional shares leading to strata differentiation and
  - Habitats on larger boulders and or riffles facilitated by the substrate size and stability pattern of the streambed.

The importance of these environmental niches within a reach is highly influenced by eco-morphological conditions. Streambeds may be straight and constricted, or meandering, or have braided channels (Frissell et al. 1986). In addition, the annual runoff volume and the effects of regular and exceptional scours after floods in turn affect the grain (substratum) size available for colonization.

Reach features (Fig. 8.4). Reach features mainly reflect channel characteristics and various seasonal aspects of water and riverine conditions for stony streams. With the winter low water period in a calcareous gravel stream (KH = limestone alps bioregion in AT) both central patches of high-flow adapted taxa and a flow related lateral algal zone with orange colours consisting of the cyanobacteria *Chamaesiphon polonicus* and *Homoeothrix varians* Geitler and bleached pustules of *Schizothrix semiglobosa* can be observed (Fig. 8.4a, b). Brownish vegetation colours of *Hydrurus foetidus* (chrysophyte) are characteristic for the clear spring phase of a high alpine glacier streams with a short spring bloom (Fig. 8.4c). This is later reduced to a remaining hue of brown from *Hydrurus* mixed with *Phaeodermatium rivulare* after spring floods (snowmelt) and as temperature increases (Fig. 8.4d), but before the water becomes turbid with the melting period of the glaciers (Fig. 8.5a). In other streams the *Hydrurus* aspects vary from dark brown to black patches within calcareous mid-altitude gravel streams (Fig. 8.4e, g) and brownish flocks on larger gravel together with other algae in later spring or early summer of lower altitude streams (calcareous, Fig. 8.4h, mixed geology peri-alpine, Fig. 8.4f). In AT and NY green tuft flocks of *Ulothrix* may also be seen (e.g. Fig. 8.4f near margin) in later spring periods with cold water. The summer aspect at cold-water calcareous stream (12–14 °C) with low runoff (Upper Lunzer Seebach) is characterized by *Gongrosira incrustans* (Reinsch) Schmidle (green) and *Schizothrix semiglobosa* (flesh-red pustules, Fig. 8.4h), associated with other algae. Intense green growths of *Klebsormidium* (before drying) are frequently observed in bedrock pools in summer in NY streams (Fig. 8.4i), as well as greenish patches in a sandy cobble streams typically colonized by cottony patches of *Vaucheria* sp. (Fig. 8.4j). The tufts of *Vaucheri*a stabilize the benthic substrata of these streams, by knitting together sand and fine silt (J. D. Wehr, personal observations).

River Type to Ecoregional Scale

River type features (Fig. 8.5). Rivers in AT (Fig. 8.5a–d)) comprise a large variety of runoff and catchment geology related types (e.g. glacial streams, crystalline and calcareous streams), with an extended altitudinal range and steep gradients. Stream and river types of southeastern NY State drain large parts of mountainous regions, although of a generally lower altitudinal range than in AT (highest peaks > 1200 m), and thus are largely located in forested areas with fairly dense canopy cover (Fig. 8.5e, f), except in agricultural and urban areas

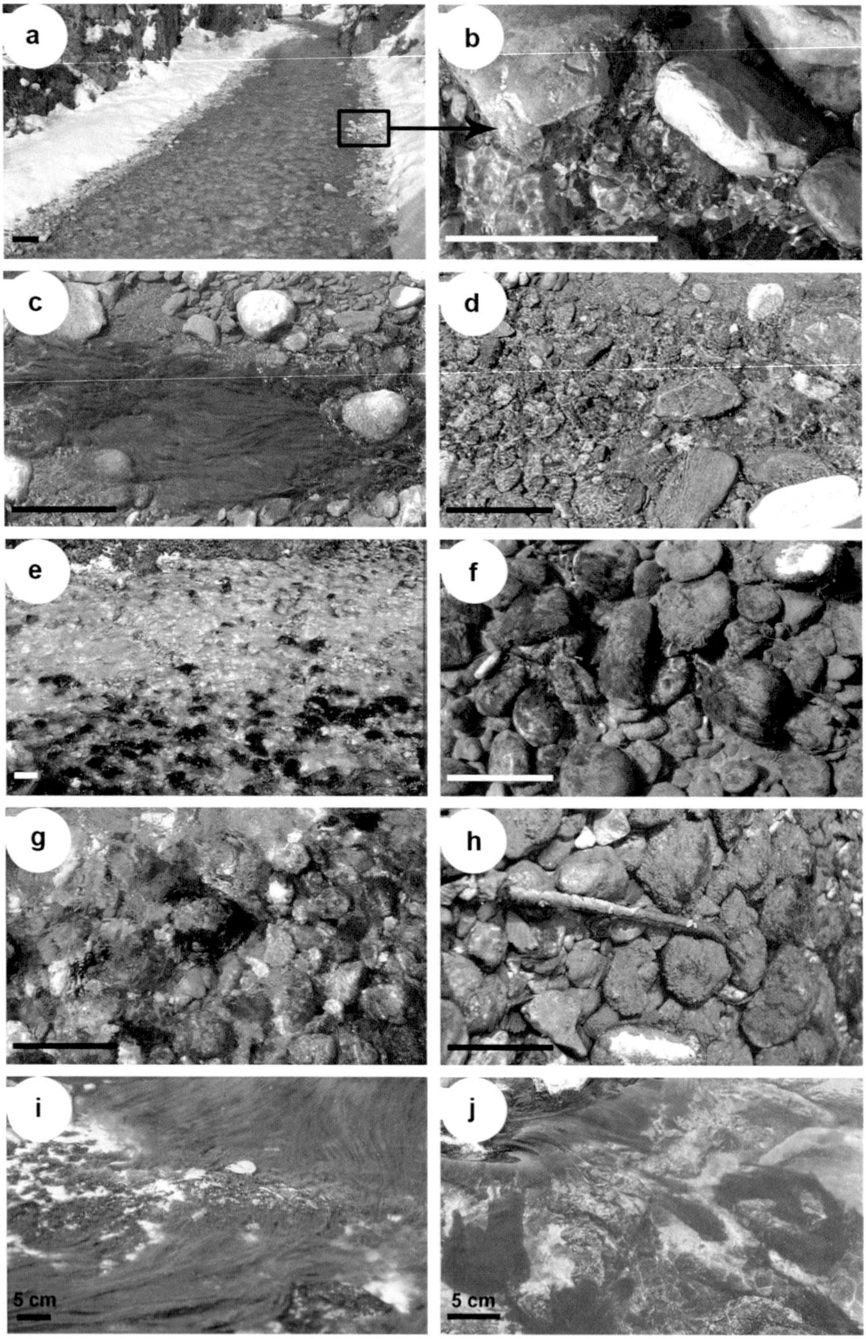

Fig. 8.4 (**a**–**j**) Field aspects of macroalgal growth in different stream types and seasons: (**a**, **b**) calcareous gravel river in winter: (**b**) close up of (**a**) as indicated. (**c**, **d**) Glacier stream in (**c**) early and (**d**) late spring. (**e**) Calcareous river in spring. (**f**) Mixed geology river in late spring. (**g**) KH river in late spring. (**h**) KV river in summer (low water period). (**i**) Isolated pool in mountain stream during summer low flow. (**j**) Spring-fed stream with large patches of *Vaucheria sp*. (**i**, **j**) from USA, all others from Austria. For KH, KV-abbreviations and location of bioregions see Fig. 8.6. Scale bars = 50 cm, when not otherwise indicated

Fig. 8.5 (**a–j**) Ecomorphological and water colour aspects of river types: (**a**, **b**) glacier river (VZA) bioregion) at (**a**) snowmelt and (**b**) with low runoff in mid-summer. (**c**) Mid-altitude stream (UZA region). (**d**) Calcareous gravel stream (KH region). (**e**) Forested mountain stream in early spring. (**f**) Large forested stream (mainly eastern Hemlock) in summer. (**g**, **h**) Two first-order tufa spring streams (AV region). (**i**) Deciduous forest river. (**j**) Brown water stream in spring. (**a–d** and **g–i**) from Austria, all others from USA. For VZA, KH, AV abbreviations and location see Fig. 8.6

increasing from NW to SE toward NY City, and coastal pine areas of Long Island. The low order streams above the timberline in the Austrian Alps streams have no canopy (Fig. 8.5a, b), and still in mid altitudes of the Alps open channels prevail (Fig. 8.5c, d). In the glaciated high central mountain areas of AT, glacial streams offer very specific and sometimes harsh conditions for macroalgal growth, due to turbidity during elevated runoff in summer (initial snowmelt and ablation period (Fig. 8.5a) and variable but sometimes low runoff during cold climate episodes within late summer (Fig. 8.5b). Central alpine rivers can become clear during longer dry periods in autumn (Fig. 8.5c, approximately 800 m a.s.l.) but have large boulders in their channels favouring growth of rheobiontic SBM on riffles (e.g. *Lemanea fluviatilis*). Mid-altitude rivers (900 m a.s.l.) in the calcareous alps in pristine pine forested catchments show high natural gravel bed dynamics and variable SBM covers, e.g. in River Isar (Fig. 8.5d) and the tributary Karwendelbach (Fig. 8.4a, b).

In the forested mountain streams and rivers of NY State, rocky substrata and canopy cover are critical factors affecting the development of SBM (Fig. 8.5e, f), comparable with a closed canopy situation in the peri-alpine areas in AT (Fig. 8.5g). Except in intensively farmed areas, riparian vegetation offers at least partial shading of most streams in southeastern NY State (Fig. 8.5f). Special stream types in AT are related to calcareous springs, e.g. near Innsbruck (Fig. 8.5g) and Lingenau (Fig. 8.5h) respectively, where tufa formation to a larger extent is attributed to both the calcification of spring moss (*Cratoneuretum*-association Fig. 8.5g) and *Oocardium, stratum* (Fig. 8.1h), a specific calcifying SBM taxon (Rott et al. 2010b, 2012). Brown water streams are generally rare and short (lowest stream orders) in AT but are somewhat more frequent in NY, especially in so-called pine barrens landscapes (e.g. pitch pine—oak forests) where in the spring, humic substance colorations are most striking (Fig. 8.5j).

Longitudinal differentiation. SBM showed interesting coincidences of changes along one continuum in different alpine situations. The study of St. Vrain River in Colorado between >3000 and 1500 m a.s.l. (Ward 1986) showed in spite of some general variability in SBM, *Hydrurus foetidus* dominance in the upper reaches, *Lemanea fluviatilis* in the middle reaches, and *Cladophora* sp. in the lower most reaches. The majority of other taxa however were found to occur along extended altitudinal ranges. Similar patterns were observed in NY streams. *Hydrurus* strictly occurs in colder mountainous locations and is never seen in lower reaches even during winter months. *Tetraspora* species also tend to be more common in upper reaches during winter months in NY. In NY *Cladophora glomerata* is the most common SBM species at lower altitudes, especially below 300 m (Table 8.2). Along River Isar, AT (originating at 1200 m a.s.l.), the crenal areas, which are richly fed by isothermal cold groundwater-fed streams are dominated by *Hydrurus* year round. This alga becomes less frequent in consecutive stretches with a peak of other SBM diversity 5–8 km downstream (see Fig. 8.9) and *Siphononema polonicum* dominates over a longer segment km 10–15 (900 m a.s.l.) in areas where the channels are more open and influenced by pasture use and urbanization. In high-alpine situations of glacier fore-field streams, the richness of SBM and microalgae was inversely correlated to the glaciation of catchment and positive correlated to conductivity, i.e. calcareous influence in part of the catchment, (see Rott et al. 2010a), whereas in subalpine and open canopy montane forest areas specific spray-zone taxa (e.g. *Trentepohlia*) are common, but fade out further downstream (Rott et al. 2006a). In contrast to alpine situations we found in AT also taxa with a preference to large rivers and low altitudes (*Bangia atropurpurea* Roth (C. Agardh) see Tables 8.1 and 8.3); this latter alga has not been observed in NY streams, although it has invaded large sections of the Laurentian Great Lakes (Graham and Graham 1984).

Spatial variability within AT. Large-scale (80000 km^2, > 1300 points) spatial distribution of SBM key species is illustrated in Fig. 8.6. Some taxa seem to be distributed over almost all areas of AT (*Cladophora glomerata* and *Hydrurus foetidus*) although both have somewhat divergent altitudinal and environmental preferences (Tables 8.1 and 8.3). Other species were more frequently recorded in either the Western Alpine (*Chamaesiphon fuscus* (Rostafinski) Hansgirg, *Lemanea fluviatilis*), Eastern Lowland (*Phormidium retzii* (Agardh) Kützing ex Gomont) or the central alpine part of AT (*Chamaesiphon geitleri*). These patterns reflect to some extent the geographic orientation and topography of the bioregions, which in fact are highly influenced by a strong altitudinal gradient decreasing from W to E as well as from SW to NE (see also related preferences of taxa for bioregions in AT, Table 8.1).

Spatial drivers for datasets from AT and USA. Multivariate analysis of both datasets indicates that spatial gradients play an important role for the location of species niches within a multidimensional space. In spite of somewhat different datasets (six variables were identical, whereas bioregional differentiation and TP data were only available for AT and canopy, land use, and TDP for USA only), the independent canonical correspondence analyses (CCA) of SBM from AT (Fig. 8.7) and USA (southeastern NY State, Fig. 8.8) respectively, showed in both cases that altitude (= a spatial gradient) was statistically the most determining variable correlated to axis 1 (explaining 20.7 % of variability for AT and 15.5 % for USA). This spatial gradient is in both cases inversely related to aqueous chemical variables; nitrate and conductivity for AT (explaining 20.4 and 14.6 %) and conductivity and nitrate for NY (explaining 14.3 and 7.9 %). Key taxa within others strongly related to the altitude axis in both regions are *Hydrurus foetidus* (HYDRFOET) and *Cladophora glomerata* (CLADGLOM) of which the first is associated with higher, and the second to lower altitudes than average (compare Figs. 8.7 and 8.8). While in NY State conductivity was the second most important variable, nitrate had this position in AT. In NY (USA) streams, the higher importance of conductivity (axis 1) is related to a stronger gradient of human-influenced sites (mostly residential, a few urban and industrial) at lower altitudes (O'Brien and Wehr 2010) than in AT. The lateral

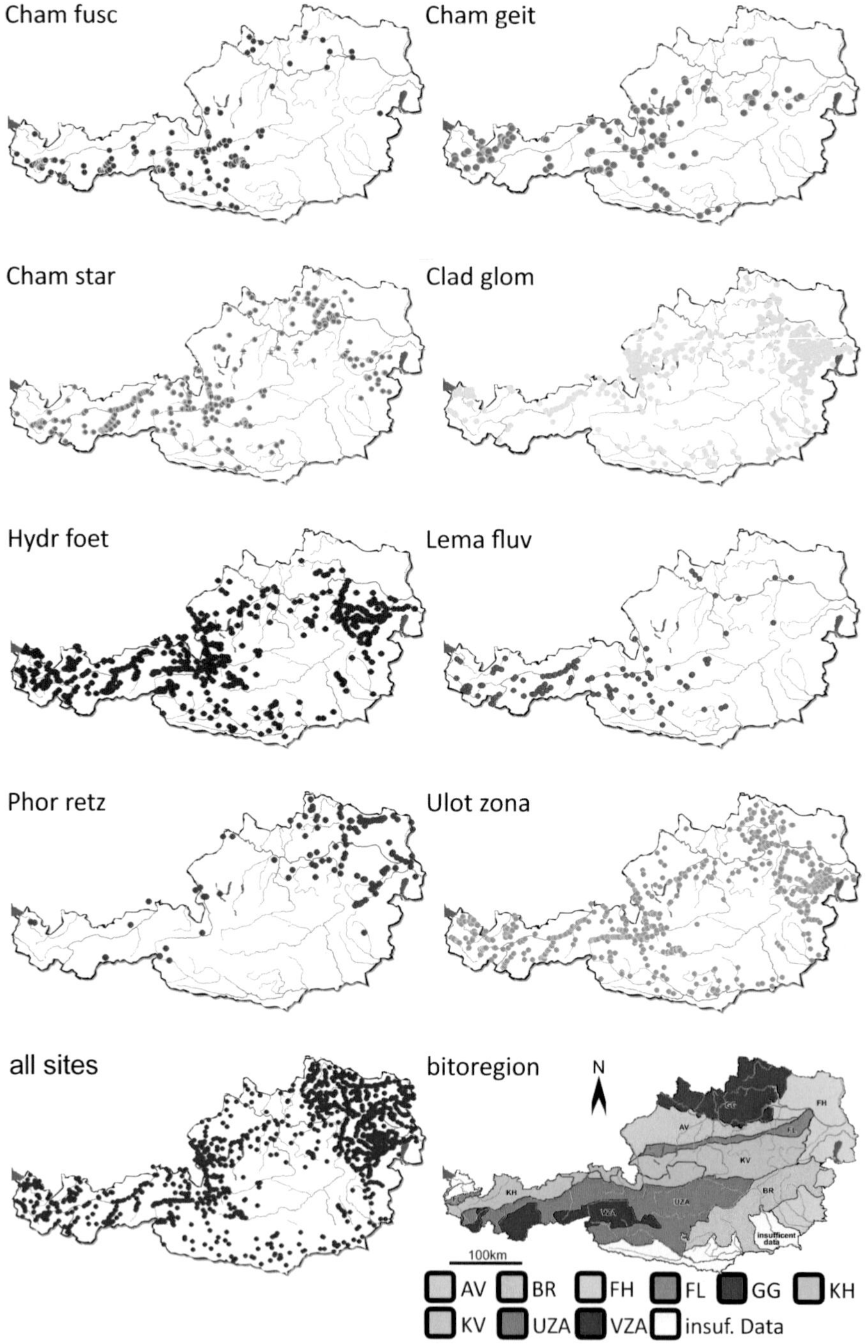

Fig. 8.6 Geographic distribution of SBM records sampled from 1345 stations across Austria within the framework of a river network and related bioregions. For taxa abbreviations the first four letters of genus and species names were used. Fluvial bioregions acc. to Moog et al. (2004): AV Bavarian Austrian Piedmont; BR Ridges and Foothills of the Crystalline Alps; FH Eastern Ridges and Lowlands; FL Flysch (Limestone); GG Granite and Gneiss Region of the Bohemian Massif; KH Limestone Alps; KV Limestone Foothills; UZA Non-glaciated Crystalline Alps; VZA Glaciated Crystalline Alps

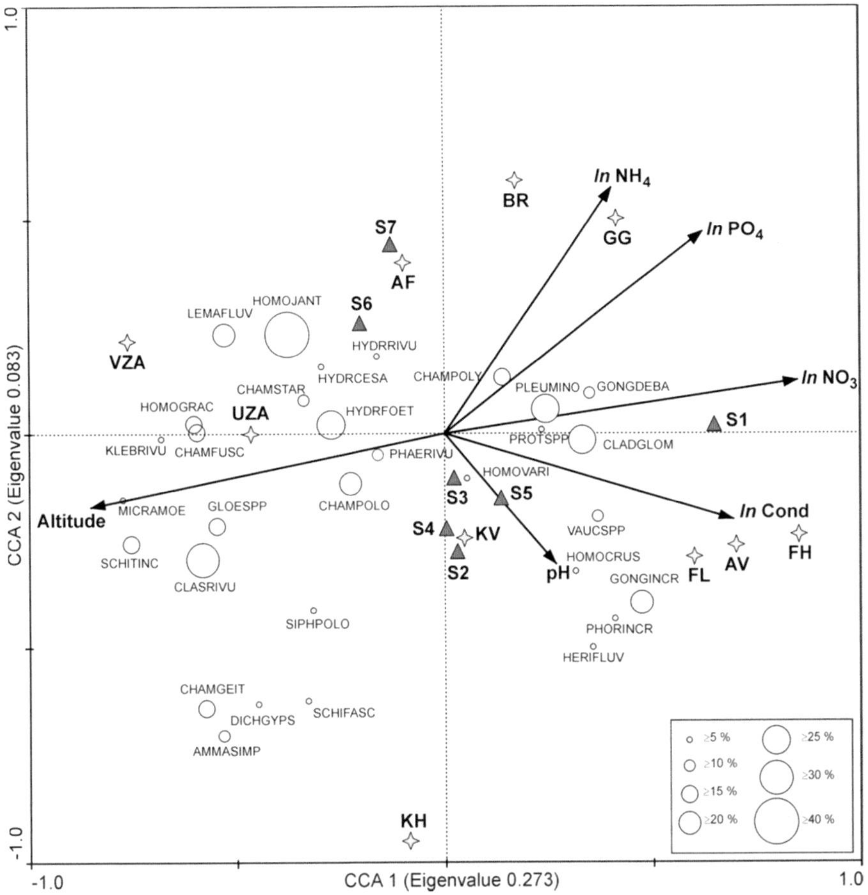

Fig. 8.7 Species niches of SBM from Austria evaluated by means of canonical correspondence analysis (CCA) based on a random selection of 30 samples from each of 10 fluvial bioregions in Austria. Main driving environmental variables used are altitude, pH, conductivity, ln-transformed nutrient concentrations, stream order (S1–S7) and bioregional affiliation. The sizes of the circles indicate the taxon fit. For taxa abbreviations the first four letters of genus and species names were used. Bioregional affiliations: AF: Alpine rivers, for all other abbreviations and location of the bioregions (see Fig. 8.6)

axis differentiation is influenced by nutrients (NH_4, SRP) and pH in AT. In the NY dataset, the gradient of canopy cover and its inverse relation to nutrients (mainly nitrogen components) appear to exert the second greatest effects on species composition. The implications of the CCA for niche differentiation of individual species are discussed in the spatio-temporal section below.

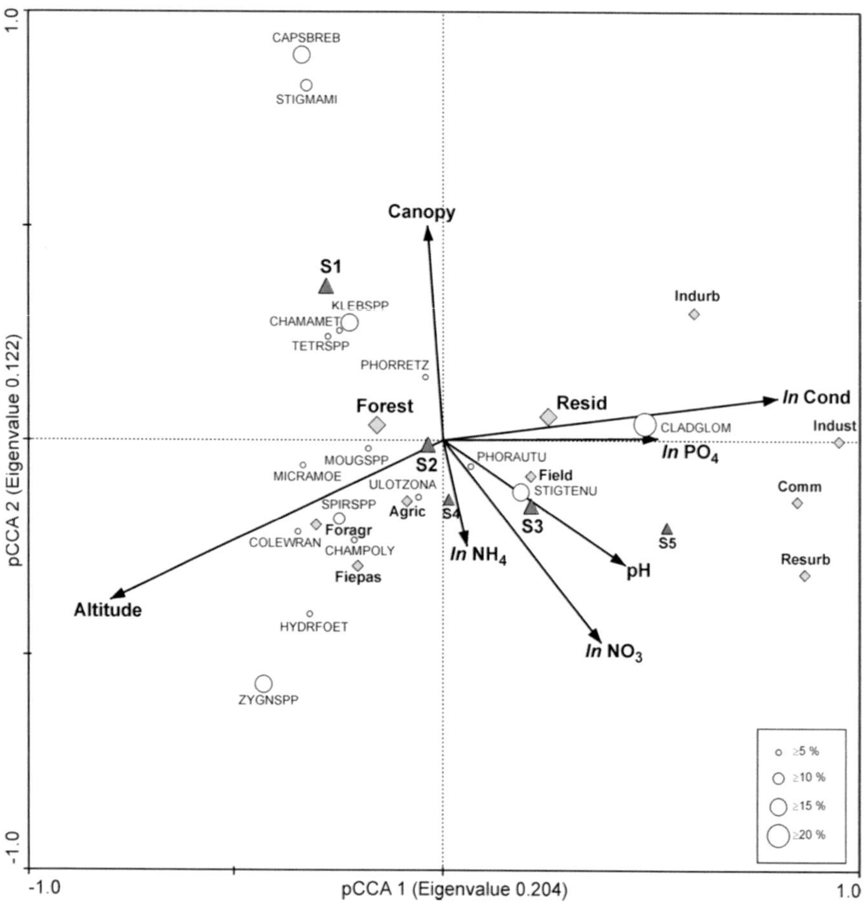

Fig. 8.8 Species niches of SBM from southeastern New York State evaluated by means of partial canonical correspondence analysis (pCCA) based on 60 stream and river sites sampled in three seasons each. Main driving environmental variables used are altitude, pH, conductivity, ln-transformed nutrients, canopy cover, stream order (S1–S5) and land use. Season was used as a co-variable. Not significant nominal variables according to Monte Carlo permutation tests (1999 permutations under reduced model, restricted to split plot-design) are indicated by smaller symbols. The sizes of the circles indicate the taxon fit. For taxa abbreviations the first four letters of genus and species names were used. Land use abbreviations: Agric: Agriculture, Comm: Commercial, Fiepas: Field/Pasture, Field: Field, Foragr: Forest/Agriculture, Forest: Forest, Indurb: Industrial/Urban, Indust: Industrial, Resid: Residential, Resurb: Residential/Urban

Temporal Aspects

Temporal Niches

The temporal duration of SBM appearance in streams is related to size, growth-form type, colonization strategy and specific ecophysiological capacity to utilize resources (including riparian influences), and resistance to major disturbances or recolonizing after disturbances (= resilience). Individual generation and doubling times of SBM can vary from a few days to weeks and months (Rott 1991). Typical time extensions of SBM niches in temperate streams are (see also Blum 1956; Rott and Pfister 1988):

- Perennial (e.g. structural perennials = calcifiers, e.g. *Rivularia*; resilient perennials, such as encrusting taxa; e.g. *Gongrosira incrustans*)
- Regular seasonal growth related to life cycle-stage variations (e.g. *Batrachospermum*, *Lemanea*, Hambrook and Sheath 1991; Sheath and Wehr 2015)
- Regular occurrence in cold seasons but related to low flow (e.g. *Bangia* in rivers in AT)
- Divergent seasonal preferences such as for *Tetraspora* spp. winter in NY and summer in AT (Tables 8.1 and 8.2).
- Short-term window growths often favoured by low flow conditions in spring or autumn.
- Periodic, but temporary and irregular growth (related to runoff history) (e.g. *Phormidium autumnale* showing irregular growth maxima during cold seasons or at other times of the year; *Nostochopsis lobata*, in large fast-flowing rivers, but expanding only during low-flow periods).

Perennials. Perennial SBM tend to be most common in headwater streams, especially in springs and spring-streams due to the stability of environmental conditions (Cantonati et al. 2012). Calcified taxa were found to persist the whole year round (12 out of 61 taxa in AT) or longer time periods, even when overgrown temporarily by other algae (e.g. in spring streams often covered by members of the Zygnematales and diatoms for some time of the year). *Rivularia* occurs in AT mainly in spring streams and shows seasonal growth by adding seasonal rings of calcite and new branches in spring and autumn. Most persisting SBM taxa tolerate being overgrown at least intermittently by other algae (diatoms and seasonal SBM), also in larger streams leading to multi-layered biofilms. *Hydrurus* may grow on top of diatom-dominated basal biofilms, and in turn, may form the substratum for smaller epiphytic algae (especially diatoms) in fast, high altitude streams. Temporal persistence can be favoured by temporary external nutrient pulses of organic P enhancing the performance of N-fixing Rivulariaceae (Whitton and Matteo, 2012), although these algae were found to be slow growing, as compared with the chrysophyte *Hydrurus foetidus* (Rott et al. 2000), and can be overgrown by *Zygnema* in summer (true in both NY and AT streams). Perennial taxa also occur in higher stream orders that experience minor seasonal variability in environmental conditions (especially discharge), and climate, as is common in the subtropics. Good examples are macroalgal ecophysiological studies conducted diurnally, which reveal variable growth capacities of gametophyte and sporophyte stages, but a large flexibility along seasonal environmental variations under subtropical conditions (Necchi and Alves 2005).

Bi-seasonal growth. Within the AT dataset (Table 8.1) many taxa exhibited two growth periods, the first in spring and the second in autumn. In some cases longer growth periods were recorded, spanning the entire cold season between autumn and spring. Such patterns are not primarily related to riparian vegetation types, although some species are specifically shade adapted (including many red algae; Bautista and Necchi 2007). For microalgae, especially diatoms, seasonality with peaks in spring and autumn or along the whole winter period is common. While closed canopy streams are a minority in the Austrian dataset, since the largest number of stream orders is between 3 and 6, canopy variability is critical and important for the NY (US) streams, with about 12 of the 41 listed SBM taxa with a shade preference (six significantly so: Fig. 8.8). Irrespective of this, in AT summer peaks of SBM growth were more often observed in the lowlands and for specific taxa, such as *Phormidium retzii* and members of the Chaetophorales. The majority of SBM in AT showed growth depression during summer in consequence of irregular and heavy rainfall peaks (or maximum ice melting period in glacier streams and alpine rivers) with persisting cold waters over summer. In contrast, many streams in southeastern NY tend to have more stable (although lower) flow during summer. In addition occasional summer rainfall likely favours expansion of many filamentous green algal taxa. In AT also in limestone areas with elevated catchment altitudes highest runoff may occur in summer by calcareous karstic phenomena (e.g. see Pfister 1993); no major karst geologies are present in the study area of southeast NY State.

Seasonal windows. For the facilitation of development of short-term windows, specific species with inherent ecophysiological capacities show the potential to extremely efficient energy utilization and fast growth capacities at low flow, such as *Hydrurus foetidus* for very short periods (6–8 weeks) in spring or autumn in glacial streams (USA: Squires et al. 1973; AT: Rott et al. 2006a, 2006b; Switzerland: Uehlinger et al. 2010). In alpine streams in Utah, macroscopic assemblages of *Hydrurus* were absent when temperatures exceeded 10 ° C (Squires et al. 1973) in European alpine streams they disappeared at > 15 °C (e.g. Cantonati et al. 2006) and they also were missing in mountainous (non-glacial) sites in NY in summer (Table 8.2). Short stable runoff periods in early spring, also favoured some taxa in large alpine rivers in AT, such as *Bangia atropurpurea* and *Ulothrix* spp., with a preference to near water level growth on boulders. Similarly in southeastern NY, *U. zonata* (Weber and Mohr) Kützing often forms macroscopic growths in early spring, especially on rocks in the splash zone of rapidly flowing mountain streams.

Phenology

The phenology of short lived (ephemeral) SBM follows a sequence from primarily exponential growth (accrual) to stagnation and decay (*sensu* Biggs 1996), but can also be related to epiphyte growth and grazing in later stages (Furey et al. 2012). Longer living SBM can be subject to temporal changes in morphology either as a part of their live cycles or in response to changes in the environment (light, temperature) or potentially (when perennial) along several consecutive phases of growth, presumably along seasonal climatic variations. Visible phenological changes are often expressed as changes in colour (pigmentation).

One of the most striking examples found in both ecoregions is *Hydrurus*, in which colours and morphological changes vary from clear brown, undefined gelatinous early stages (in low flow) (Fig. 8.2a), or thin orange brown hues together with *Phaeodermatium rivulare* (Fig. 8.4d), to the typical extended consistent brown gelatinous masses in moderate flow (Figs. 8.2b and 8.4c). But in rapid current and when exposed to sunlight, pigmentation can become dark brown to black in both AT and NY calcareous streams (Fig. 8.4e). In non-favourable conditions *Hydrurus* masses can become bleached, with the mucilage loose and sponge-like (partly seen in Fig. 8.4c). Another example is *Cladophora* filaments, which appear macroscopically bright green early in the season, yellowish in mid-summer, and then turn rusty-red in NY streams late in the year, with heavy epiphyte cover by the diatom *Cocconeis*. In contrast dark green *Cladophora* tufts were found even during the coldest periods of the year on riffles in mid altitude streams in AT.

Furthermore, phenological aspects of SBM can also be caused by desiccation, if a taxon tolerates drying in a vegetative status. One example is the weakly calcified cyanobacterium *Schizothrix semiglobosa*, growing within the wetted perimeter of KH streams in AT. It starts with small isolated clear brown pustules (Fig. 8.3g), becoming dense (Fig. 8.1g) and finally forming closed darker pigmented covers (Fig. 8.3h), which may be easily removed by strong currents. Single pustules of this taxon can also dry and bleach completely in shallow areas (Fig. 8.4b), but may re-establish from minute resting stages quickly after rewetting. The other example is the green alga *Klebsormidium rivulare* (Kützing) M.O. Morison and Sheath, from low-nutrient streams in NY (Fig. 8.4i), which during summer low-flow periods forms patches of filaments accumulating in isolated pools. The filaments become gradually macroscopically darker in colour, develop thickened cell walls, shrink (in diameter), and may fragment into shorter sections. Variation in colouration is typical for *Siphononema polonicum* in calcareous streams in AT varies from blood red/violet (observed in an alpine spring area) to red (Figs. 8.1b and 8.4g) or orange. Another of example of this and morphological variability is shown for the red alga *Hildenbrandia*, forming circular dark red patches in mountain rivers (GG ecoregion) (Fig. 8.3e), and thin even dark red non-calcified crusts in deep areas of a mountain river (Fig. 8.1c).

Long-Term Variation

In the study of the river Atna catchment in Norway, a study over 12 years (Lindstrøm et al. 2004), the majority of taxa within one catchment/river network was found to be resilient over long-term in respect to species composition and richness in spite of high and recurrent seasonal variability and a slight general increasing trend in richness observed in downstream areas. Eventual human influence in the lower catchment (possibly in combination with global change) was the reason that the lowland segments became slightly more species rich. The case of River Atna, however, was not subject to airborne acidification, since the catchment is a non-acidified, well-buffered system, whereas in the larger majority of other South Norwegian catchments SBM changes were related to long-term environmental change with rather slow recovery in recent years (see Schneider and Lindstrøm 2009).

Spatio-Temporal Aspects

Spatio-Temporal Niche Model

In clear streams, high-frequency phenological observations reveal a bridge between temporal and spatial aspects of SBM growth in streams. For this purpose a spatio-temporal niche model of growth related to time and floods for a reach in a calcareous gravel stream (KH bioregion at 900 m a.s.l.) was developed (see also Pfister 1993). Growth patterns were mapped across a 1 m long and 11 m wide permanent transect, observing presence within 25 20 × 20 cm squares per m^2, over 14 months. From relative changes in areas, where SBM were recorded, accrual and loss rates between sampling dates were estimated (Fig. 8.9). The most rapidly growing taxa in this river transect were *Schizothrix semiglobosa* (orange) in spring and autumn, *Siphononema* (red) in late summer/early autumn, and *Phormidium autmnale* (blue) in autumn. Compared to these three taxa, the expansion of *Hydrurus foetidus* was found to be much slower, with a peak in late autumn, a preference for large boulders, and microhabitats receiving maximum exposure to the current.

Within this reach, the total SBM cover had peaks in spring and autumn, and two minima: one in summer (peak flow time) and one in winter, after an exceptional scouring in February. The detailed growth pattern (Fig. 8.10) suggests strong spatio-temporal partitioning, since growth maxima (darkest coloured areas) rarely overlap. This is especially true for *Chamaesiphon geitleri*, which occupied shallow right bank areas, while during the same period two co-occurring taxa, *Schizothrix* and *Siphononema,* predominated in the central channel. *Phormidium autumnale* shared a more opportunistic (i.e. irregular) type of growth pattern in this reach, with *Hydrurus* colonizing central channel areas in autumn. *Phormidium autumnale* was almost absent during winter in the first year, but showed high growth in autumn and in spring following a spate in the second year (Fig. 8.10).

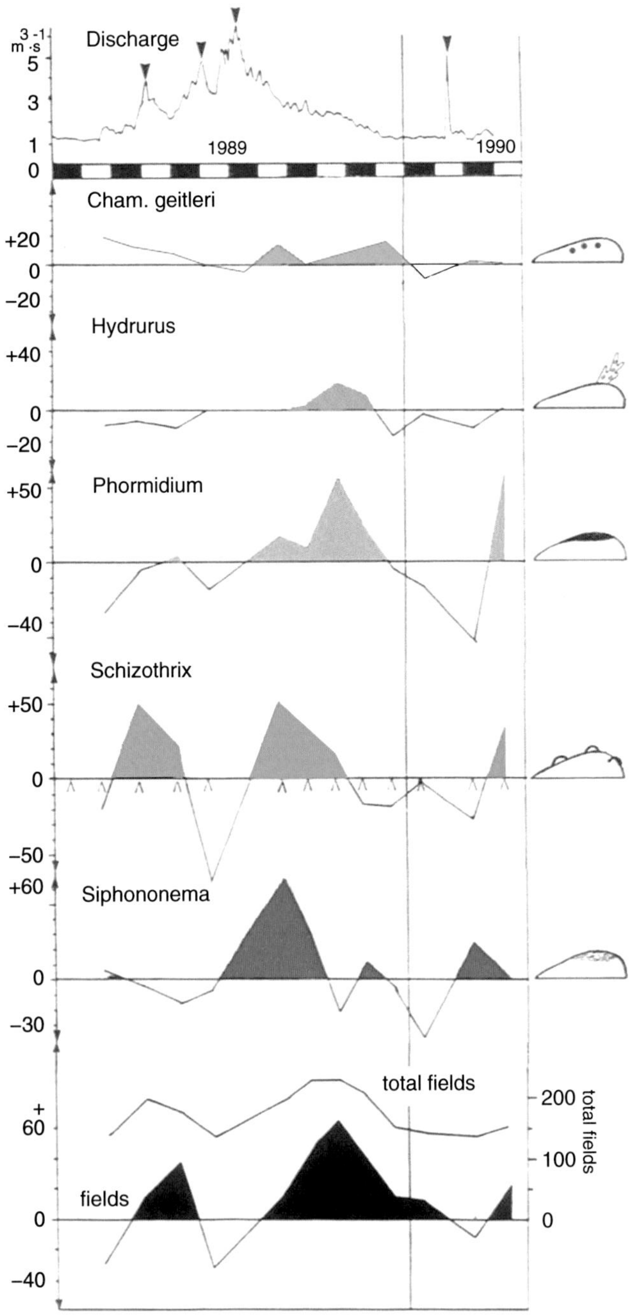

Fig. 8.9 Model graph showing seasonal changes for five SBM (Figs. 8.1a, b, d, g) in terms of changes of covered fields between two assessments (*left lower scales*) in relation to discharge line and total cover within a 1 m broad and 11 m wide permanent transect mapped 13 times between January 1989 and April 1990 in a calcareous mountain river. In discharge graph (uppermost) inverted triangles indicate runoff peaks (for related details on spatial-temporal pattern see Fig. 8.10)

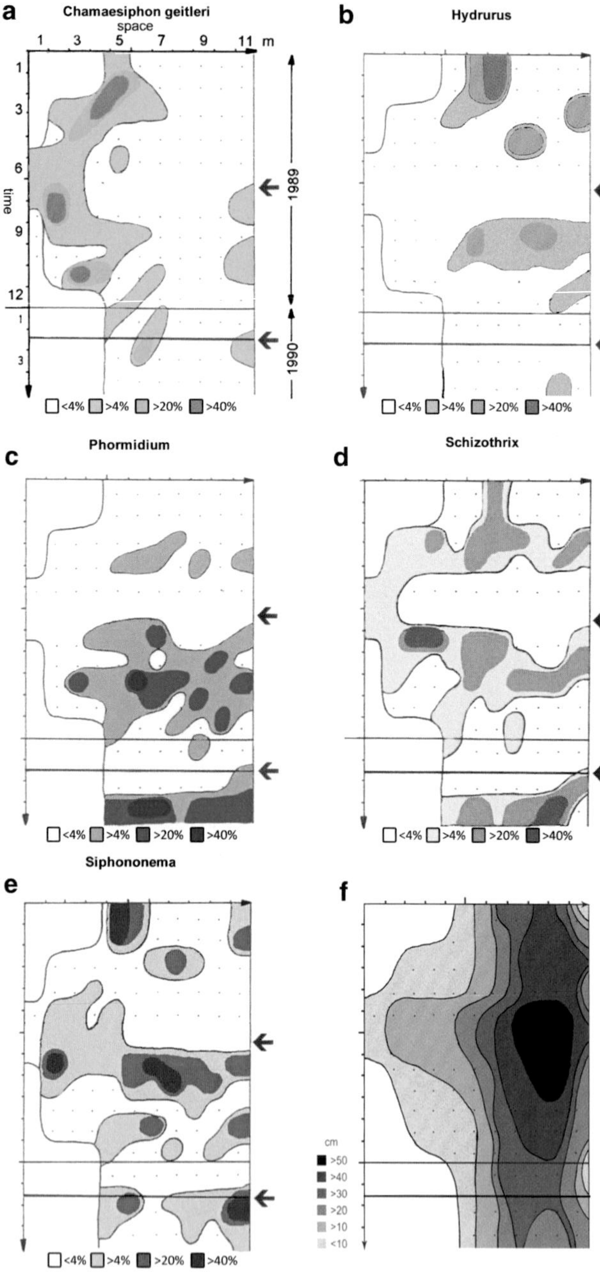

Fig. 8.10 Model graph of spatio-temporal growth pattern of five macroalgae: (**a**–**e**) same taxa as in Fig. 8.9); (**f**) water depths expressed as net changes in species presence and/or water depths in 25, 20 × 20 cm squares per m^2. Horizontal axis: distance from right bank, vertical axis: temporal distance based on 13 field assessments between January 1989 and April 1990

Complex Species Preferences

Species niches encompass spatial and temporal effects and related changes of environmental variables in a complex manner becoming evident from multivariate analysis (CCA) as follows:

- In both datasets the multivariate evaluation of the SBM by CCA relating species presence with water chemical data, altitude, stream order and other environmental data indicates the influence of a larger number of individual species on the overall distributions (Figs. 8.7 and 8.8) (Tables 8.3 and 8.4).
- Between the two CCAs the positioning of specific single species (statistical importance indicated by the size of the circles) varies to some extent, although in both cases the third most common shared species, *Cladophora glomerata,* is related to low altitudes and increased conductivity and nutrients (SRP in NY and nitrate in AT).
- In the CCA from AT the statistically most important taxon *Homoeothrix janthina* (Bornet and Flahault) Starmach (Fig. 8.7) is positioned together with *Lemanea fluviatilis,* both characterizing soft water streams at higher altitudes. *Clastidium setigerum* Kirchner is associated with the preference of highest altitudes (see Table 8.3) and also very low conductivity and low nutrients, whereas *Hydrurus foetidus* is positioned more to the centroid of the axes since it is widely distributed over almost all altitudes (except < 300 m) and wide ranges of environmental conditions and seasons. The positioning of several, but statistically less important, taxa in relation to bioregions confirms many earlier findings of bioregion-specific reference taxa within AT (e.g. *Ammatoidea simplex* Woronichin, *Dichothrix gypsophila* (Kützing) Bornet and Flahault (KH), *Gongrosira incrustans* and *Heribaudiella fluviatilis* (Areschoug) Svedelius (AV); see Pfister and Pipp 2013).
- In the NY dataset (Fig. 8.8) *Zygnema spp.* were characteristic for the highest pristine mountainous sites, while both urban and agricultural nutrient-rich streams typically have extensive covers of *Cladophora*, the statistically most important taxon together with *Stigeoclonium,* or occasionally *Phormidium autumnale*. Differences in shading preferences become evident from the arrangement of taxa along the canopy axis for *Capsosira brebissonii* Kützing ex Bornet and Flahault and *Stigonema mamillosum* C. Agardh ex Bornet and Flahault (related to dense canopy expressed by extreme position within positive sector) contrasting to *Klebsormidium* spp. (intermediate) and *Stigeoclonium tenue* (C. Agardh) Kützing (open canopy, negative sector).

Additional findings integrating several variables are the following:

- There were a few widely distributed taxa shared in both datasets, although none of them was found everywhere (= ubiquitous). The most frequent was *Phormidium autumnale,* a potential mix of genotypes with similar morphology (see Loza et al. 2013)—reaching 67 % in AT and 58 % in NY, USA. In both regions this taxon was evidently characterized by a broad ecological

Table 8.3 Habitat preferences of SBM from Austria ($n > 10$) expressed by medians of altitude and water chemical parameters (i.e. geochemical variables and nutrients) tested for significance by means of Wilcoxon-Mann–Whitney-tests of samples with taxon presence versus samples where taxon was absent

Species name (abbrev.)	n	pH Median	Cond. Median	TP Median	SRP Median	NH_4 Median	NO_3 Median	Altitude Median
AMMASIMP*	14	8.19	303	14	2	5	562	881
AUDOHERM	299	7.93	271	33	20	65	1299	549
AUDOPYGM	62	8.20	503	45	31	35	1900	305
BANGATRO	58	8.04	344	24	16	59	885	420
BATRGELA	23	8.10	453	25	13	38	1610	420
CALOSPP	33	8.10	337	19	5	20	700	462
CHAESPP	63	8.10	469	176	118	120	2000	315
CHAMFUSC	91	7.90	206	11	5	10	500	927
CHAMGEIT	45	8.17	285	9	4	10	510	870
CHAMPOLO	333	8.06	303	17	9	20	620	595
CHAMPOLY	500	8.00	331	37	21	61	1132	468
CHAMSTAR	175	7.93	209	24	15	25	660	708
CHLOMICR	49	8.10	368	35	26	69	1200	450
CLADGLOM	343	8.10	452	33	20	38	1400	393
CLASRIVU	121	7.97	232	11	5	10	450	940
CLASSETI*	15	7.93	191	5	2	10	472	1068
DICHGYPS*	16	8.25	295	8	2	9	456	927
GLOESPP	56	8.00	284	11	5	15	570	840
GONGDEBA	211	8.00	358	50	30	147	2200	420
GONGINCR	135	8.24	566	34	21	24	1700	280
HERIFLUV*	19	8.20	417	19	10	13	1150	357
HILDRIVU	60	8.01	415	71	42	62	2373	409
HOMOCRUS	40	8.30	443	19	11	19	1165	353
HOMOGRAC	81	7.77	190	12	5	10	410	870
HOMOJANT	322	7.90	252	22	12	32	600	640
HOMOVARI	460	8.10	350	26	16	30	988	488
HYDRCESA	76	8.00	287	14	11	13	567	620
HYDRRIVU	136	8.00	311	19	10	20	745	570
HYDRHOMO	25	8.00	319	13	5	8	520	755
HYDRFOET	420	8.04	311	20	12	25	695	590
KLEBRIVU*	18	7.51	90	9	1	16	322	905
LEMAFLUV	96	7.89	187	16	9	14	485	845
LEPTFOVE	166	8.10	401	47	28	57	1305	390
MICRAMOE	26	7.97	284	14	4	12	500	818
OSCILIMO	59	8.00	731	212	152	380	2300	220
PHAERIVU	457	8.02	307	20	12	25	720	570
PHORAUTU	715	8.00	350	36	20	50	1130	465
PHORINCR	50	8.30	468	31	18	25	1309	339
PHORRETZ	87	8.03	450	114	62	97	2000	270
PHORSUBF	45	8.00	360	34	14	91	900	406
PLEUAURA	47	8.03	315	21	16	70	630	580
PLEUMINO	653	8.04	402	50	30	67	1600	408
PROTSPP	179	8.10	396	38	21	52	1144	420
SCHIFASC	20	8.31	313	11	6	10	600	835
SCHITINC	43	8.00	191	9	5	10	370	927
SIPHPOLO	55	8.02	332	17	8	25	810	600
SPHAFLUV	109	8.00	490	78	55	190	2900	310
STIGTENU	106	8.00	483	125	80	145	1254	394
TETRSPP*	11	8.10	802	125	79	26	2400	182
TRENAURE	22	7.97	318	14	2	7	755	426
ULOTZONA	492	8.01	319	27	16	32	900	510
VAUCSPP	176	8.01	450	57	37	53	2200	357
XENOKERN	28	8.30	423	23	13	25	1130	370
All samples	**1064**	**8.00**	**405**	**48**	**28**	**63**	**1356**	**420**

Statistically significant medians ($p < 0.001$) highlighted: median of samples with taxon significantly lower: ■ for pH/altitude, ■ for conductivity and nutrients; significantly higher: ■ for pH/altitude, ■ for conductivity and nutrients). For taxa abbreviations the first four letters of genus and species names were used ([a]not stastistically tested, $n > 10 < 20$)

Table 8.4 Habitat preferences of SBM from southeaster New York State ($n > 10$) expressed by medians of altitude and water chemical parameters (i.e. geochemical variables and nutrients) tested for significance by means of Wilcoxon-Mann–Whitney-tests of samples with taxon presence versus samples where taxon was absent

Species name (abbrev.)	n	pH Median	Cond. Median	TDP Median	SRP Median	NH_4 Median	NO_3 Median	Altitude Median
AUDOHERM	102	7.50	92	7	7	29	241	221
BATRANAT	6	6.85	156	10	7	32	1291	7
BATRGELA	5	7.11	174	5	8	50	31	255
BATRHELM	6	6.97	131	7	7	33	669	7
CAPSBREB	7	6.60	44	5	2	18	17	224
CHAMAMET	10	7.08	68	4	6	17	237	316
CHAMCURV	8	6.58	54	8	6	20	158	328
CHAMPOLO	2	7.65	61	3	3	4	77	250
CHAMPOLY	19	7.32	52	3	3	19	384	396
CHLORIVU	8	7.38	171	8	8	33	788	97
CHRYMAXI	3	6.56	52	3	6	41	171	337
CLADGLOM	56	8.00	287	19	14	32	508	74
COLEWRAN	11	7.41	69	5	7	26	240	425
DRAPACUT	4	6.85	112	15	14	39	207	132
GONGFLUM	7	7.80	163	15	13	23	247	203
GONGINCR	22	7.90	105	9	11	25	249	250
HOMOVARI	58	7.41	75	6	7	25	229	265
HYDRFOET	10	7.14	38	1	4	13	523	388
KLEBSPP	30	7.00	53	5	6	19	75	274
LEMAFLUV	36	7.74	144	10	7	28	240	203
LEPTFOVE	24	7.44	88	9	7	33	169	218
MICRAMOE	16	7.28	69	4	5	15	152	319
MOUGSPP	41	7.19	77	6	7	30	124	319
NOSTLOBA	2	7.80	117	17	17	25	163	58
OEDOSPP	70	7.39	93	10	10	31	214	312
PHORAUTU	99	7.50	117	9	10	32	281	255
PHORRETZ	57	7.30	82	7	7	27	186	254
RHIZHIER	9	7.53	95	13	13	20	123	354
SIROSPP	5	7.63	95	10	5	43	349	274
SPIRSPP	25	7.26	71	6	7	43	200	354
STIGMAMI	6	6.90	39	5	2	15	28	180
STIGTENU	62	7.70	152	8	8	37	307	156
TETRSPP	16	7.23	67	7	8	25	90	367
TRIBAFFI	6	7.64	103	4	8	189	62	377
TRIBVULG	3	7.28	74	10	10	12	214	457
ULOTAEQU	12	7.27	70	8	7	58	180	85
ULOTVARI	20	7.67	182	7	6	32	379	139
ULOTZONA	57	7.60	78	6	7	22	274	340
VAUCSPP	28	7.20	80	9	8	23	284	298
ZYGNSPP	6	7.42	38	2	6	17	458	388
All samples	**172**	**7.50**	**95**	**8**	**8**	**28**	**240**	**253**

Statistically significant medians ($p < 0.001$) highlighted: median of samples with taxon significantly lower: ■ for pH/altitude, ■ for conductivity and nutrients; significantly higher: ■ for pH/ altitude, ■ for conductivity and nutrients). For taxa abbreviations the first four letters of genus and species names were used niche, but associated with statistically high nitrate preferences in the NY dataset only (Table 8.4). Next was *Audouinella hermannii* (59 % in USA), with a tendency to lower pH, conductivity and dissolved phosphorus compounds in both datasets (Tables 8.3 and 8.4). The niche preferences of *Cladophora glomerata* (found in 32 % of samples in both datasets) indicate statistically higher pH and conductivity medians than average in both regions, whereas higher nutrient preferences (for TDP, NO_3) were only statistical significant for the US dataset.

- Several additional shared taxa (>10) were also ecologically specified, including the exclusive running water representatives (mainly *Hydrurus foetidus*, *Lemanea fluviatilis*), and seasonal, water-level fluctuation zone taxa (e.g. *Ulothrix zonata*) (see Tables 8.1, 8.2, 8.3, and 8.4).

Functional Species Groups

Especially in fast-flowing, clear mountain streams, SBM follow specific adaptive strategies, which allow them to grow in specific niches, including flow adaptation, desiccation resistance, adaptation to overcome nutrient shortage (N_2 fixers), use nutrient pulses (hair-forming taxa), and strategies to cope with excess $CaCO_3$ (calcifiers). Although taxa following the same strategy would be categorized within functional groups (e.g. rheobiontic taxa, calcifiers, N_2 fixers, Tables 8.1 and 8.2), in some taxa more than one of these adaptive capacities can be expressed.

The importance of flow adaptions of SBM in streams becomes evident from the high number of species categorized as rheobiontic and rheophilous in both ecoregions (36 % AT, 46 % USA) with some extremely flow resistant taxa such as *Lemanea fluviatilis* (resistant to 4 m s^{-1} of extreme drag; Sheath and Hambrook 1988, 1990). Taxa specifically adapted to drying are more important in warm and dry summers in some North-American streams: e.g. *Klebsormidium* forms mats that become isolated during prolonged desiccation periods in late summer. Filaments of this species form thickened walls, accumulate lipids and starch, and exhibit reduced pigmentation (Morrison and Sheath 1985). Filaments also fragment to form akinete-like structures that serve as propagules that germinate and recolonize streams later in the year when stream flow increases. The fact that a considerable number of species (approximately 20 % in both datasets), is capable to form hairs (sites of enhanced phosphatase activity, Livingstone and Whitton 1984; Gibson and Whitton 1987; Whitton et al. 1991) may indicate temporal shortage of SRP or potential pulses of organic phosphates washed out periodically from soils in the catchment (e.g. Whitton and Neal 2011; Whitton and Matteo 2012). The hair-forming taxa, as a functional group span multiple taxonomic affiliations. In NY streams and rivers these include the cyanobacteria *Coleodesmium wrangelii* and *Homoeothrix varians*, green algae such as *Draparnaldia acuta* and *Stigeoclonium tenue,* as well as several rhodophyte *Batrachospermum* species (Table 8.2). More cyanobacterial taxa with hairs were observed in AT streams (8), but the same two green algal species as recorded in NY (Table 8.1). In the AT dataset a larger number of hair-forming taxa were related to low SRP (as a group statistically highly significant lower medians than other algae) and mostly a preference for higher altitude alpine areas (UZA, VZA) with, e.g., *Homoeothrix janthina* identified as statistically most representative taxon (Fig. 8.7). Within both ecoregions, taxa capable of forming both hairs and heterocytes were represented by four cyanobacteria, which are likely N_2 fixers. This functional group (especially *Capsosira brebissonii, Stigonema. mamillosum*) was in the CCA, clearly separated from other taxa, strongly associated with high canopy cover in the US dataset (Fig. 8.8), but this was not the case for *Coleodesmium wrangelii* (with heterocysts), a taxon associated with lower altitude sites and higher NO_3. A taxonomically related, but oligotraphentic and presumably geographically restricted taxon, *Capsosira lowei* Casamatta et al. 2006, occurs in high gradient streams with low nutrients in the Smoky Mountains. However, this double-functional group was not so well represented within the Austria dataset and normally related to more oligotrophic higher altitude alpine areas or associated with clean calcareous spring streams (e.g. *Rivularia* spp., see Cantonati et al. 2012) not represented in the CCA.

General Remarks

Using two temperate stream datasets, we demonstrated a wide diversity of spatial and temporal growth strategies in soft-bodied macroalgae. These spatial and temporal patterns were some cases identical in the two regions, in others complementary, or were contradictory, between Austria and NY State (USA). Factors determining these traits and strategies were in most cases based on species-specific ecological preferences and limitations, as well as extrinsic (environmental) and occasionally local factors that shape the phenology and growth of SBM at a certain point in time. The analyses identified a number of SBM species with specific and statistically significant ecological deviations from each regional median (i.e. specialized niches; Tables 8.3 and 8.4). However, some specific regional preferences differed between the AT and NY datasets, presumably due to different ecological and geomorphological regional settings.. Nonetheless, the first four most important driving variables for each region were identical for both (altitude, nitrate, conductivity, SRP) datasets, and the six common shared taxa (*Cladophora glomerata, Stigeoclonium tenue, Phormidium autumnale, Hydrurus foetidus, Ulothrix zonata, and Lemanea fluviatilis* see also Figs. 8.7 and 8.8) followed a clear trophic gradient (in respect to both N- and P-components). This pattern is based on more widespread taxa, whereas there were also striking biogeographical differences of SBM between the two datasets. For example, *Capsosira brebissonii* and *Nostochopsis lobata* was observed only in NY, and *Chamaesiphon geitleri* and *Siphononema polonicum* only in AT. Some of these taxa in particular were observed in specific habitat/stream types (e.g. humic streams in NY, calcareous gravel streams in AT characterized by specific taxa). We would suggest that our data, which attempted to bridge local features to regional scales, confirm that at least for temperate clear mountain stream and rivers of orders 2–7, the multivariate-based niche concept is useful to understand SBM growth and phenology for both widespread and regional important taxa. As recently shown in the comparison of SBM preferences between Norway and AT (Rott and Schneider 2014) the considerable number of shared SBM species despite fundamental differences in resource levels, supports the idea that the more widespread taxa that colonize streams in temperate North America and Central Europe follow similar, generalizable mechanisms (i.e. ranking according to longitudinal gradients). In both region, *Cladophora* preferred lower altitudes, *Hydrurus* mid to higher altitudes, while *Lemanea* was ranked between the two. Our data are in agreement with patterns observed in the UK. A detailed, km-by-km survey of the length of the River Tweed in Scotland (156 km long) documented that *C. glomerata* was absent from the upper 25 km of the river's length, and became abundant only in lower reaches (Holmes and Whitton 1981). The apparent preference of this alga for low-altitude sites in NY and the UK may be the result, in part, of both the negative effects of scouring and decreases in photosynthetic rates at higher current velocities, as well as elevated nutrients (Dodds and Gudder 1992). In contrast, *Hydrurus* has a demonstrated preference for colder temperatures and greater current velocity (Klaveness and Lindstrøm 2011), conditions common at higher altitudes and in most cold headwaters, facilitating free CO_2 supply. We also documented specific resilience strategies by certain macroalgal taxa developed against desiccation, such as *Klebsormidium*, *Trentepohlia* and *Gloeocapsa,* although each had its own specific adaptations.

By the specific nature of the present data analysis (spatially extended, but not temporally dense), we may have overlooked key variables that may play an important role for species niche delimitations. In particular, these data did not assess effects of physical disturbance (frequency), although it is generally elevated in fast-flowing, clear mountain streams, the target stream type selected in this comparative study. Harsh physical conditions were found to be favourable for several SBM, and may in fact contribute to maintain the niche mosaic of five spatially varying common taxa (Fig. 8.10).

Because our study comparing AT and NY datasets emphasized spatial differentiation of SBM niches, we may also have underestimated the importance of temporal factors regulating potential seasonal succession. These patterns may follow different rules than periphyton in general, when studied as communities of micro- and macroalgae together. We did not obtain evidence, which would strongly support seasonal succession in our temperate study streams, a situation that is evident in frequently disturbed desert streams, (Grimm and Fisher 1989; Peterson and Grimm 1992), and Mediterranean-type streams with alternating species of the dominant SBM (*Cladophora* vs. *Nostoc*, see Power 1992). Under desert situations, taxa capable of extensive growth and high growth rates (short doubling times) have an advantage over slow-growing taxa and follow each other quickly. A classical pattern of succession of taxonomic group dominances from diatoms to green algae and cyanobacteria with regard to canopy cover was also observed along a longitudinal gradient in a Colorado spring-brook system (Ward and Dufford 1979).

In contrast to our primary theory that SBM would mostly be restricted in space and especially in time (often referred to as seasonal patterns), we found at least for AT and a multiannual dataset (1984–2001) that the majority of the taxa occurred along extended time periods or variable seasons, while a minority of taxa exhibited a clear seasonal preference, especially with regard to warm summer conditions. This is in contrast to the paradox of the plankton, where in most cases a specific seasonal succession (e.g. rapid species turnover), with a small spatio-temporal niche, is thought to be common. Although the broad picture of seasonality in our datasets show minor differentiations, especially for temporal succession of SBM species in AT may occur in many sites, but not along consistent succession rules. Such results do not necessarily involve competitive exclusion of taxa, and instead, many coexist. For example, short seasonal pulses of some SBM taxa can co-occur with perennial taxa in the same reach, although a reach-level map of SBM assemblages in a calcareous gravel stream showed a minor overlap of five co-occurring taxa (Fig. 8.9). There is some convincing evidence based on expert studies in the field (e.g. Lindstrøm et al. 2004; Pfister 1993) that benthic macroalgae are in fact conservative elements of stream biota (perhaps in contrast to micro species and many diatoms), which are extremely resilient to physical environmental changes based (scours in general and even catastrophic events). A special strategy of persistence of macroalgal taxa within a habitat may be related to an ability to quickly form resting stages for effective re-colonization, or undergo complex life cycle adaptations best studied for red algae until now (e.g. some species of *Batrachospermum*, Minckley and Tindall 1963; Necchi 1997; Necchi and Vis 2005). In addition, widespread, eutraphentic macroalgal species can have species-specific growth capacities that hinder seasonal succession or replacement by other less well-adapted species, as demonstrated by the greater photosynthetic plasticity in *Cladophora glomerata* versus *Vaucheria* sp. in one headwater stream (Ensminger et al. 2005).

Conclusion

The significance of spatial regulation of benthic algae on large, inter-regional and global scales are currently focused mainly on in situ studies of morphology and in some cases the genetic identity of taxa, as opposed to experiments with cultures and molecular based lab studies. We recommend that a link should be made between the ecological understandings and the genetic variation (and molecular ecological adaptation capacities), especially along the two lines:

- On the microhabitat scale SBM offer specific and delimited niches in clear near-natural streams closely linked to other biota (e.g. bacteria, ciliates, insects) and potentially triggered by trace substances of the algae and potential co-evolutionary effects.

- On the bioregional level, long-term studies of defined reference key species of SBM (such as in headwater streams or springs) should improve our understanding of wider environmental changes such as that caused by remote air pollution and climate change.

Acknowledgements Support for the NY portion of this work was provided by a grant from the NY State Biodiversity Research Institute to JDW.

References

Allen JD (1995) Stream ecology. Structure and function of running waters. Chapman and Hall, London
Algarte VM, Siqueira NS, Rodrigues L (2013) Desiccation and recovery of periphyton biomass and density in a subtropical lentic ecosystem. Acta Sci Biol Sci 35:311–318
Bautista ANI, Necchi O Jr (2007) Photoacclimation in three species of red algae. Brazil J Plant Physiol 19:23–34
Becker B, Marin B (2009) Steptophyte algae and the origin of embryophytes. Ann Bot 103:999–1004
Berrendero E, Perona E, Matteo P (2008) Genetic and morphological characterization of *Rivularia* and *Calothrix* (Nostocales, Cyanobacteria) from running water. Int J Syst Evol Microbiol 58:447–460
Biggs BJF (1996) Patterns in benthic algae of streams. In: Stevenson RJ, Bothwell ML, Lowe RL (eds) Algal ecology: freshwater benthic ecosystems. Academic, San Diego, pp 31–56
Biggs BJF, Stevenson RJ, Lowe RL (1998) A habitat matrix conceptual model of periphyton. Arch Hydrobiol 143:21–56
Blum JL (1956) The ecology of river algae. Bot Rev 22:291–341
Cantonati M, Gerecke R, Bertuzzi E (2006) Springs of the Alps—sensitive ecosystems to environmental change: from biodiversity assessments to long-term studies. Hydrobiologia 562:59–96
Cantonati M, Rott E, Spitale D et al (2012) Are benthic algae related to spring types? Freshw Sci 31:481–498
Caron DA, Countway PD (2009) Hypotheses on the role of the protistan rare biosphere in a changing world. Aquat Microb Ecol 57:227–238
Casamatta DA, Gomez SR, Johansen JR (2006) *Rexia erecta* gen. et sp. nov. and *Capsosira lowei* sp. nov., two newly described cyanobacterial taxa from the Great Smoky Mountains National Park (USA). Hydrobiologia 185:13–26
Castenholz RW, Garcia-Pichel F (2012) Cyanobacterial responses to UV radiation. In: Whitton BA (ed) Ecology of Cyanobacteria II. Their diversity in space and time. Springer, New York, pp 562–591
DeNicola DN, Hoagland KD (1996) Effects of solar spectral irradiance (visible to UV) on a prairie stream epilithic community. J N Am Benthol Soc 15:155–169
Dodds WK, Gudder DA (1992) The ecology of *Cladophora*. J Phycol 28:415–426
Eloranta P, Kwandrans J (2012) Illustrated guidebook to common freshwater red algae. Polish Academy of Sciences, Krakow, Poland
Ensminger I, Förster J, Hagen C et al (2005) Plasticity and acclimation to light reflected in temporal and spatial changes of small-scale macroalgal distribution in a stream. J Exp Bot 56:2047–2058
Entwisle TJ, Sonneman J, Lewis SH (1997) Freshwater algae in Australia: a guide to conspicuous genera. Sainty & Associates, Potts Point
Fenchel T, Finlay BJ (2004) The ubiquity of small species: patterns of local and global diversity. Bioscience 54:777–784

Förster J, Gutowski A, Schaumburg J (2004) Defining types of running waters in Germany using benthic algae: a prerequisite for monitoring according to the Water Framework Directive. J Appl Phycol 16:407–418
Frissell CA, Liss WJ, Warren CE et al (1986) A hierarchical framework for stream habitat classification: viewing streams in a watershed context. Environ Manage 10:199–214
Furey PC, Lowe RL, Power ME et al (2012) Midges, *Cladophora*, and epiphytes: shifting interactions through succession. Freshw Sci 31:93–107
Garcia-Pichel F, Castenholz RW (1991) Characterization and biological implications of scytonemin, a cyanobacterial sheath pigment. J Phycol 27:395–409
Gibson MT, Whitton BA (1987) Hairs, phosphatase activity and environmental chemistry in *Stigeoclonium*, *Chaetophora* and *Draparnaldia* (Chaetophorales). Brit Phycol J 22:11–22
Graham JM, Graham LE (1984) Growth and reproduction of *Bangia atropurpurea* (Roth) C. Ag. (Rhodophyta) from the Laurentian Great Lakes. Aquat Bot 28:317–331
Grime JP (1977) Evidence for the existence of three primary strategies in plants and its relevance to ecological and evolutionary theory. Am Nat 111:1169–1194
Grimm N (1995) Why link species to ecosystems? In: Jones CG, Lawton JH (eds) Linking species and ecosystems. Chapman and Hall, New York, pp 5–15
Grimm NB, Fisher SG (1989) Stability of periphyton and macroinvertebrates to disturbance by flash floods in a desert stream. J N Am Benthol Soc 8:293–307
Gutowski A, Förster J (2009) Benthische Algen ohne Diatomeen und Characeen. Bestimmungshilfe Landesamt für Natur, Umwelt und Verbraucherschutz-Arbeitsblatt 9. Nordrhein-Westfahlen, Recklingshausen.
Hambrook JA, Sheath RG (1991) Reproductive ecology of the fresh water red alga Batrachospermum boryanum Sirodot in a temperate headwater stream. Hydrobiologia 218:233–246
Hieber M, Robinson CT, Pushforth SR et al (2001) Algal communities associated with different alpine stream types. Arct Antarct Alp Res 33:447–456
Holmes NTH, Whitton BA (1981) Phytobenthos of the River Tees and its tributaries. Freshw Biol 11:139–163
John D, Whitton BA, Brook AJ (2011) The freshwater algal flora of the British Isles, 2nd edn. Cambridge Univ Press, Cambridge
Kawecka B (1980) Sessile algae in European mountain streams. 1. The ecological characteristics of communities. Acta Hydrobiol 22:361–420
Kawecka B (1981) Sessile algae in European mountain streams. 2. Taxonomy and autecology. Acta Hydrobiol 23:17–46
Klaveness D, Lindstrøm EA (2011) *Hydrurus foetidus* (Chromista, Chrysophyceae): a large freshwater chromophyte alga in laboratory culture. Phycol Res 59:105–112112
Komárek J (2013) Cyanoprokaryota III: Heterocytous genera, Süsswasserflora von Mitteleuropa, 19/3. Springer Spektrum, Heidelberg
Komárek J, Anagnostidis K (1998) Cyanoprokaryota I: Chroococcales, Süsswasserflora von Mitteleuropa, 19/1. G. Fischer Verlag, Stuttgart
Komárek J, Anagnostidis K (2005) Cyanoprokaryota II: Oscillatoriales, Süsswasserflora von Mitteleuropa, 19/2. G. Fischer Verlag, Stuttgart
Lindstrøm EA, Johansen SW, Saloranta T (2004) Periphyton in running waters—long-term studies of natural variation. Hydrobiologia 521:63–86
Livingstone DA, Whitton BA (1984) Water chemistry and phosphatase activity of the blue-green alga *Rivularia* in upper Teesdale streams. J Ecol 72:405–421
Loza V, Perona E, Mateo P (2013) Molecular fingerprinting of cyanobacteria from river biofilms as a water quality monitoring tool. Appl Environ Microbiol 79:1459–1472
McNeely C, Power ME (2007) Spatial variation in caddisfly grazing regimes within a northern California watershed. Ecology 88:2609–2619
Minckley WL, Tindall DR (1963) Ecology of *Batrachospermum* sp. (Rhodophyta) in Doe Run, Meade County, Kentucky. B Torrey Bot Club 90:391–400

Moog O, Schmidt-Kloiber A, Ofenböck T et al (2004) Does the ecoregion approach support the typological demands of the EU Water Framework Directive? Hydrobiologia 516:21–33
Morrison MO, Sheath RG (1985) Responses to desiccation stress by *Klebsormidium rivulare* (Ulotrichales, Chlorophyta) from a Rhode Island stream. Phycologia 24:129–145
Necchi O Jr (1997) Microhabitat and plant structure of *Batrachospermum* (Batrachospermales, Rhodophyta) populations in four streams of São Paulo State, southeastern Brazil. Phycol Res 45:39–45
Necchi O Jr, Alves AHS (2005) Photosynthetic characteristics of the freshwater red alga *Batrachospermum delicatulum* (Skuja) Necchi & Entwisle. Acta Bot Brasil 19:125–137
Necchi O Jr, Vis ML (2005) Reproductive ecology of the freshwater red alga *Batrachospermum delicatulum* (Batrachospermales, Rhodophyta) in three tropical streams. Phycol Res 53:194–200
OBrien PJ, Wehr JD (2010) Periphyton biomass and ecological stoichiometry in streams within an urban to rural land-use gradient. Hydrobiologia 657:89–105
Ott DW, Brown RM Jr (1974) Developmental cytology of the genus *Vaucheria* I. organisation of the vegetative filament. Brit Phycol J 9:111–126
Parker BC, Samsel GE, Prescott GW (1973) Comparison of microhabitats of macroscopic subalpine stream algae. Am Midl Nat 90:143–153
Peterson CG, Grimm NB (1992) Temporal variation in enrichment effects during periphyton succession in a nitrogen-limited desert stream ecosystem. J N Am Benthol Soc 11:20–36
Pfister P (1993) Seasonality of macroalgal distribution pattern within the reach of a gravel stream (Isar, Tyrol, Austria). Arch Hydrobiol 80:39–51
Pfister P, Pipp E (2013) Part A3, Phytobenthos. In: Mauthner-Weber R (ed) Guidance on the monitoring of the biological quality elements. Federal Ministry of Agriculture, Environment and Water Management, Vienna, pp 1–92
Power ME (1992) Hydrologic and trophic controls of seasonal algal blooms in northern California rivers. Arch Hydrobiol 125:385–410
Pringle CM (1996) Atyid shrimps (Decapoda: Atyidae) influence the spatial heterogeneity of algal communities over different scales in tropical montane streams, Puerto Rico. Freshw Biol 35:125–140
Rott E (1991) Methodological aspects and perspectives of the use of periphyton for monitoring and protecting rivers. In: Whitton BA, Friedrich G, Rott E (eds) The use of algae for monitoring rivers. Inst Bot Univ Innsbruck, Innsbruck, pp 9–16
Rott E, Pfister P (1988) Natural epilithic algal communities in fast-flowing mountain streams and rivers and some man-induced changes. Verh Int Ver Limnol 23:1320–1324
Rott E, Schneider SC (2014) A comparison of ecological optima of soft-bodied benthic algae in Norwegian and Austrian rivers and consequences for river monitoring in Europe. Sci Total Environ 475:180–186
Rott E, Walser L, Kegele M (2000) Ecophysiological aspects of macroalgal seasonality in a gravel stream in the Alps (River Isar, Austria). Verh Int Ver Limnol 27:1622–1625
Rott E, Cantonati M, Füreder L et al (2006a) Benthic algae in high altitude streams of the Alps—a neglected component of aquatic biota. Hydrobiologia 562:195–216
Rott E, Füreder L, Schütz C et al (2006b) A conceptual model for niche separation of biota within an extreme stream microhabitat. Verh Int Ver Limnol 29:2321–2323
Rott E, Gesierich D, Binder N (2010a) Lebensraumtypen und Diversitätsgradienten lotischer Algen in einem Gletschereinzugsgebiet. In: Koch EM, Erschbamer B (eds) Glaziale und periglaziale Lebensräume im Raum Obergurgl. Innsbruck Univ Press, Innsbruck, pp 203–212, 287–296, (In German with English abstract)
Rott E, Holzinger A, Gesierich D et al (2010b) Cell morphology, ultrastructure and calcification pattern of *Oocardium stratum*, a peculiar lotic desmid. Protoplasma 243:39–50
Rott E, Hotzy R, Cantonati M et al (2012) Calcification types of *Oocardium stratum* Nägeli and microhabitat conditions in springs of the Alps. Freshw Sci 31:610–624

Schneider S, Lindstrøm EA (2009) Bioindication in Norwegian rivers using non-diatomaceous benthic algae: the Acidification Index Periphyton (AIP). Ecol Indic 9:1201–1211
Sheath RG, Cole KM (1992) Biogeography of stream macroalgae in North America. J Phycol 28:448–460
Sheath RG, Hambrook JA (1988) Mechanical adaptations to flow in freshwater red algae. J Phycol 24:107–111
Sheath RG, Hambrook JA (1990) Freshwater ecology. In: Cole KM, Sheath RG (eds) Biology of the red algae. Cambridge Univ Press, Cambridge, pp 423–453
Sheath RG, Vis ML (2013) Biogeography of Freshwater Algae. eLS. John Wiley & Sons Ltd, Chichester: 0.1002/9780470015902.a0003279.pub3.
Sheath RG, Wehr JD (2015) Introduction to freshwater algae. In: Wehr JD, Sheath RG, Kociolek JP (eds) Freshwater algae of North America: ecology and classification. Academic, San Diego, pp 1–11
Sobczyk A, Schyns G, Tandeau de Marsac N, Houmard J (1993) Transduction of the light signal during complementary chromatic adaptation in the cyanobacterium *Calothrix* sp. PCC 7601: DNA-binding proteins and modulation by phosphorylation. EMBO J 12:997–1004
Squires LE, Rushforth SR, Endsley CJ (1973) An ecological survey of the algae of Huntington Canyon, Utah. Brigham Young U Sci Bull 18:1–87
Steinman AD, Mulholland PJ, Hill WR (1992) Functional responses associated with growth form in stream algae. J N Am Benthol Soc 11:229–243
Tank JL, Rosi-Marshall EJ, Griffiths NA et al (2010) A review of allochthonous organic matter dynamics and metabolism in streams. J N Am Benthol Soc 29:118–146
Tonetto AF, Branco CCZ, Peres CK (2012) Effects of irradiance and spectral composition on the establishment of macroalgae in streams in southern Brazil. Int J Limnol 48:363–370
Uehlinger U, Robinson CT, Hieber M et al (2010) The physico-chemical habitat template for periphyton in alpine glacial streams under a changing climate. Hydrobiologia 657:107–121
Vannote RL, Minshall GW, Cummins KW et al (1980) The river continuum concept. Can J Fish Aquat Sci 37:130–137
Vavilova VV, Lewis WM Jr (1999) Temporal and altitudinal variations in the attached algae of mountain streams in Colorado. Hydrobiologia 390:99–106
Ward JV (1986) Altitudinal zonation in a Rocky Mountain stream. Arch Hydrobiol Suppl 74:133–199
Ward JV, Dufford RG (1979) Longitudinal and seasonal distribution of macroinvertebrates and epilithic algae in a Colorado springbrook-pond system. Arch Hydrobiol 86:284–321
Wehr JD, Sheath RG, Kociolek JP (2015) Freshwater algae of North America: ecology and classification, 2nd edn. Academic, San Diego
Wehr JD, Sheath RG (2015) Habitats of freshwater algae. In: Wehr JD, Sheath RG, Kociolek JP (eds) Freshwater algae of North America, ecology and classification. Academic Press, San Diego, pp 13–74
Whitford LA, Schumacher GJ (1963) Communities of algae in North Carolina streams and their seasonal relations. Hydrobiologia 22:133–196
Whitton BA (1970) Biology of *Cladophora* in freshwaters. Water Res 4:457–476
Whitton BA, Matteo P (2012) Rivulariaceae. In: Whitton BA (ed) Ecology of cyanobacteria II. Their diversity in space and time. Springer, Drodrecht, Netherlands, pp 562–591
Whitton BA, Neal C (2011) Organic phosphate in UK rivers and its relevance to algal and bryophyte surveys. Int J Limnol 47:3–10
Whitton BA, Grainger SL, Hawley GR et al (1991) Cell-bound and extracellular phosphatase activities of cyanobacterial isolates. Microb Ecol 21:85–98
Winter JG, Duthie HC (2000) Stream biomonitoring at an agricultural test site using benthic algae. Can J Bot 78:1319–1325

Chapter 9
Ecophysiology of River Algae

Sergi Sabater, Joan Artigas, Natàlia Corcoll, Lorenzo Proia, Xisca Timoner, and Elisabet Tornés

Abstract Algae in rivers are affected by light, water turbulence, and nutrient availability. These environmental factors ultimately affect algae according to their habitat, growth form, and specific physiological abilities. Water flow imposes limitations in the diffusion and availability of gases and resources, also in relation to algal size and growth form. Algae adapt physiologically to light scarcity or excess via photosynthetic mechanisms, as well as by modifying their pigment composition. The algal ability to obtain and keep resources is mediated by enzymes, and its ability to use and store materials is specific of the different algal groups. Toxicants impose a limit to algal performance and may affect photosynthesis as well as nutrient uptake, amongst other effects on algal cells.

Keywords Algae • Biofilm • Cyanobacteria • Light • Nutrient uptake • Periphyton • Photosynthesis • River • Waterflow

The Mechanisms of Photosynthesis

Photosynthesis is the main metabolic function of algae, and provides the necessary energy for their maintenance and growth. River algae (including cyanobacteria) perform oxygenic photosynthesis in which the water is the electron donor and through its hydrolysis release oxygen. The process of photosynthesis involves the

S. Sabater (✉) • X. Timoner • E. Tornés
Catalan Institute for Water Research (ICRA), Scientific and Technologic Park of the UdG, Emili Grahit, 101-17003 Girona, Spain

Institute of Aquatic Ecology, Faculty of Sciences, University of Girona, Campus Montilivi, 17071 Girona, Spain
e-mail: sergi.sabater@udg.edu

J. Artigas
Laboratoire Microorganismes, Génome et Environnement, UMR 6023 CNRS, Université Blaise-Pascal, 63177 Aubière Cedex, France

N. Corcoll • L. Proia
Catalan Institute for Water Research (ICRA), Scientific and Technologic Park of the UdG, Emili Grahit, 101-17003 Girona, Spain

O. Necchi Jr. (ed.), *River Algae*, DOI 10.1007/978-3-319-31984-1_9

coordination of a large number of reactions, which are separated both spatially and temporally in the algal organelles. Photosynthesis begins in the thylakoid membrane with the absorption of light by antenna pigments, photochemical reactions in the pigments center and the synthesis of ATP, and terminates in the cytoplasm with carbon fixation and their distribution to the various metabolic pathways of the organism. The photosynthesis performed in algae has similar functioning than in cyanobacteria; the main difference resides in the site of photosynthesis, located in the chloroplasts in eukaryotic photosynthetic organisms (algae), or in the plasma membranes in the prokaryotes (cyanobacteria).

Evidences exist that biological and environmental factors impact on primary production. Steinman et al. (1992) showed that photosynthesis is related to the growth form of lotic algae, and specifically to the alga's surface-to-volume ratio, being highest in branched filaments and lowest in gelatinous forms. Algal mat thickness also plays an important role on photosynthesis performance, since they might interfere with light attenuation within the biofilm matrix (Dodds et al. 1999), and define a negative relationship between photosynthetic processes and algal thickness (Enríquez et al. 1996; Dodds et al. 1999). Experimental data have shown that the initial slope of the irradiance-production response curve (α) is predominantly controlled by light history, whereas the light-saturated rate of photosynthesis (P_{max}) is controlled by temperature (Harrison and Platt 1986). Temperature is also relevant to the carbon fixation rate (Falk et al. 1999). Maxwell et al. (1994) reported that growth of *Chlorella vulgaris* at 5 °C compared to growth at 27 °C resulted in a marked increase in the chlorophyll a/b-ratio and a concomitant decrease in the content of light harvesting pigments.

The diffusional resistance of CO_2 in water is four orders of magnitude higher than in air (Keely and Sandquist 1992). The greater viscosity of water creates a stagnant boundary layer around benthic algae, restricting the rate of CO_2 or HCO_3^- diffusion to the cell, thereby representing a major rate-limiting step in photosynthesis (Smith and Walker 1980). Higher resistance to diffusion of dissolved inorganic carbon has been observed in benthic algae than in phytoplankton (France 1995). Phytoplankton cells have a thinner boundary layer than that around benthic algae, and can be therefore more easily supplied with renewed inorganic carbon at higher rates (France 1995). The boundary layer developed over benthic algae is a key regulation factor of their community metabolism, and promotes internal recycling of nutrients (Riber and Wetzel 1987) and carbon (France 1995). This has been highlighted by means of carbon stable isotope ratios ($\partial^{13}C$) studies (France 1995; Finlay 2001; Hill et al. 2008).

Measuring the in vivo photosynthetic performance of algal cells can be made by means of the pulse-amplitude-modulated (PAM) fluorometry. This methodology is useful to estimate the "health" status of photosynthetic organisms. This technique is based on the principle that light energy absorbed by PSII pigments can either drive the photochemical energy conversion at PSII reaction centers (which allow photosynthesis activity), or be dissipated into heat (non-photochemical energy), or instead emitted in the form of chlorophyll fluorescence. As these three pathways of energy conversion are complementary, the fluorescence yield may serve as a convenient indicator of time- and state-dependent changes in the relative rates of photosynthesis and heat dissipation (Schreiber et al. 1986). The PAM fluorescence parameters are based on measurements obtained from the light adapted (Fs, Fm′) and the dark adapted (F0 and Fm). The steady-state fluorescence level (Fs) reflects the redox state of plastoquinone (QA) and Fm′ represents fully reduced PSII. The minimal fluorescence yield (F0) reflects the Chl-a fluorescence emission of all open reaction centers in a non-excited status. Hence, F0 could be used as surrogate of algal biomass since chlorophyll fluorescence is proportional to total chlorophyll content (Serôdio et al. 1997). The maximum fluorescence (Fm) reflects the fluorescence emission of all closed reaction centers. Deriving from these light and dark fluorescence measurements, a large number of different coefficients can be used to quantify photochemical and non-photochemical processes. One of the mostly commonly used photochemical parameters is the effective PSII quantum yield (Φ'PSII), which represents the algal capacity to convert photo energy into chemical energy once steady-state electron transport has been achieved. This parameter is commonly used as an indicator of the physiological state of algae. Amongst the non-photochemical parameters, the non-photochemical quenching (NPQ) represents the excess of light energy arriving to PSII and dissipated in non-radiative processes (Bilger and Bjorkman 1990). This parameter is commonly used as an indicator of physiological stress caused by light, pollutants, or other environmental stressors (Corcoll et al. 2012).

Photodamage, Photoprotection, and Light Adaptation in River Algae

The non-radiative dissipation of excess energy is a short-term process for the photoprotection of PSII against light-induced damage. Photoprotective dissipation is attributed to rapid modifications within the light-harvesting complex (LHC) of the PSII, leading to non-photochemical Chl-a fluorescence quenching (NPQ). The NPQ is induced by the formation of a proton gradient across the thylakoid membrane (DpH), and is associated with the operation of a xanthophyll cycle, which converts epoxidized to deepoxidized forms of xanthophylls. In green algae the xanthophyll cycle occurs via the deepoxidation of violaxanthin to anteroxanthin/zeaxanthin, and in diatoms by the deepoxidation of diadinoxanthin to diatoxanthin (Müller et al. 2001). Antioxidant enzymes such as catalase (CAT), ascorbate peroxidase (APX), superoxide dismutase (SOD), or glutathione reductase (GR) are the first barrier to cope with oxidative stress.

These enzymes participate in scavenging ROS and in avoiding their accumulation and the resulting oxidative stress. These photoprotective mechanisms have been observed also recently in stream biofilms (e.g., Bonnineau et al. 2012). Longer term adaptations and responses involve quantity and quality of algal pigments. Pigments are the main photosynthetic molecules responsible to capture light energy from solar radiation. Pigments are localized within the algal cell in association with the photosynthetic or thylakoid membranes. Pigment molecules are bound to peptides forming the light-harvesting protein complexes, which surround the photosynthetic reaction center where the photosynthetic process occurs. Algae and cyanobacteria contain pigments in an up to 5 % or more of its dry weight.

Algal pigments are divided in three major classes (chlorophylls, carotenoids, and phycobilins) depending on their chemical structure. Chlorophylls and carotenoids are water insoluble, whereas phycobilins are water soluble. Chlorophylls are composed of a phorphyrin ring system with a central magnesium atom. There are four types of chlorophylls: chlorophyll-a, -b, -c (-c1, -c2, and -c3), and -d. While chlorophyll-a (Chl-a) occurs in all photosynthetic organisms as the primary photosynthetic pigment (the photosynthetic process does not occur without it), the other chlorophylls function as accessory pigments. As such, absorb the light energy from wavelengths that Chl-a cannot, and transfer this energy to the Chl-a, ensuring a major capacity for the light-harvesting complexes.

Chlorophylls other than Chl-a are distinctively distributed within the different algal groups (Table 9.1). Carotenoids are long-chain molecules and include carotenes and xanthophylls. There is a wide diversity of carotenoids that function both

Table 9.1 Distribution of chlorophylls and carotenoids in the different algal groups from various sources (Rowan 1989; Jeffrey et al. 1997; Buchaca 2009)

	Bacillariophyta (Diatoms)	Chlorophyta (Green-algae)	Cyanophyta (Cyanobacteria)	Cryptophyta	Chrysophyta (Yellow-green algae)	Euglenophyta	Rhodophyta (Red algae)
Chlorophylls							
Chl-a	x	x	x	x	x	x	x
Chl-b		x				x	
Chl-c1	x				–		
Chl-c2	x			x	x		
Chl-c3					x		
Carotenoids							
Alloxanthin				x			
β, ε-carotene		x		x	–		x
β, β-carotene	x	x	x	–	x	x	x
Cantaxanthin			x				
Diadinoxanthin	x					x	
Diatoxanthin	x					x	
Echinenone			x				
Scytonemin			x				
Fucoxanthin	x				–		
Lutein		x					–
Myxoxanthophyll			x				
Neoxanthin	–	x	–	–	–	–	
Violaxanthin	–	x					x
Zeaxanthin	x		x	–	x	x	x

The different symbols indicate the pigment abundance (x, abundant; –, minoritary). Green algae have Chl-b and lutein as the primary photosynthetic pigments, diatoms Chl-c1 and c2 and fucoxanthin, cyanobacteria have myxoxanthophyll and zeaxanthin, and red algae violaxanthin and zeaxanthin as accessory pigments as well as protecting agents against photooxidation (Porra et al. 1997). Carotenoids are also characteristic of the different algal groups. Finally, phycobilins consist on a tetrapyrrole molecule bounded to a non-pigmented protein and function as accessory pigments. Phycobilins only occur in cyanobacteria, red algae, and some cryptophytes. It is therefore obvious that pigments composition is indicative of a given algal group, and can be used as a tool in ecological studies. The so-called chemotaxonomic approach is widely used in planktonic communities (e.g., Jeffrey et al. 1997; Brotas and Plante-Cuny 2003; Schlüter et al. 2006) but of limited use in river algae.

Pigment analysis also gives valuable information about the physiological state of the algal community. Living algae mainly contains undegraded chlorophyll but chlorophyll is highly susceptible to destruction once protecting pigments disappear or cells become senescent. Chlorophyllides, phaeophytins and phaeophorbides are the main chlorophyll degradation products. Phaeophytins occur when the magnesium atom of the chlorophyll molecule is lost; the loss of the phytol tail from the D ring results in chlorophyllides. Further degradation of phaeophytins or chlorophyllides produce phaeophorbides. The phaeophytization index summarizes the physiological state of algae by relating all the chlorophyll derivates (CD) to the a-phorbines (CD/a-phorb). The a-phorbines (a-phorb) is the sum of Chl-a and its degradation products. A CD/a-phorb ratio of 1 or around 1 indicates a complete degradation of the Chl-a (adapted from Moss 1968), while low CD/a-phorb values indicate a good physiological state of algae.

Maximizing photosynthesis and adapting the photosynthetic apparatus to actual light conditions is made by adjusting their intracellular pigment content. Shade and sun-adapted algae presents different physiological adaptations to optimise the use of light under predominant light conditions. Usually algae adapted to shade conditions are more light-efficient at low irradiances than sun-adapted algae. Shade-adapted algae present a higher ratio of Chl-a per cell than those that are sun adapted ("the greening effect"). Chl-a synthesis increases to maintain an efficient conversion of light energy to chemical energy. Sun-adapted algae show higher efficiency in the use of light at high irradiances and higher values of light saturation (Ik) and corresponding maximum productivity (P_{max}) (Hill and Boston 1991) (Fig. 9.1). Because different algal groups show specific abilities, green algae occur in open, non-limited light conditions (Guasch and Sabater 1995; Ensminger et al. 2000), while diatoms and red algae usually dominate under shaded conditions probably due to their higher efficiency in low-light irradiances (Hill and Boston 1991). However, some red algae are also able to acclimate to high-light conditions (Bautista and Necchi 2007). Generally, algae subjected to high irradiances have higher photoinhibition and recovery capacity than those inhabiting shaded sites, which tend to present chronic photoinhibition (Figueroa and Viñegla 2001). Other adaptive mechanisms of stream biofilms against light excess include downward micro migration of motile diatoms (Waring et al. 2007).

In some highly intense light conditions, photooxidation may occur. In these situations, algae reduce their photosynthetic efficiency, Chl-a concentration decreases, and usually carotenoids accumulate. Two types of carotenoids are produced to

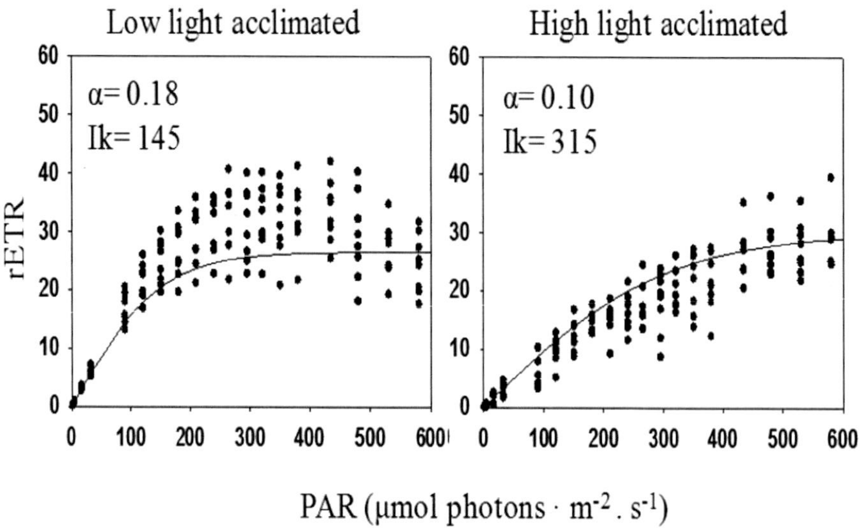

Fig. 9.1 Relationship between relative electron transport (rETR), *closed circles*, and irradiance (μmol photons m^{-2} s^{-1}) for stream biofilm acclimated to low-light (25 μmol photons m^{-2} s^{-1}) and to high-light (500 μmols photons $m^{-2}s^{-1}$) conditions. rETR-I curves (*solid lines*) were obtained by fitting the model of Eilers and Peeters (1988). The derived photosynthetic parameters from rETR-I curves are also presented: the initial slope (α) that indicates the efficiency in the use of light and the semi saturation irradiance (Ik). Adapted from Corcoll et al. (2012)

protect algae from UV degradation. Scytonemin occurs in the extracellular polysaccharide sheaths of Cyanobacteria (Garcia-Pichel and Castenholz 1991). Others accumulate intracellularly and protect cells from lethal singlet oxygen generated by UV or other stress situations (Karsten et al. 1998). Amongst these are echinenone, canthaxanthin, β-carotene, lutein, zeaxanthin, and myxoxanthophyll that capture free radicals generated by UV penetration within the cell (Adams et al. 1993). The synthesis of carotenoids as a mechanism of cell protection has been observed under high light irradiances, but also in algae exposed to metal toxicity (e.g., Corcoll et al. 2012). These mechanisms are unleashed in response to flow intermittency (e.g., Timoner et al. 2014), nutrient concentration (e.g., Devesa-Rey et al. 2009), and even pollution (e.g. Corcoll et al. 2012).

Gases and Nutrient Requirements and Limitations in River Algae

River algae, as many other microorganisms, live within a world of low Reynolds numbers. Viscosity forces prevail, and algae compensate such a harsh environment through movement and physiological adaptation. The properties inherent to this small scale has particular implications for nutrient uptake, heat exchange, and gases

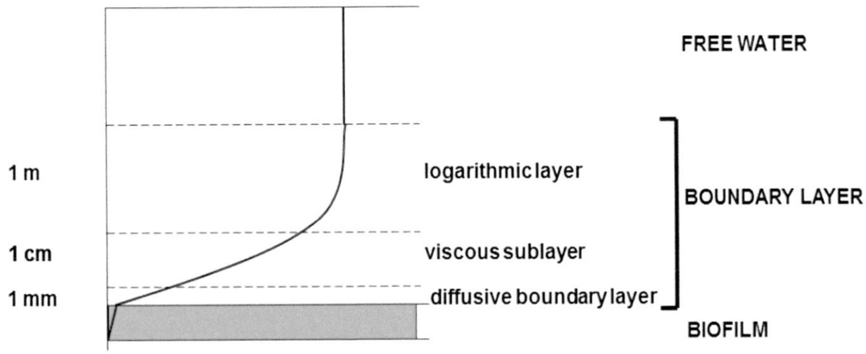

Fig. 9.2 Scheme of the boundary layer and corresponding sub-layers, with indication of the change of water velocity in the different sublayers and the scale at which changes take place

transfer depending on whether algae develop under a turbulent or non-turbulent environment.

The benthic boundary layer represents a hot spot not only for the dissipation of kinetic energy, but also for the biological activity involved in all biogeochemical transformation processes (Boudreau and Jorgensen 2001). Of special relevance is the diffusive boundary layer (DBL, where the exchange processes occur), at the scale of mm and in the vicinity of substrata (Fig. 9.2).

Phytoplankters live in a stirred environment with thin diffusive boundary layers, where they grow fast and have high nutrient requirements, but their biomass is often nutrient limited. However, the diffusive boundary layer is thicker in benthic microalgae, therefore reducing the exchange of gases and nutrients (Sand-Jensen and Borum 1991). As a result, the vertical transport of dissolved gases, solids, solutes, and heat in the immediate vicinity of the substratum surface of benthic algae is reduced to their respective molecular levels. At the top of the diffusive boundary layer, the concentration gradient gradually decreases to zero. Within the biofilm or periphytic mat, where the DBL penetrates (Fig. 9.2), the molecular diffusivity is further conditioned by the porosity, the topography, the meiofauna bioturbation, and the biogeochemical production and loss processes. As a result, oxygen diffusion shows a parabolic profile, with zero values within the sediment and the biofilm mat.

The limitations for the external supply of resources (gases, nutrients, organic carbon) may determine the productivity of these communities, as well as the internal biofilm chemical composition, and this progress more steeply when biofilm thickness increases. One obvious strategy is movement, since breaking the exhaustion of nutrients surrounding the cells can be achieved by motility (passive or active). Steinman et al. (1992) showed that algal growth forms (i.e., filamentous vs. prostrate gelatinous) played an important role on flow attenuation and nutrient uptake capacity. The transfer of nutrients through the boundary layer which surrounds the algal cells to their interior is biologically mediated by means of substance-specific membrane transport systems. These work against a concentration gradient, are site-specific, and require investing energy in terms of ATPs. The uptake of phosphorus (mostly in the forms of H_2PO4^- and $HPO4^{2-}$) by algae is mediated by an active transport system, which includes transport across the plasma membrane and incorporation into organic matter by means of phosphorylation processes.

Main Nutrient Requirements in River Algae

Every replication of an algal cell requires the uptake and assimilation of a given amount of inorganic nutrients. This amount needs to be at least equal to that of the mother cell (termed as the nutrient quota), if the daughter cells are to have the same composition than the parent cell. This type of rule applies to all elements which are essential for the growth, maintenance, and reproduction of the cells. In higher amounts, carbon, oxygen, hydrogen, and nitrogen; in smaller amounts, phosphorus, silicon, sulfur, chlorine, calcium, magnesium, and potassium; and even in trace amounts, a long list of essential elements such as silicon, iron, manganese, molybdenum, copper, vanadium, cobalt, zinc, or boron: All of them follow a relation between demand and supply, and when the balance is negative any of these elements may be limiting for the growth and survival of the algal cells.

Phosphorus has been often considered as the main limiting nutrient in freshwater systems, even though large-scale meta-analysis has questioned the uniqueness of this paradigm (Elser et al. 2007). Anyhow, phosphorus is accumulated in the cell so long as the uptake of another element (as nitrogen) is controlling the rate of the deployment of the two in the structure of new cell material. The phosphorus cell quota may range between the 0.2 and 0.4 % of the dry weight, but this can amount up to 3 % in repleted cells. Therefore, as a result of the luxury uptake, the cell may contain 8–16 times the minimum P quota that can sustain up to 3–4 cell doublings. The percent of nitrogen in the cell may account for 7–8.5 % (Reynolds 2006).

When the concentration of P in water exceeds a characteristic threshold value, phosphate is stored in the form of polyphosphate granules. Experiments using ^{32}P reveal that phytoplankton communities adapt their physiology according to their pre-exposure history to environmental phosphate fluctuations (Aubirot et al. 2011). However, nitrogen assimilation consists on the uptake of inorganic nitrogen (in the forms of nitrate, nitrite, and ammonium) from river water and translocation across the plasma membrane, followed by the reduction of oxidized nitrogen and the incorporation into amino acids (Cai et al. 2013). The nutrient uptake efficiency depends on the lifestyle of algal cells; planktonic cells are more efficient than benthic algae in terms of nutrient uptake (Price and Carrick 2011). Also depending on the size, smaller cells have low surface to volume ratios, and have faster nutrients uptake than larger cells. The nutrient uptake by stream algae increases with higher water N and P availability, but is also limited by the kinetic uptake capacity of river biota (Dodds et al. 2002). Thus, algal biomass accrual in biofilms determines nutrient uptake. Light and nutrient availability are tightly linked to the nutrient content of algal cells (Fanta et al. 2010). The light:nutrient hypothesis states that lower light availability (relative to nutrients) produces nutrient-rich algae (Sterner et al. 1997), and vice versa.

The biofilm architecture may difficult algal nutrient uptake even when they are placed at the most external cell layer (Kühl and Polerecky 2008). Adnate growth forms can create boundary layers between the cells and the environment, reducing external nutrient uptake from the flowing water but promoting internal nutrient recycling (Price and Carrick 2011). Nutrient uptake rates may not only may be enhanced by water movement. Grazing and physical losses may modify the thickness of the diffusive boundary layer. Algal communities located in the middle river channel (greater current speed/thinner biofilms) uptake nutrients faster than those located at the littoral zone (lower water velocity/thicker biofilm). The presence of cell-free pores and channels in the biofilm can eventually facilitate nutrient diffusion. Battin et al. (2003) observed that surface sinuosity and biofilm fragmentation increased with thickness, and that these changes likely reduced resistance to the mass transfer of solutes from the water column into the biofilms. The physical structure of the EPS matrix can both interfere on the ion flux between periphyton communities and water (Stevenson and Glover 1993), but also permits the binding and concentrating of organic molecules and ions (including nutrients) close to the cells (Decho 2000).

Benthic Algae Stoichiometry

Phototrophic organisms can produce biomass at extremely low nutrient content (high C:nutrient ratio) and support a wide range of C:N:P molar ratios (Sterner and Elser 2002). Despite this high plasticity, algal species distribution responds often to nutrient supply gradients in aquatic environments (Tilman et al. 1986).

The concept of resource limitation has shifted over the past two decades from an earlier paradigm of single-resource limitation (Liebig's law of the minimum) towards concepts of co-limitation by multiple resources (Elser et al. 2007). As an example, the study of Harpole et al. (2011) evidenced that more than half of the factorial studies on N and P limitation in primary producers including algae (over a dataset of 641 studies in freshwater, marine and terrestrial ecosystems) displayed some type of synergistic response to N and P addition. The type of nutrient co-limitation (i.e., simultaneous or independent) appears to be sensitive on the length of the experiment, total N and P concentration, and changes in the nutrient molar ratios.

Nutrient limitation in algae can be derived from both total nutrient concentration and nutrient ratios, though nutrient ratios mostly suggest potential (i.e., not real) limitation. Deviation from the Redfield ratio (C106:N16:P1) has been used as an indication of which nutrient is limiting for algal growth (Redfield et al. 1963). Specifically, the N:P ratio of 16:1 is used as a benchmark for differentiating P-limitation (N:P > 16) from N-limitation (N:P < 16), though this critical N:P ratio value may have variations among algal and cyanobacterial groups (Geider and La Roche 2002).

Water N:P molar ratios may control population dynamics and species co-existence in river algal communities. This relationship confirms the ecological stoichiometry (Sterner and Elser 2002) that exists between the lotic benthic microbial communities and their corresponding nutritional sources (Artigas et al. 2008). Requirements for specific elements, such as silicon, have been described in diatoms, where it accounts for an average ratio of ca. 1:1 with dissolved inorganic nitrogen (Redfield et al. 1963).

When the N:P ratios are imbalanced relevant physiological changes may occur. Under low P in relation to N supply (high water N:P) algal growth slows down due to reduced cellular RNA (P-rich) molecules and increased storage of carbohydrates (Healey and Hendzel 1979). Surplus of carbohydrates is released outside from the cell and stored in the EPS matrix. Stelzer and Lamberti (2001) concluded that the relative abundance of 9 of the 11 most common algal taxa was affected by N:P ratio, but also by total nutrient concentration. Total nutrient concentration (especially N) may increase the algal cell biovolume in early stages of exposition. The nutrient ratio preferentially favors one or another species or group of species. The diatoms *Achnanthidium minutissima, Amphipleura pellucida*, and *Cymbella affinis* are sensitive to the N:P whereas *Gomphonema* sp. is sensitive to total nutrient concentration (Stelzer and Lamberti 2001). Hill and Knight (1988) determined that the prostrate diatoms *Achnanthes lanceolata* and *Cocconeis placentula* displayed the most consistent positive responses to nitrate enrichment.

Since N and P are precursors of a wide range of organic molecules (including nucleic acids, lipids, proteins, and energy transfer molecules such as ADP and ATP) required for algal growth, nutrient requirements essentially vary between fast- and slow-growing algal species. Larger and morphologically complex algae, characterized by slower growth, have the capacity to allocate nutrients from supportive tissue to metabolically active tissue thus reducing their nutrient demand. However, algae without tissue differentiation must have a complete physiological apparatus, and should therefore contain high levels of N and P- organic rich compounds. Fast-growing species (i.e., diatoms) have higher N requirements per unit of biomass and unit of time than slow-growing taxa (i.e., cyanobacteria). Consequently, fast-growing species are more sensitive to nutrient limitation than slow-growing algae (Pedersen and Borum 1996).

It might be said that the distribution of phototrophic organisms depends on the trophic status of freshwater environments. While some nanoflagellate taxa belonging to chrysophyceae and cryptophyta dominate in oligotrophic environments, diatoms, and green algae often proliferate in environments with moderate-to-high nutrient availability. Cyanobacteria can fulfill nitrogen requirements easier than most algae because they can fix inorganic N from the atmosphere. Overall, conditions of light, nutrients, growth rate, as well as species and physiological diversity influence the C:N:P ratios of phototrophic communities in river ecosystems (Sterner and Elser 2002).

pH Requirements and Osmotic Stress in River Algae

Algal communities differ spatially and temporally mostly because of their preferences to specific water chemistry and geomorphological characteristics (Stevenson et al 1999). Salinity and major ions affect algal distribution (Cholnoky 1968; Potapova and Charles 2003), but ultimately depend on climate, geology, topology, and other physiographical features (Potapova and Charles 2003). Effects of soil erosion, irrigation, or inputs from wastewaters often increase total mineral content or the concentration of individual ions in river water (Meybeck and Helmer 1989). This has effects on the osmotic equilibrium as well as on the pH equilibrium of algal cells.

Salt Stress and Osmotic Effects

The osmotic stress associated to high salt content may also occur as a water deficit without a direct role of sodium ions (Munns 2002). Stream seasonal drought offer an example of this possibility. Desiccation, and the resulting reduction in water activity and increase in salinity, causes osmotic stress on microorganisms and ultimately decreases of cytoplasmic volume, damage to membranes, proteins and nucleic acids, and cellular lysis (Potts 1999). Under prolonged salt stress, cells respond to hyper-ionic and hyper-osmotic stress in addition to dehydration stress (Chaves et al. 2009). Some algae have developed thickened cell walls to reduce water loss in response to desiccation stress (Morison and Sheath 1985), while others secrete mucilaginous substances that act as a source of water and thus reduce osmotic stress during drought (Shephard 1987) (Fig. 9.3).

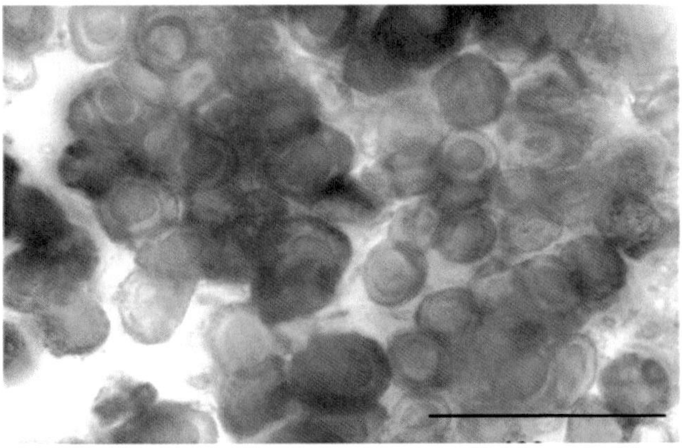

Fig. 9.3 Algal biofilm under drought conditions in a Mediterranean stream develops thickened cell walls to reduce osmotic stress. Scale bar = 100 μm

Algae and cyanobacteria inhabit in a wide range of salt concentrations or changing salinities, from freshwater and marine, and hypersaline environments. However, they differ in salinity preferences (halophilic or non-halophilic) or tolerances (stenohaline or euryhaline), and their distribution is according to these preferences (Table 9.2). Algal cells adapt to osmotic stress through different mechanisms

Table 9.2 Principal algal groups and genera with their main preferences to salinity and pH. Different sources were consulted, and some references are given

	Salinity	pH		
	Halophilous	Acidophilous	Alkalophilous	References
Cyanobacteria				
Calothrix			x	Freytet and Verrecchia (1998)
Gloeocapsa			x	Freytet and Verrecchia (1998)
Homoeothrix			x	Freytet and Verrecchia (1998)
Nostoc			x	Freytet and Verrecchia (1998)
Phormidium			x	Freytet and Verrecchia (1998)
Rivularia			x	Freytet and Verrecchia (1998)
Schizothrix			x	Freytet and Verrecchia (1998)
Spirulina	x			Sudhir and Murthy (2004)
Synechocystis	x			Allakhverdiev et al. (1999), Sudhir and Murthy (2004)
Green algae				
Chamydomonas	x	x		Raven (1995), Gross (2000), Ahmad and Hellebust (1986)
Chlorella			x	Yates and Robbins (1998)
Chlorococcum			x	Yates and Robbins (1998)
Cosmarium		x	x	Negro et al. (2003), Greenwood and Lowe (2006)
Closterium		x	x	Negro et al. (2003, Greenwood and Lowe (2006)
Dunaliella	x	x		Gross (2000), Goyal (2007)
Gongrosira			x	Freytet and Verrecchia (1998), Arp et al. (2010)

(continued)

Table 9.2 (continued)

	Salinity	pH		
	Halophilous	Acidophilous	Alkalophilous	References
Mougeotia		x	x	Sabater et al. (2000), Greenwood and Lowe (2006)
Spirogyra		x	x	Sabater et al. (2000), Greenwood and Lowe (2006)
Zygnema			x	Sabater et al. (2000)
Red algae				
Cyanidium		x		Gross (2000)
Diatoms				
Amphora	x		x	Potapova and Charles (2003), Rothschild and Mancinelli (2001), Arp et al. (2010)
Cymbella			x	Freytet and Verrecchia (1998)
Diploneis	x			Potapova and Charles (2003)
Eunotia		x		Winterbound et al. (1992), Negro et al. (2003)
Gomphonema			x	Freytet and Verrecchia (1998)
Gyrosigma	x			Potapova and Charles (2003)
Mastogloia	x			Potapova and Charles (2003)
Nitzschia	x			Potapova and Charles (2003), Rothschild and Mancinelli (2001)
Pinnularia		x		Sabater et al. (2003)

(e.g., morphological, physiological, biochemical changes) (Bohnert et al. 1995). Several metabolic processes are affected by changes in salinity. Osmotic stress inhibits algal photosynthesis by damaging the Photosystem (PS) II reaction centers (Vonshak and Torzillo 2004), and enhance respiration and electron transport activity of PS I (Endo et al. 1995). The PS I has a role in the protection of PS II from excessive excitation energy under salt stress, and contributes to the synthesis of organic osmolytes to maintain the osmotic balance (Vonshak and Torzillo 2004). Algae accumulate osmoregulatory substances in response to an increase in salinity or osmotic pressure of the environment, such as polyols (e.g., glycerol, mannitol, galactitol, sorbitol) and derivatives (e.g., glycerol galactoside), and sugars (e.g., sucrose and trehalose) (Sudhir and Murthy 2004; Yancey 2005). Halophilic green algae of the genus *Dunaliella* accumulate very large amounts of intracellular glycerol (Goyal 2007). Moreover, the formation or degradation of glycerol in *Dunaliella* follows the osmotic changes of the medium (Ben-Amotz and Avron 1973). The change in the intracellular lipid content is also a response to salt stress. The unsaturation of fatty acids is important in the tolerance of the photosynthetic machinery to salt stress. The unsaturation of fatty acids in membrane lipids reverses the suppressed activity and synthesis of the Na^+/H^+ antiporter system under salinity stress (Allakhverdiev et al. 1999).

Hydrogen Ion Concentration Effects

Most temperate river systems are characterized by neutral or slightly alkaline conditions. Hydrogen ions concentration affects the availability of inorganic ions and metabolites to the algal cells, and ultimately determines the ability of algae to live in an environment. One of the most important effects of pH is its role on the carbonate-bicarbonate buffering system. Algae occur in a wide range of pH (Table 9.2). Acidophiles live in extremely acid waters below 2 and alkaliphiles in waters with pH exceeding 10.

Highly acidic environments result from natural processes (e.g., volcanic activities, hot springs, acidic ponds) as well as from anthropogenic activities (e.g., acid mine drainage, alterations in land use, acid deposition) (Planas 1996). These environments are usually characterized also by high concentrations of iron, aluminum, nickel, and other heavy metals, which may be toxic and may affect the biota (Sabater et al. 2003). Thus, highly acidic streams are extreme environments and only few organisms like bacteria, fungi and algae are able to cope with these conditions (Gross 2000; Sabater et al. 2003). Algae have developed different mechanisms to live under low pH; algal cell membranes become temporarily impermeable for protons and use active pumps to prevent H^+ ions to enter in the cell (Seckbach and Oren 2007). High or low pH also affects the availability and form of carbon dioxide. At pH below 4 almost all dissolved inorganic carbon (DIC) is present in the form of CO_2, and the usually high exchange rate between the dissolved CO_2 and atmospheric CO_2 compensates the depletion of CO_2 pool during photosynthesis (Gross 2000). Under this acidic conditions, CO_2 entering the cell is converted to HCO_3^- in the cytosol which has a neutral pH (Gehl and Colman 1985). In river waters of neutral or basic pH dissolved CO_2 is mainly present in the form of bicarbonate. When concentration of calcium ions in these systems is high, calcium carbonate precipitation occurs and forms carbonate deposits such as tufa and travertines (Freytet and Plet 1991). Stromatolites are also characteristic of these systems, which are carbonated and laminated structures as the product of the microbial activity (Sabater et al. 2000).

Calcium carbonate precipitation will only occur if both an appropriate saturation index and nucleation sites are present. Photosynthetic induced calcification is considered the most common form of microbiological carbonate precipitation (Hammes and Verstraete 2002). Cyanobacteria have been reported to be the most abundant primary producers in these formations, which provide nucleation sites for mineral formation (Merz-Preiß and Riding 1999). Several species of the genera *Gloeocapsa*, *Calothrix*, *Homoeothrix*, *Nostoc*, *Phormidium*, *Rivularia*, and *Schizothrix* include most abundant calcifying cyanobacteria (Freytet and Verrecchia 1998). Diatoms are the dominant eukaryotic microalgal group in these formations. Calcium carbonate precipitation by diatoms is associated more directly with cell products (e.g., exopolymer stalks) than with their photosynthetic activity (Freytet and Verrecchia 1998; Golubić et al. 2008). Diatoms of the genera *Amphora, Cymbella*, and *Gomphonema* are those associated with calcium carbonate precipitation around their stalks (Freytet and Verrecchia 1998).

Pollution Effects on River Algae

A large number of toxicants, including organic compounds and metals, are released into fluvial ecosystems, and have potential adverse effects on algal communities. Chemical toxicity to algae depends on the mode of action of each toxicant, the time and concentration of exposure, as well as the sensitivity of organisms (Escher and Hermens 2002). Organic pollutants with mode of action (MoA) interfering with the PSII are mostly organic herbicides, some of the most toxic pollutants for algae. The active constituents of PSII inhibitors bind with a protein called the D1 protein, block the photosynthetic electron transport at the plastoquinone pool via Q_A and Q_B, and thus inhibit photosynthesis activity (Faust et al. 2001). Herbicides such as diuron, atrazine, prometryn, and/or isoproturon have relevant effects on algal photosynthesis (e.g., Ricart et al. 2009). Herbicides can also affect the absorption of nutrients (NO_3, NO_2, and Si) by algae (Debenest et al. 2010). In particular photosynthesis-inhibitor herbicides (*i.e.*, triazines) reduce the capacity of microalgae to uptake nitrate, nitrite, and silica by disrupting the absorption of nutrients by cells in both phytoplankton and biofilm communities.

Some heavy metals directly act on the photosynthetic process. Copper in excess act as a PSII inhibitor by blocking the electron transport an altering the structure of the antenna size (Dewez et al. 2005). Mercury exhibits a high toxicity to photosynthesis, including both light and dark reactions, because inhibit reaction centers of photosystems I and II, and the metabolism of CO_2 fixation, respectively (Juneau et al. 2001). Other metals such as zinc, cadmium, selenate, and dichromate, which do not have an MoA directly on PSII may also inhibit photosynthesis at determined concentrations (e.g. Corcoll et al. 2011). Photosynthesis has been shown to be also sensitive to emerging toxicants as the bactericide triclosan (e.g., Proia et al. 2013), pharmaceuticals as β-blockers (Bonnineau et al. 2010), or antibiotics (Liu et al. 2011).

The algal capacity to uptake nutrients may be altered by the toxicants presence. Significant and persistent reduction of the biofilms capacity to uptake phosphorus occurs after 48-h exposure to the bactericide Triclosan (TCS; Proia et al. 2011). Damaged *Spirogyra* sp. reduced chlorophyll concentration, and increased diatom and bacteria mortality occurred in communities exposed to TCS. The reduced uptake capacity of biofilms was consequence of negative effects on autotrophs, probably as a result of the TCS capacity to block fatty acid synthesis (McMurry et al. 1998) compromising permeability-barrier functions (Phan and Marquis 2006) and destabilizing cell membranes (Villalaín et al. 2001).

Other toxicants may directly affect the physiological capacity of algal cells to uptake nutrients by interfering with specific membrane receptor or metabolic uptake ways. This is the case of arsenic (As) and phosphorus (P), which have analogous chemical behavior including their respective incorporation into organic molecules (Anderson and Bruland 1991). The toxicity of As(V) to algae is related to the similarity between arsenic and phosphate, with consequent interference of phosphate metabolism (Sanders 1979). Different algae in marine and freshwater environments have shown increased growth in response to the addition of As(V) under P-limited conditions, and are therefore more susceptible than P-replete cells to experience toxicity (Thiel 1988). Moreover, freshwater algae have been shown to possess mechanisms that uptake and detoxify As(V), by reduction to As(III), which is actively transported out of the cell, by production of methylation and organoarsenicals (Wood 1974).

Acknowledgements This chapter benefitted from funding of the projects SCARCE (CONSOLIDER-INGENIO CSD2009-00065), and CARBONET (CGL2011-30474-C02-01) of the Spanish Ministry of Science and Innovation.

References

Adams WW, Demmig-Adams B, Lange OL (1993) Carotenoid composition and metabolism in green and blue-green algal lichens in the field. Oecologia 94:576–584

Ahmad I, Hellebust JA (1986) The role of glycerol and inorganic ions in osmoregulatory responses of the euryhaline flagellate Chlamydomonas pulsatilla Wollenweber. Plant Physiol 82: 406–410

Allakhverdiev SI, Nishiyama Y, Suzuki I et al (1999) Genetic engineering of the unsaturation of fatty acids in membrane lipids alters the tolerance of *Synechocystis* to salt stress. Proc Natl Acad Sci U S A 96:5862–5867

Anderson LCD, Bruland KW (1991) Biogeochemistry of Arsenic in natural waters: the importance of methylated species. Environ Sci Technol 25:420–427

Arp G, Bissett A, Brinkmann N et al (2010) Tufa-forming biofilms of German karstwater streams: microorganisms, exopolymers, hydrochemistry and calcification. In: HM Pedley, Rogerson M (eds) Tufas and speleothems: unravelling the microbial and physical controls. Geological Society, London, Special Publications 336, pp 83–118.

Artigas J, Romani AM, Sabater S (2008) Relating nutrient molar ratios of microbial attached communities to organic matter utilization in a forested stream. Fund Appl Limnol 173:255–264

Aubirot L, Bonilla S, Flakner G (2011) Adaptive phosphate uptake behavior of phytoplankton to environmental phosphate fluctuations. FEMS Microbiol Ecol 77:1–16

Battin TJ, Kaplan LA, Newbold JT et al (2003) Effects of current velocity on the nascent architecture of stream microbial biofilms. Appl Environ Microbiol 69:5443–5452

Bautista AIN, Necchi O Jr (2007) Photoacclimation in three species of freshwater red algae. Braz J Plant Physiol 19:23–24

Ben-Amotz A, Avron M (1973) The role of glycerol in the osmotic regulation of the halophilic alga *Dunaliella parva*. Plant Physiol 51:875–878

Bilger W, Bjorkman O (1990) Role of the xanthophyll cycle in photoprotection elucidated by measurements of light-induced absorbance changes, fluorescence and photosynthesis in leaves of *Hedera canariensis*. Photosynth Res 25:73–185

Bohnert HJ, Nelson DE, Jensen RG (1995) Adaptations to environmental stresses. Plant Cell 7:1099–1111

Bonnineau C, Guasch H, Proia L et al (2010) Fluvial biofilms: A pertinent tool to assess beta-blockers toxicity. Aquat Toxicol 96:225–233

Bonnineau C, Gallardo-Sague I, Urrea G et al (2012) Light history modulates antioxidant and photosynthetic responses of biofilms to both natural (light) and chemical (herbicides) stressors. Ecotoxicology 21:1208–1224

Boudreau BP, Jorgensen BB (2001) The benthic boundary layer: transport processes and biogeochemistry. Oxford University Press, Oxford

Brotas V, Plante-Cuny M-R (2003) The use of HPLC pigment analysis to study microphytobenthos communities. Acta Oecol 24:109–115

Buchaca T (2009) Pigments indicadors: estudi del senyal en estanys dels Pirineus i de la seva aplicació en paleolimnologia. PhD Thesis, Arxius de les Seccions de Ciències 142, Institut d'Estudis Catalans, Barcelona

Cai T, Park SY, Li Y (2013) Nutrient recovery from wastewater streams by microalgae. Renew Sust Energ Rev 19:360–369

Chaves M, Flexas J, Pinheiro C (2009) Photosynthesis under drought and salt stress: regulation mechanisms from whole plant to cell. Ann Bot 103:551–560

Cholnoky BJ (1968) Die Ökologie der Diatomeen in Binnengewässern. J. Cramer, Lehre

Corcoll N, Bonet B, Leira M et al (2011) Chl-a fluorescence parameters as biomarkers of metal toxicity in fluvial biofilms: an experimental study. Hydrobiologia 673:119–136

Corcoll N, Bonet B, Leira M et al (2012) Light history influences the response of fluvial biofilms to Zn exposure. J Phycol 48:1411–1423

Debenest T, Silvestre J, Coste M et al (2010) Effects of pesticides on freshwater diatoms. In: Whitacre DM (ed) Reviews of Environmental Contamination and Toxicology 203, pp 87–103

Decho AW (2000) Microbial biofilms in intertidal systems: an overview. Cont Shelf Res 20:1257–1273

Devesa-Rey R, Models AB, Díaz-Fierros F (2009) Study of phytopigments in river bed sediments: effects of the organic matter, nutrients and metal composition. Environ Monit Assess 153:147–159

Dewez D, Geoffroy L, Vernet G et al (2005) Determination of photosynthetic and enzymatic biomarkers sensitivity used to evaluate toxic effects of copper and fludioxonil in alga *Scenedesmus obliquus*. Aquat Toxicol 74:150–159

Dodds WK, Biggs BJF, Lowe RL (1999) Photosynthesis-irradiance patterns in benthic microalgae: variations as a function of assemblage thickness and community structure. J Phycol 35:42–53

Dodds WK, Smith VH, Lohman K (2002) Nitrogen and phosphorus relationships to benthic algal biomass in temperate streams. Can J Fish Aquat Sci 59:865–874

Eilers PHC, Peeters JCH (1988) Model for the relationship between light intensity and the rate of photosynthesis in phytoplankton. Ecol Modell 42:199–215

Elser JJ, Bracken MES, Cleland EE et al (2007) Global analysis of nitrogen and phosphorus limitation of primary producers in freshwater, marine and terrestrial ecosystems. Ecol Lett 10:1135–1142

Endo T, Schreiber U, Asada K (1995) Suppression of quantum yield of photosystem II by hyperosmotic stress in *Chlamydomonas reinhardtii*. Plant Cell Physiol 36:1253–1258

Enríquez SC, Duarte M, Sand-Jensen K et al (1996) Broad-scale comparison of photosynthetic rates across phototrophic organisms. Oecologia 108:197–206

Ensminger I, Xyländer M, Hagen C et al (2000) Strategies providing success in a variable habitat: II. Ecophysiology of photosynthesis of *Cladophora glomerata*. Plant Cell Environ 23: 1129–1136

Escher BI, Hermens JLM (2002) Critical review modes of action in ecotoxicology: their role in body burdens, species sensitivity, QSARs, and mixture effects. Environ Sci Technol 36:4201–4217
Falk S, Maxwell DP, Laudenbach DE et al (1999) Photosynthetic adjustment to temperature. In: Baker NR (ed) Photosyhtesis and the environment, vol. 5. Kluwer Academic Press, New York, pp 367–384.
Fanta SE, Hill WR, Smith TB et al (2010) Applying the light: nutrient hypothesis to stream periphyton. Freshwater Biol 55:931–940
Faust M, Altenburger R, Backhaus T et al (2001) Predicting the joint algal toxicity of multicomponent s-triazine mixtures at low-effect concentrations of individual toxicants. Aquat Toxicol 56:13–32
Figueroa FL, Viñegla B (2001) Effects of solar UV radiation on photosynthesis and enzyme activities (carbonic anhydrase and nitrate reductase) in marine macroalgae from southern Spain. Rev Chil Hist Nat 74:237–249
Finlay JC (2001) Stable-carbon-isotope ratios of river biota: implications for energy flow in lotic food webs. Ecology 82:1052–1064
France RL (1995) Carbon-13 enrichment in benthic compared to planktonic algae: foodweb implications. Mar Ecol-Progr Series 124:307–312
Freytet P, Plet A (1991) Les formations stromatolitiques (tufs calcaires) récentes de la région de Tournus (Saône et Loire). Geobios 24:123–139
Freytet P, Verrecchia EP (1998) Freshwater organisms that build stromatolites: a synopsis of biocrystallization by prokaryotic and eukaryotic algae. Sedimentology 45:535–563
Garcia-Pichel F, Castenholz RW (1991) Characterization and biological implications of scytonemin, a cyanobacterial sheath pigment. J Phycol 27:395–409
Gehl KA, Colman B (1985) Effect of external pH on the internal pH of *Chlorella saccharophila*. Plant Physiol 77:917–921
Geider R, La Roche J (2002) Redfield revisited: variability of C:N:P in marine microalgae and its biochemical basis. Eur J Phycol 37:1–17
Golubić S, Crescenzo V, Plenković-Moraj A et al (2008) Travertines and calcareous tufa deposits: an insight into diagenesis. Geolog Croat 61:363–378
Goyal A (2007) Osmoregulation in *Dunaliella*, part II: photosynthesis and starch contribute carbon for glycerol synthesis during a salt stress in *Dunaliella tertiolecta*. Plant Physiol Biochem 45:705–710
Greenwood JL, Lowe RL (2006) The effects of pH on a periphyton community in an acidic wetland, USA. Hydrobiologia 561:71–82
Gross W (2000) Ecophysiology of algae living in highly acidic environments. Hydrobiologia 433:31–37
Guasch H, Sabater S (1995) Seasonal variations in photosynthesis-irradiance responses by biofilms in Mediterranean streams. J Phycol 31:727–735
Hammes F, Verstraete W (2002) Key roles of pH and calcium metabolism in microbial carbonate precipitation. Rev Environ Sci Biotechnol 1:3–7
Harpole WS, Ngai JT, Cleland EE et al (2011) Resource co-limitation of primary producer communities. Ecol Lett 14:852–862
Harrison WG, Platt T (1986) Photosynthesis-irradiance relationships in polar and temperate phytoplankton populations. Polar Biol 5:153–164
Healey FP, Hendzel LL (1979) Indicators of phosphorus and nitrogen deficiency in five algae in culture. J Fish Res Board Canada 36:1364–1369
Hill WR, Boston HL (1991) Effects of community development on photosynthesis-irradiance relations in stream periphyton. Limnol Oceanogr 36:375–389
Hill WR, Knight AW (1988) Nutrient and light limitation of algae in two northern California streams. J Phycol 24:125–132
Hill WR, Fanta SE, Roberts BJ (2008) 13C dynamics in benthic algae: effects of light, phosphorus, and biomass development. Limnol Oceanogr 53:1217–1226

Jeffrey SW, Mantoura RFC, Wright SW (1997) Phytoplankton pigments in oceanography: guidelines to modern methods. UNESCO, Paris
Juneau P, Dewez D, Matsu S et al (2001) Evaluation of different algal species sensitivity to mercury and metolachlor by PAM-fluorometry. Chemosphere 45:589–598
Karsten U, Maie J, Garcia-Pichel F (1998) Seasonality in UV-absorbing compounds of cyanobacterial mat communities from an intertidal mangrove flat. Aquatic Microb Ecol 16:37–44
Keely JE, Sandquist DR (1992) Carbon: freshwater plants. Plant Cell Environ 15:1021–1035
Kühl M, Polerecky L (2008) Functional and structural imaging of phototrophic microbial communities and symbioses. Aquat Microb Ecol 53:99–118
Liu B, Weiqiu N, Xiangping G et al (2011) Growth response and toxic effects of three antibiotics on Selenastrum capricornutum evaluated by photosynthetic rate and chlorophyll biosynthesis. J Environ Sci 23:1558–1563
Maxwell DP, Falk S, Trick CG et al (1994) Growth at low temperature mimics high light acclimation in *Chlorella vulgaris*. Plant Physiol 105:535–543
McMurry LM, Oethinge M, Levy SB (1998) Triclosan targets lipid synthesis. Nature 394:531–532
Merz-Preiß M, Riding R (1999) Cyanobacterial tufa calcification in two freshwater streams: ambient environment, chemical thresholds and biological processes. Sediment Geol 126:103–124
Meybeck M, Helmer R (1989) The quality of rivers: from pristine stage to global pollution. Palaeogeogr Palaeocl 75:283–309
Morison MO, Sheath RG (1985) Responses to desiccation stress by *Klebsormidium rivulare* (Ulotrichales, Chlorophyta) from a Rhode Island stream. Phycologia 24:129–145
Moss B (1968) Studies on the degradation of chlorophyll a and carotenoids in freshwaters. New Phytol 67:49–59
Müller P, Li XP, Niyogi KK (2001) Non-photochemical quenching. A response to excess light energy. Plant Physiol 125:1558–1566
Munns R (2002) Comparative physiology of salt and water stress. Plant Cell Environ 25:239–250
Negro A, De Hoyos C, Aldasoro J (2003) Diatom and desmid relationships with the environment in mountain lakes and mires of NW Spain. Hydrobiologia 505:1–13
Pedersen MF, Borum J (1996) Nutrient control of algal growth in estuarine waters. Nutrient limitation and the importance of nitrogen requeriments and nitrogen storage among phytoplankton species of macroalgae. Mar Ecol-Progr Ser 142:261–272
Phan T, Marquis RE (2006) Triclosan inhibition of membrane enzymes and glycolysis of Streptococcus mutans in suspension and biofilm. Can J Microbiol 52:977–983
Planas D (1996) Acidification effects. In: Stevenson RJ, Bothwell ML, Lowe RL (eds) Algal ecology: freshwater benthic ecosystems. Academic, San Diego, pp 497–530
Porra RJ, Pfündel EE, Engel N (1997) Metabolism and function of photosynthetic pigments. In: Jeffrey SW, Mantoura RFC, Wright SW (eds) Phytoplankton pigments in oceanography: guidelines to modern methods. UNESCO, Paris, pp 85–126
Potapova MG, Charles DF (2003) Distribution of benthic diatoms in U.S. rivers in relation to conductivity and ionic composition. Freshwater Biol 48:1311–1328
Potts M (1999) Mechanisms of desiccation tolerance in cyanobacteria. Eur J Phycol 34:319–328
Price KJ, Carrick HJ (2011) Meta-analytical approach to explain variation in microbial phosphorus uptake rates in aquatic ecosystems. Aquat Microb Ecol 65:89–102
Proia L, Morin S, Peipoch M et al (2011) Resistance and recovery of river biofilms receiving short pulses of triclosan and diuron. Sci Total Environ 409:3129–3137
Proia L, Vilches C, Boninneau C et al (2013) Drought episode modulate biofilm response to pulses of Triclosan. Aquat Toxicol 127:36–45
Raven J (1995) Costs and benefits of low intracellular osmolarity in cells of freshwater algae. Funct Ecol 9:701–707
Redfield AC, Ketchum BH, Richards FA (1963) The influence of organisms on the composition of seawater. In: Hill MN (ed) The sea, vol 2. John Wiley, New York, pp 26–77
Reynolds CS (2006) The ecology of phytoplankton. Cambridge University Press, Cambridge

Riber HH, Wetzel RG (1987) Boundary-layer and internal diffusion effects on phosphorus fluxes in lake periphyton. Limnol Oceanogr 32:1181–1119
Ricart M, Barceló D, Geiszinge A et al (2009) Effects of low concentrations of the phenylurea herbicide diuron on biofilm algae and bacteria. Chemosphere 76:1392–1401
Rothschild LJ, Mancinelli RL (2001) Life in extreme environments. Nature 409:1092–1101
Rowan KS (1989) Photosynthetic pigments of algae. Cambridge University Press, Cambridge
Sabater S, Guasch H, Romaní A et al (2000) Stromatolitic communities in Mediterranean streams: adaptations to a changing environment. Biodivers Conserv 9:379–392
Sabater S, Buchaca T, Cambra J et al (2003) Structure and function of benthic algal communities in an extremely acid river. J Phycol 39:481–489
Sand-Jensen K, Borum J (1991) Interaction between phytoplankton, periphyton and macrophytes in temperate freshwaters and estuaries. Aquat Bot 41:137–176
Sanders, JG (1979). Effects of arsenic speciation and phosphate concentration on arsenic inhibition of Skeletonema costatum (Bacillariophyceae). J Phycol 15:424–428
Schlüter L, Lauridsen TL, Krogh G et al (2006) Identification and quantification of phytoplankton groups in lakes using new pigment ratios—a comparison between pigment analysis by HPLC and microscopy. Freshwater Biol 51:1474–1485
Schreiber U, Schliwaand U, Bilge W (1986) Continuous recording of photochemical and non-photochemical chlorophyll fluorescence quenching with a new type of modulation fluorometer. Photosynth Res 10:51–62
Seckbach J, Oren A (2007) Oxygenic photosynthetic microorganisms in extreme environments: possibilities and limitations. In: Seckbach J (ed) Algae and cyanobacteria in extreme environments. Springer, Dordrecht
Serôdio J, Marques da Silva J, Catarino F (1997) Nondestructive tracing of migratory rhythms of intertidal benthic microalgae using in vivo chlorophyll a fluorescence. J Phycol 33:542–553
Shephard K (1987) Evaporation of water from the mucilage of a gelatinous algal community. Br Phycol J 22:181–185
Smith FR, Walker NA (1980) Photosynthesis by aquatic plants effects of unstirred layers in relation to assimilation of CO_2 and HCO_3^- and to carbon isotopic discrimination. New Phytol 86:245–259
Steinman AD, Mulholland PJ, Hill WR (1992) Functional responses associated with growth from in stream algae. J North Am Benthol Soc 11:229–243
Stelzer RS, Lamberti GA (2001) Effects of N: P ratio and total nutrient concentration on stream periphyton community structure, biomass, and elemental composition. Limnol Oceanogr 46:356–367
Sterner RW, Elser JJ (2002) Ecological stoichiometry: the biology of elements from molecules to the biosphere. Princeton University Press, Princeton
Sterner RW, Else JJ, Fee EJ et al (1997) The light: nutrient ratio in lakes: the balance of energy and materials affects ecosystem structure and process. Am Nat 150:663–684
Stevenson RJ, Glover R (1993) Effects of algal density and current on ion transport through periphyton communities. Limnol Oceanogr 38:1276–1281
Stevenson RJ, Pan Y, van Dam H (1999) Assessing environmental conditions in rivers and streams using diatoms. In: Stoermer EF, Smol JP (eds) The diatoms: application for the environmental and earth sciences. Cambridge University Press, Cambridge, pp 57–85
Sudhir P, Murthy S (2004) Effects of salt stress on basic processes of photosynthesis. Photosynthetica 42:481–486
Thiel T (1988) Phosphate transport and arsenate resistance in the cyanobacterium *Anabaena variabilis*. J Bacteriol 170:1143–1147
Tilman D, Kiesling R, Sterne R et al (1986) Green, bluegreen and diatom algae: taxonomic differences in competitive ability for phosphorus, silicon and nitrogen. Arch Hydrobiol 106:473–485
Timoner X, Buchaca T, Acuña V et al (2014) Photosynthetic pigment changes and adaptations in biofilms in response to flow intermittency. Aquatic Sci 76:565–578
Villalaín J, Reyes Mateo C, Arand FJ et al (2001) Membranotropic effects of the antibacterial agent triclosan. Arch Biochem Biophys 390:128–136

Vonshak A, Torzillo G (2004) Environmental stress physiology. In: Richmond A (ed) Handbook of microalgal culture: biotechnology and applied phycology. Blackwell Publishing, Oxford, pp 57–82

Waring J, Baker NR, Underwood GJC (2007) Responses of estuarine intertidal microphytobenthic algal assemblages to enhanced ultraviolet B radiation. Glob Change Biol 13:1398–1413

Winterbound MJ, Hildrew AG, Orton S (1992) Nutrients, algae and grazers in some British streams of contrasting pH. Freshwater Biol 28:173–182

Wood JM (1974) Biological cycles for toxic elements in the environment. Science 183: 1049–1052

Yancey PH (2005) Organic osmolytes as compatible, metabolic and counteracting cytoprotectants in high osmolarity and other stresses. J Exp Biol 208:2819–2830

Yates KK, Robbins LL (1998) Production of carbonate sediments by a unicellular green alga. Am Mineral 83:1503–1509

Chapter 10
Biogeography of River Algae

Morgan L. Vis

Abstract Biogeography, or the study of the distribution of organisms in time and space, has a rich history. New molecular tools are providing another line of evidence in understanding distributions: in some instances confirming previous morphological data, while in other cases providing new insights. There have been few studies devoted to the biogeography of river algae, but much information can be gleaned from floristic surveys and systematic research. Water, wind, animals, and humans have played an important role as dispersal agents and may affect river algal biogeography. There is convincing evidence that some species of river algae are specialists and others generalists. There have been various invasions of freshwater rivers by marine algae and some of these events have raised awareness of and studies on possible transport mechanisms of microorganisms. Many recent studies of river algae have shown similar results to other microorganisms that many species are not ubiquitous, but are more geographically restricted than was previously reported. This certainly seems to be the case for freshwater Rhodophyta for which new molecular data are being amassed. Lastly, there is a need for easily accessed repositories of biogeographic data to better understand these organisms and their global distributions.

Keywords Biogeography • Chlorophyta • Cyanobacteria • Diatom • Dispersal • Invasive • Phylogeography • Rhodophyta • River

Introduction

Biogeography is the study of the distribution of organisms in geographic space and through geological time as well as the potential related patterns among floras/faunas (Lomolino et al. 2010). This area of research has fascinated biologists for centuries: understanding why an organism may or may not occur in a suitable habitat can provide insights into its evolutionary history, reproductive biology, and dispersal capabilities. With the advent of molecular tools, a new area of biogeography

M.L. Vis (✉)
Department of Environmental and Plant Biology, Ohio University,
Porter Hall 315, Athens, OH 45701, USA
e-mail: vis-chia@ohio.edu

O. Necchi Jr. (ed.), *River Algae*, DOI 10.1007/978-3-319-31984-1_10

research, phylogeography, emerged in the 1990s (Avise 2000). Phylogeography integrates genetic studies and biogeography to investigate the evolutionary history of a species or closely related species geographic distribution. This research area has provided much understanding of the distribution, evolutionary history, and potential barriers to dispersal for numerous organisms. However, there have been few studies on freshwater algae and even fewer focused on river-dwelling algae. Nevertheless, numerous insights have been gained and are discussed in this chapter.

For microorganisms, such as river algae, in the past there has been the prevalent idea that "everything is everywhere and the environment selects" (Finlay 2002). However, this idea has been challenged and many times it has been shown that microorganisms harbor cryptic diversity, which is often specific to particular geographic areas (Casamatta et al. 2003; Stancheva et al. 2013). Therefore, the current thinking is that freshwater algal species are more geographically restricted than previously thought. For example, in groups like diatoms and desmids that might be readily dispersed, the majority of taxa appear not to be cosmopolitan (Coesel and Krienitz 2008; Kociolek and Spaulding 2000).

Biogeographic trends in freshwater river algae have been studied based primarily on morphological identifications and molecular tools (Sheath and Cole 1992; Moreno et al. 2012; Necchi et al. 2013; Stancheva et al. 2013). With the advent of molecular data, some of the biogeographic trends, originally based on morphological information, have been revised or questioned. For example, the red alga *Batrachospermum gelatinosum* (Linnaeus) DeCandolle was thought to be a single cosmopolitan taxon, occurring in both the Northern and Southern Hemispheres based on morphological identifications (Entwisle 1989; Entwisle and Foard 1997). However, analyses of molecular data indicated that the Southern Hemisphere specimens were not closely related and constituted cryptic variation with a new species subsequently described (Vis and Entwisle 2000). Likewise, the cosmopolitan cyanobacterium, *Phormidium retzii* Kützing ex Gomont, has been shown to most likely harbor cryptic species and that these species distributions are more localized (Casamatta et al. 2003). A biogeographic pattern correlating genetic and geographic distance as well as cryptic species were demonstrated in the cosmopolitan diatom, *Gomphonema parvulum* (Kützing) Kützing (Abarca et al. 2014). Therefore, biogeographic literature without molecular data is discussed in this chapter, but caution is recommended in the conclusions drawn.

There has been little research on the biogeography of microscopic (single-celled) river algae with the exception of diatoms. River diatom floras have been studied at various geographic levels from small regional areas or islands, to countries (e.g., Potapova and Charles 2002; Juttner et al. 2010; Antoniades et al. 2009; Morales et al. 2009; Chattova et al. 2014). These diatom floras have been well studied due to their usefulness as bioindicators of river water quality in many biomonitoring programs (Rimet 2012). Since water quality is the focus of these studies, there is chemical and physical data associated with the algal records that can provide important information on the ecological niche of individual species, which may help explain species distributions (Potapova and Charles 2003). However, there are potential pitfalls in making sweeping biogeographical conclusions because diatoms as a group are extremely species rich and the taxonomy is highly complex, with taxonomies varying among regions or studies (Edlund and Jahn 2001; Potapova and Charles 2002; Soininen 2007).

In contrast to river habitats, there has been far more research on the biogeography of planktonic microalgae in lakes and reservoirs. Biogeographic patterns of large areas within a continent have been studied (i.e. Furlotte et al. 2000; Bouchard et al. 2004; Stomp et al. 2011). As well, the biogeography of particular algal groups such as the scaled chrysophytes have been documented (Kristiansen 2001; Siver and Lott 2012). Although many studies have employed only morphological tools, phytoplankton research is now applying both morphological and molecular data for a better understanding of biogeography. An elegant study in the genus *Synura* has revealed both cosmopolitan and geographically restricted taxa based on the combination of molecular and morphological data (Boo et al. 2010). In a study of *Microcystis* from reservoirs in northern Ethiopia, researchers noted low genetic diversity, which was unlike studies from other areas of the world such as Europe (van Gremberghe et al. 2011). They concluded that the local environment and geographic position of the reservoirs most likely influenced the genetic diversity. Recently, a volume of *Hydrobiologia* was devoted to the biogeography and spatial patterns of biodiversity of freshwater phytoplankton (Naselli-Flores and Padisák 2016). Although lentic habitats are not the focus of this chapter, the rich literature on freshwater phytoplankton could inform future biogeographic studies of river microorganisms.

The biogeography of freshwater macroalgae from both lentic and lotic habitats has been studied. However, there appear to be more studies of stream/river habitats than lakes/reservoirs. One notable recent study from lake habits was of *Aegagropila linnaei* Kützing and produced interesting results that suggested that this taxon is widespread, yet rare, probably due to low dispersal capabilities (Boedeker et al. 2010). River macroalgae primarily in the cyanobacteria, Chlorophyta and Rhodophyta have been the subject of much study using morphological data alone (i.e., Sheath and Cole 1992; Moreno et al. 2012) and more recently combined with molecular data (i.e., Casamatta et al. 2003; Strunecky et al. 2012; Necchi et al. 2013). These studies have provided considerable information on both species communities as well as individual taxa and will be highlighted along with other studies later in the chapter.

In this chapter, the current understanding of biogeography of benthic river-dwelling freshwater algae, both micro- and macro-algae, is explored. The chapter has been divided into three sections: the study of the dispersal of freshwater algae and the understanding of the floras of large regions, ecosystems, or continents; and the study of individual taxa, determining endemism, rarity, and cosmopolitanism. Lastly, a case study of freshwater red algae of the Batrachospermales is presented and future research needs assessed.

Dispersal of Freshwater Organisms

The biogeography of freshwater organisms is strongly linked to the dispersal capabilities and mechanisms of individual taxa. Since many species of river algae are sessile, and those that are motile (i.e., raphid diatoms, flagellates) have limited movement capabilities within or among rivers, nearly all rely on external methods for their dispersal. River algae, like other freshwater organisms, can be thought of having four potential modes of dispersal—water, air, other motile organisms, and humans (Kristiansen 1996). In this section, I will provide examples from the literature summarized by Kristiansen (1996), but primarily focus on research that has occurred since this review. As most studies do not focus on specific habitats, this section will consider both lentic and lotic habitats.

Water is an obvious mechanism for dispersal of aquatic organisms especially given the unidirectional flow of stream and river ecosystems. This unidirectional flow would be advantageous to spread propagules downstream and throughout a watershed. There is substantial literature on metacommunity structure in periphytic diatoms that has examined the question of environment versus spatial structuring of communities (i.e. Heino et al. 2010; Liu et al. 2013; Bottin et al. 2014). Many of the studies have shown that spatial partitioning can be at least as influential as environmental parameters and sometimes more important (Heino et al. 2010). Overland versus hydrological distances were tested within a large river basin and it was observed that hydrological distance better explained the observed metacommunity pattern (Liu et al. 2013). These results suggest that diatoms may be confined by their hydrological network and that water dispersal and flow direction may play a significant role in regional distribution of river diatoms. Nevertheless, downstream dispersal may be disadvantageous, if downstream habitat is unsuitable for a particular taxon. For example, it has been hypothesized that the chantransia stage of freshwater red algae is an adaptation to unidirectional flow and the need to maintain a population in a favorable habitat (Sheath 1984).

There has been a rich history of studying airborne (wind) dispersal of algae (Kristiansen 1996). As early as the mid-1800s, dust samples collected by Darwin on the H.M.S. Beagle were studied by Ehrenberg (1849) with numerous diatom taxa identified, but their viability was not assessed (Kristiansen 1996). van Overeem (1937) studied filtered samples from an airplane and was able to demonstrate that there were viable chlorophytes and cyanobacteria present. In the 1960s, there were numerous studies of airborne dust and the taxonomic richness and abundance assessed (i.e., Schlichting 1961; Brown et al. 1964; Geissler and Gerloff 1965). These works confirmed that a variety of chlorophytes, cyanobacteria, and diatoms as well as chrysophytes and euglenoid from terrestrial and freshwater habits could be dispersed via this manner.

In recent years, there continues to be research on airborne algae. Reviews by Sharma et al. (2007), Després et al. (2012), and Sahu and Tangutur (2015) provide much information regarding mechanisms for dispersal via this method, taxa that have been found to be viable, and suggestions for further research. Sharma et al.

(2007) summarized the known species of cyanobacteria, chlorophytes, tribophytes, and diatoms recovered from six biogeographic regions. They also suggest numerous methods for suspending airborne algae including farming practices, heavy rainfall and gusts of wind. They suggest in the future that there is a need for controlled laboratory experiments on algal tolerance to radiation and desiccation. Després et al. (2012) examined the wide literature on biological particles including cyanobacteria and algae. Sahu and Tangutur (2015) review the literature from the perspective of human health/allergens, deterioration of structures and contamination of drinking water. In future research, the authors of all three of these reviews advocate for both traditional methods of culturing and light microscopy as well as the incorporation of molecular identification techniques.

Some of the more recent research has revealed more information about the likelihood of dispersal, important meteorological conditions and further characterized the algae that can be airborne. Hall (1998) found a single colony of *Pseudopediastrum boryanum* (Turpin) E. Hegewald (as *Pediastrum boryanum*) in a pollen trap 42 km from the likely freshwater source and concluded that although the rate of dispersal was low (0.8 cm^2 per year), there is the potential for this taxon to be dispersed long distance via this method. Whole communities of microorganisms that colonized water containers were identified with morphological and molecular data (Genitsaris et al. 2011, 2014). These researchers showed that a wide variety of heterotrophic and autotrophic organisms were dispersed via wind with Chlorophyta being one of the dominant lineages. In addition, rainfall was the most important meteorological parameter among other parameters such as sunshine, wind speed, and relative humidity in a multivariate analysis contributing to the deposition of many of the organisms in the water containers (Genitsaris et al. 2014). Recent research has built on the study by Brown (1971) in which he documented the diversity of algae and fern spores across the island of Oahu, Hawaii. The new research investigated airborne algae along the same transect and documented the diversity with morphological observations and molecular analyses (Sherwood et al. 2014b, c). These researchers found that 97 % of taxa collected had previously been unrecorded in the Hawaiian Islands. It would appear that there is still much to learn about airborne algae and new molecular tools will be helpful in understanding the taxonomic richness and relatedness of these algae to those in nearby and more distant terrestrial and freshwater habitats.

Along with prevailing wind currents, other less frequent events such as volcano eruptions may disperse microalgae via the atmosphere. A study of the Oruanui eruption of Taupo volcano, New Zealand, showed that freshwater diatoms were dispersed more than 100 km downwind of the eruption, much farther than the 30 km dispersal of the lacustrine sediments (Van Eaton et al. 2013). These researchers hypothesize that both ancient and modern eruptions may disperse a variety of microorganisms to other habitats.

Numerous aquatic organisms may aid in the dispersal of freshwater macro- and microalgae. Within a drainage basin, it is likely that various aquatic animals including fish, macroinvertebrates, and crayfish may transport algae both short and long distances (i.e., Velasquez 1940; Hambrook and Sheath 1987; Fuelling et al. 2012).

It has been shown that a number of algal taxa can pass through fish and macroinvertebrate guts and still be intact or viable (Velasquez 1940; Hambrook and Sheath 1987). In addition, algae such as *Thorea* may adorn the antennae of crayfish and be carried up or down stream (Fuelling et al. 2012). Other aquatic organisms may transport algae from one water body to another (see Kristiansen 1996 for references).

Various mammals that might spend time near water have been investigated for their potential to disperse algae. To date, minks, muskrats, and raccoons have been shown to transport algae either internally or externally (Maguire 1963; Roscher 1967; Leone et al. 2014). In recent research, minks were investigated as possible wildlife vectors for the spread of the invasive freshwater alga *Didymosphenia geminata* (Lyngbye) M. Schmidt (Leone et al. 2014). These researchers observed that these animals could have a large numbers of cells on their fur, averaging more than 700 cells from a large bloom area. These animals travel up and downstream, potentially as far as 10 km and therefore it is likely they could disperse *Didymosphenia*, as well as other diatoms, to adjacent watersheds.

The animal dispersal vectors that have received the most attention are birds, especially wading waterfowl, which have the potential to transport algae and their propagules long distances (Kristiansen 1996 and references therein). There were numerous studies showing that viable algae can be recovered from fecal pellets along with being washed off feet (Schlichting 1960; Proctor 1963). In other studies, algae were fed to waterfowl and showed that viable cells were obtained, but not always (i.e. Atkinson 1970, 1971). From a search of the recent literature (1996–2015), only a single experimental study on the dispersal of algae by waterfowl was found. In this study, fecal pellets from three waterfowl species were examined and found that 7 % of the pellets examined had oogonia from charophytes, but the viability of the propagules was not determined (Charalambidou and Santamaria 2005). In addition, the pellets were not examined for other algal taxa. It is surprising this area of research has not expanded, as there are many studies of plant and invertebrate propagules (i.e., Frisch et al. 2007; Green et al. 2008). Numerous molecular algal studies find evidence of long-distance dispersal and hypothesize waterfowl as a possible dispersal mechanism (Meiers et al. 1999; Oberholster et al. 2005; Lam et al. 2012). But more specific studies of flyways and birds that visit smaller first- and second-order rivers where benthic algae are abundant would greatly enhance our understanding of potential mechanisms of long distance dispersal. In a recent thorough review on the dispersal of aquatic organisms by waterfowl, numerous research priorities such as studies of the characteristics and digestibility of propagules, retention time, and better tracking of bird movements were expressed (Figuerola and Green 2002). If these studies come to fruition, they could greatly augment our knowledge of dispersal capabilities of birds as well as specific algal taxa with particular propagules.

Human transport of freshwater organisms has been widely recorded over many decades. There have been numerous introductions into the Laurentian Great Lakes and surrounding rivers through ballast water from primarily European lakes and rivers (Mills et al. 1993). For example, *Bangia atropurpurea* (Mertens ex Roth) C. Agardh has been documented as a nonnative alga transported via shipping from large rivers and lakes of Europe (Sheath 1987; Müller et al. 2003; Shea et al. 2014). Likewise, *Nitellopsis obtusa* (N.A. Desvaux) J. Groves was found to be abundant in the St. Lawrence River and the St. Clair-Detroit River system in the 1980s and hypothesized to have been transported from Europe via shipping (Geis et al. 1981; Schloesser et al. 1986). Numerous brackish water/marine taxa of diatoms and *Enteromorpha* spp. have been reported to be potentially introduced in the Great Lakes Region via ballast water (Mills et al. 1993 and the references therein). Indoor and outdoor aquaria have been known to house exotic algal species that have typically been transported with aquatic plants and numerous species such as *Rhizoclonium fractiflexum* Gardavsky have been described from these artificial habitats (Gardavsky 1993). The home aquarium trade may also play a role in dispersing freshwater algae such as *Compsopogon caeruleus* (Balbis ex C. Agardh) Montagne to suitable natural habitats, but empirical data are lacking (Necchi et al. 2013). Aquaria can house numerous freshwater taxa and our laboratory alone has collected the freshwater red algae *Compsopogon, Kumanoa*, and *Thorea* from home aquaria (unpublished data); this does not mean that these taxa would necessarily be dispersed inadvertently to suitable natural habitats, but given the number of freshwater aquarium enthusiasts worldwide, it is a distinct possibility. Lastly, there is the example of *Didymosphenia geminata* being dispersed via fishing gear and recreational boating (Kilroy and Unwin 2011). Due to bloom formation of this taxon, its spread has been carefully documented, but it is likely that other freshwater algae that are not invasive have been spread short and longer distances through recreational use of river and lake habitats.

Studies of Continents or Regions

Biogeographic studies and survey research of river algae either whole communities or particular taxonomic groups are daunting. A researcher has to make numerous decisions about the depth and breadth of the study. For example, how many rivers to sample within a given geographic area? Should rivers be resampled seasonally? What taxonomic groups to be included? Should identification be based solely on morphological data or molecular data generated as well? Answers to each of these questions shape a study and the biogeographic knowledge gained. In this section I will present example studies, admittedly biases towards river macroalgae, which have provided biogeographic data for specific regions, certain algal morphologies or species-rich groups. When appropriate I will use the studies to illustrate particular difficulties in collecting biogeographic data. Lastly, I will highlight the few studies that have synthesized river algal data across continents.

The river macroalgal flora of the Hawaiian Islands is one of the better studied floras. There had been sporadic collection of freshwater specimens in the islands for decades (Sherwood 2006). Starting in the early 1990s, the river macroalgal flora from rivers (accessible and remote) has been studied on several islands (Vis et al. 1994; Filkin et al. 2003; Sherwood 2006). These studies revealed a freshwater macroalgal flora belonging to the Chlorophyta, cyanobacteria, Rhodophyta, Bacillariophyta, and Tribophyta. As well, a difference between the flora of Hawai'i the youngest island and others was detected with fewer of the widespread taxa being present (Sherwood 2006). More recently, molecular tools have been employed to study the red algal flora to better understand species present and their biogeographic affinities (Chiasson et al. 2007; Carlile and Sherwood 2013). These studies linked the cryptic life history stage (chantransia) to members of the Thoreales and Batrachospermales, but the origin of most taxa were difficult to discern due to lack of molecular data from potential dispersal areas (Chiasson et al. 2007; Carlile and Sherwood 2013). Of those taxa for which the biogeographic pattern could be deduced, it would appear that they showed a western Pacific origin (Carlile and Sherwood 2013). Most recently, a large study of freshwater and terrestrial algae was conducted and assessed biogeographic affinities using molecular data (Sherwood et al. 2014a). Like the previous studies, there was no clear biogeographic pattern discerned but a little more than 27 % and 11 % of the flora was cosmopolitan or endemic, respectively (Sherwood et al. 2014a).

Rivers in southeastern Brazil have been well studied for river macroalgae (i.e., Branco and Necchi 1996; Necchi et al. 1997, 1999). In the eastern Atlantic Rain Forest region of southeastern Brazil, it was shown that the flora was composed of cosmopolitan species reported frequently on other continents such as *Chaetophora elegans* (Roth) C. Agardh and *Draparnaldia glomerata* (Vaucher) C. Agardh, as well as numerous chlorophyte, rhodophyte, and cyanobacterial species with tropical distributions (Branco and Necchi 1996). A study of another area within southeastern Brazil (northwestern São Paulo State), the flora was shown to be a mix of taxa reported for temperate and tropical areas of other continents and the percentage of Chlorophyta (45 %), cyanobacteria (33.5 %), Rhodophyta (14.5 %), and Chrysophyta (7 %) species was similar to surveys of other regions of the world (Necchi et al. 1997). These surveys provide good baseline data for southeastern Brazil with morphological identifications of taxa. With the advent of molecular tools, it may be possible to determine if the species collected represent cosmopolitan taxa as the morphology would suggest or more geographically restricted taxa. For example, recent molecular studies of red algae indicate that some taxa may be more geographically restricted such as *Sirodotia delicatula* Skuja commonly reported from Brazil, but probably is a cryptic species restricted to South America rather than *S. delicatula* originally described from Indonesia (Johnston et al. 2014). Certainly, there are endemic taxa to various regions of Brazil such as *Kumanoa amazonensis* Necchi and ML Vis and *Batrachospermum orthostichum* Skuja such that other cryptic endemics may be present (Necchi 1990; Necchi et al. 2010).

There have been a few floristic studies of the river macroalgae of North American arctic rivers (Sheath et al. 1996; Sheath and Muller 1997). These rivers have the same taxonomic groups, cyanobacteria, Chlorophyta, Chrysophyta, and Rhodophyta, represented as other regions of the continent, but the flora contains fewer species than other biomes (Sheath and Cole 1992; Sheath et al. 1996). As well, there are few endemics and many of the taxa just represent an extension of the boreal forest flora (Sheath et al. 1996). There has been little molecular data collected for these taxa so it is difficult to be certain these do not represent cryptic species. At least in one case, *Batrachospermum gelatinosum* that is widespread in arctic is confirmed via molecular data to be cosmopolitan in North America and Europe, boreal and temperate regions (House et al. 2010; Keil et al. 2015).

In many cases, research in understudied areas of the globe provides new biogeographic information. For example, a small number of rivers were studied in Bolivia, but yielded 35 new records for South America and 46 new for Bolivia (McClintic et al. 2003). These numbers undoubtedly represent the lack of information for this country. Necchi et al. (1997) did a similar size study and only reported four species to be new to Brazil most likely because this is a much better studied country. In a survey of French Guiana, many of the non-Rhodophyta taxa had not been previously noted from that country, but most had been reported from better studied areas of South America (Vis et al. 2004). Under-sampled areas of the world represent a major challenge to understanding the biogeography of a species or group of taxa. These areas may provide key information such as the new information on *Aegagropila* in Kazakhstan filled in a disjunct distribution (Boedeker and Sviridenko 2012).

Intensive surveys have yielded rich floras not previously recognized for an area. For example, unique habitats such as springs have shown to be biodiverse. In the southeastern Alps, springs of various sizes and physical and chemical parameters were sampled from which 120 taxa (exclusive of diatoms) have been identified (Cantonati et al. 2012). Likewise, a study of springs in the Pyrenees Mountains showed that diatom affinities were similar to habitats elsewhere with the same chemistry and that these springs had a high number of taxa with poorly known distributions (Sabater and Roca 1992). In addition to unique habitats such as springs, studies of many rivers in small geographic areas can yield a surprising number of taxa such as a checklist of the algal flora of southeastern Ohio for rivers and other habitats reporting over 1700 infrageneric algal taxa (Verb et al. 2006). These geographically small-scale, yet intensive, surveys can provide a wealth of information for larger scale biogeography studies. Although it cannot be known with complete assurance that a taxon is truly absent from an area, these in-depth studies, which sample the majority of flowing water in a region, provide more confidence in stating a taxon is absent in comparison to a broad study of fewer sites.

In-depth surveys of a number of rivers in a geographical area can yield new insights by finding new taxa and confirming older reports. A study of *Spirogyra* and *Zygnema* from primarily the South Island of New Zealand produced ten new species (Novis 2004). Likewise, studies of Zynematales from California have yielded new species and insights into the biogeography of taxa previously unknown from California (Stancheva et al. 2012, 2013). Two new *Spirogyra* species were described and pointed to the possibility of many morphologically cryptic species (Stancheva et al. 2013). This insight regarding possible cryptic species will potentially have biogeographical implications as previously cosmopolitan species may be separated into more geographically restricted species.

Studies of continents and descriptions of continental floras for river-dwelling algae are scarce. The study of 1000 river segments from tundra to tropical biomes in North America by Sheath and Cole (1992) stands alone. Two hundred fifty-nine infragenic taxa were identified with the most widespread species being *Phormidium retzii*, *P. subfuscum* Kützing ex Gomont, *Cladophora glomerata* (Linnaeus) Kützing, *Batrachospermum gelatinosum*, and *Audouinella hermannii* Roth (Duby) (Sheath and Cole 1992). All identifications were based on morphology. Interestingly, *Phormidium retzii* has subsequently been shown to harbor cryptic variation maybe suggesting that it is not as widespread (Casamatta et al. 2003). In contrast, studies of *Batrachospermum gelatinosum* have confirmed it to be cosmopolitan in North America with little genetic variation (Vis and Sheath 1997; House et al. 2010). The other three taxa have not had enough molecular research conducted to state whether they are truly cosmopolitan species or composed of multiple cryptic species with regional distributions. For other continents, there are no single studies, but distributional data have been collated. Entwisle (2007) summarized the state of the knowledge on freshwater macroalgae for Australia providing information on regional endemism, rarity, and origins of the flora. However, he noted that much more research is needed and distributional data are "generally meager." The Süßwasserflora von Mitteleuropa series has documented the freshwater flora by taxonomic group, but the only volume that is specific to rivers would be that of the Rhodophyta because they are primarily restricted to flowing waters (Eloranta et al. 2011).

Specialist Taxa

Specialist taxa have restricted ranges for a variety of reasons (Sheath and Vis 2013). They may be restricted to a certain geographic area or a specific habitat that is widespread, but is rare. In addition, they may be only able to survive under particular ecological conditions. The ability to recognize a taxon as a specialist or endemic can be difficult. Tyler (1996) refers to "flagship" taxa that are so showy that they would be difficult to overlook and therefore one could be fairly certain if it were widespread or endemic to a certain habitat or region. A taxon might be thought to be endemic to an isolated habitat such as a spring-fed river in an arid region. *Sheathia involuta* (M.L. Vis and Sheath) Salomaki and M.L. Vis was thought to be endemic to the San Marcos River system, but recent molecular data have shown this taxon to be widespread in springs and rivers in North America (Salomaki et al. 2014). Although some taxa have been shown to be more widespread than previously believed such as the example of *S. involuta*, there are truly rare taxa. The rare, yet widespread brown algal taxon, *Pleurocladia lacustris* A. Braun, may be an example of a specialist taxon. It appears to be restricted to calcareous rivers when growing in freshwaters (Wehr et al. 2013). However, the authors noted more research on various aspects of its biology is needed to be certain of its status as a specialist. The desmid, *Oocardium stratum* Nägeli appears to be a true river-dwelling specialist (Rott et al. 2010). It is uncertain whether this species is common or rare, but it is only present in calcareous springs and potentially may be restricted to a certain area of the spring, in which there are particular chemical characteristics, primarily excess

CO_2 and very low phosphorus (Rott et al. 2012). Research is being conducted to parameterize the ecological niche of this taxon for potential habitat protection (Rott et al. 2012). Numerous other taxa, primarily cyanobacteria have been shown to be indicative of particular spring types, but may also occur in rivers with similar conditions (Cantonati et al. 2012). For example, *Chamaesiphon starmachii* Kann and *Pleurocapsa aurantiaca* Geitler are known to be rheophilic and two closely related taxa *Tapinothrix varians* (Geitler) Komárek and *T. janthina* (Starmach) Komárek have been shown to occur on carbonate and siliceous substrata, respectively (Cantonati et al. 2012). These are a few of the taxa that appear to be good examples of specialists.

Generalist Taxa

Typically, generalist taxa can be thought of as those species that have a wide ecological niche, and inhabiting rivers from various biomes and/or rivers with diverse chemical and physical attributes (Sheath and Vis 2013). The list of generalist or cosmopolitan taxa has been shrinking in the recent past due primarily to molecular studies that have revealed cryptic diversity within these species, such as *Phormidium retzii* (Casamatta et al. 2003). Nevertheless, there are still cosmopolitan taxa that occur in river habits over a wide geographic area.

The macroalga, *Cladophora glomerata*, is probably the most well-known example of a generalist from streams and rivers. It is also very abundant in the Great Lakes as well as other small and large lakes globally (Higgins et al. 2008). A small molecular data set has shown low sequence variation across continents suggesting that it is a single cosmopolitan species (Marks and Cummings 1996). *Cladophora glomerata* has been reported from geographically wide range of rivers with concomitant range in physical and chemical habitat (Dodds and Gudder 1992). Its ecology has been well studied especially in the Great Lakes due to large blooms that can form (Higgins et al. 2008). Some blooms were correlated with phosphorus addition, but more recent blooms appear to be associated with increased water clarity, sloughing from substratum and phosphorus recycling from the benthos (Higgins et al. 2008). Some researchers consider *Cladophora* and this species in particular to be an ecological engineer, having considerable influence on the microorganisms that colonize it and thereby influences the carbon cycle and cycling of other elements (Zulkifly et al. 2013).

Compsopogon caeruleus is a generalist and cosmopolitan species, which has been sampled from locations in North America, South America, Australia, Europe, and Pacific Islands with little molecular sequence variation over this very large geographic range (Necchi et al. 2013). Despite low molecular variation, the phenotypic plasticity of this species is great with numerous morphological oddities that have led researchers in the past to name new species (Kumano 2002). The range in physical and chemical parameters, in which this species can be found or cultured, is wide (Necchi et al. 2013). In fact, this alga is known commonly in the aquarium trade as "Staghorn" algae and there are numerous websites devoted to how to get rid of it, if one enters the search term staghorn algae. It would seem that *Compsopogon caeruleus* is very much a generalist and may owe its worldwide distribution to its ability to tolerate a wide range of conditions and freshwater aquarium enthusiasts.

The freshwater brown alga, *Heribaudiella fluviatilis* (Areschoug) Svedelius, is another cosmopolitan taxon once thought to be rare (Wehr 2015). This species is known from several hundred locations worldwide. In North America, there have been at least 30 reported locations, but these rivers can be hundreds to thousands of km from next nearest location (Wehr 2015). Unlike *Compsopogon* with its large three-dimensional thallus, *Heribaudiella fluviatilis* is a crust, which may be overlooked in general river surveys. Therefore, it may be that this species is abundant in between the known sites and as more researchers look for it, the more it will be found (Wehr and Perrone 2003).

The red alga, *Thorea hispida* (Thore) Desvaux, is also considered a generalist taxon. This alga appears to be similar to *Heribaudiella fluviatilis*, having been reported from locations worldwide and the records are geographically widespread (Sheath et al. 1993). Although geographically widespread, it appears to be distributed in isolated and scattered populations from springs, streams and large rivers in temperate and tropical areas. Ecological preference seems to be for neutral to basic pH and high conductivity (Sheath et al. 1993). This taxon must have a wide ecological niche as both the chantransia phase and gametophytes have been observed in home aquaria in addition to natural conditions (Vis, personal observation).

"True" Disjunct Taxa

It is difficult to know if a taxon is truly disjunct, as reporting and identification methods may affect this interpretation. For example, it may be that it has a morphology that is makes it difficult to distinguish from another widespread species or it might be a rare species that is just collected infrequently. It might be in a taxonomic group that has few experts worldwide who actively seek it out; *Batrachospermum helminthosum* is quite common in Ohio, but had never been reported in Ohio prior to those of my laboratory (House et al. 2008). Likewise, a new species might have been described that is subsequently discovered to be an ecological variant of a more widespread species when molecular tools were applied; such was the case of a species described from Kazakhstan found to be conspecific with *Aegagropila linnaei* and filled in the disjunction between Europe and Japan (Boedeker and Sviridenko 2012). In surveying the literature, there were examples with molecular data of disjuncts from lentic systems such as *Lychnothamnus barbatus* (Meyen) Leonhardi from Taiwan, Australia and Europe or *Desmodesmus pirkollei* E. Hegewald from North America and Bali (Wang and Chou 2004; Johnson et al. 2007). However, there are few from lotic habitats.

There are a few examples of disjunction in river-dwelling freshwater red algae. *Kumanoa mahlacensis* (Kumano and W.A. Bowden-Kerby) M.L. Vis, Necchi,

W.B. Chiasson and Entwisle, was first described from Micronesia (Kumano and Bowden-Kerby 1986). Molecular data for specimens attributable to this species have been recently produced. Specimens from southwestern Spain and New Mexico, USA, have been shown to have identical sequence for the *rbc*L gene (García-Fernández et al. 2015). Regardless if future molecular data show the Micronesian specimen not to be closely related, the disjunct geography between the European and North American specimens is noteworthy. A specimen of *Sheathia arcuata* (Kylin) Salomaki and M.L. Vis from New Zealand had identical sequence to specimens from Washington State, USA (Vis et al. 2010). Likewise, *Sirodotia suecica* Kylin is common in Europe and North America, but also has disjunct populations in New Zealand and Australia (Vis and Entwisle 2000; Lam et al. 2012). For *Sheathia arcuata* and *Sirodotia suecica*, long-distance dispersal by waterfowl is a likely explanation, but the dispersal mechanism for *Kumanoa mahlacensis* is less clear.

Nonnative Invasives

Both marine and freshwater invasive taxa have been reported from river habitats. In this section, there will be only examples of marine algae that invaded, as there is another chapter devoted to invasive algae. Undoubtedly there are probably more taxa that have gone unnoticed for a variety of reasons such as the small size of many algae, lack of previous distribution data, or cryptic species.

Marine macroalgal taxa that have a wide salinity tolerance and physiological mechanisms to cope with varying salinity may be found in freshwater streams and rivers. For example, two red algal marine genera, *Caloglossa* and *Polysiphonia*, have been shown to be in strictly freshwater rivers (Sheath and Cole 1992; Sheath et al. 1993 and references therein). The conspecificity of *Polysiphonia subtilissima* individuals from freshwaters in North America and Europe with marine populations was confirmed using *rbc*L gene data (Lam et al. 2013). Few *Caloglossa* specimens from freshwaters have been sequenced, but those from Malaysian freshwater rivers appear to be *Caloglossa beccarii*, a taxon known primarily from mangroves (Johnston et al. 2014). Other taxa within the Ceramiales have been reported from freshwaters using morphological identification (Sheath et al. 1993 and the references therein). However, it should be noted that subsequent molecular analyses have been applied and the *Ballia* species have been transferred to the freshwater genus *Balliopsis* as well as the freshwater representatives of *Ptilothamnion* determined to be the chantransia phase of Batrachospermales (Saunders and Necchi 2002; Vis et al. 2006).

The diatom, *Thalassiosira lacustris* (Grunow) Hasle, is an example of a marine alga that has invaded freshwater rivers (Smucker et al. 2008). This taxon is euryhaline occurring in estuarine habitats and has been verified in river from Texas, Louisiana, Maryland, and Ohio (Smucker et al. 2008 and references therein). In some, but not all, cases the rivers have high conductivity that may provide a suitable habitat. In all cases, this species was reported as part of the diatom flora and did not reach nuisance abundances.

The marine haptophyte species, *Prymnesium parvum* Carter, can be a toxin producer and has been implicated in many freshwater fish kills not only in lakes and reservoirs, but also streams and rivers (Zamor et al. 2014). The persistent problem of blooms of this taxon appears to be in lakes and reservoirs of Texas and Oklahoma. However, it has been implicated as the cause of a massive fish kill in a West Virginia/Pennsylvania river as well (West Virginia DEP website accessed 23 August 2014). Research on its biogeography and recent (since 2001) range expansion suggest that it is linked to changes in salinity in freshwater systems, but not necessarily clearly associated with other water quality changes (Patino et al. 2014). Interestingly, it may be dispersed to a water body and take quite some time before toxic blooms develop (Patino et al. 2014).

For a long time, the marine red alga, *Bangia fuscopurpurea* (Dillwyn) Lyngbye, was thought to have invaded the St. Lawrence Seaway and subsequently the Laurentian Great Lakes (Sheath 1987). However, it has been shown using molecular data that the *Bangia* found in North America is a freshwater species, *Bangia atropurpurea* (Mertens ex Roth) C. Agardh from European Lakes and large rivers which was most likely transported via shipping (Müller et al. 2003; Shea et al. 2014). Therefore, it can be considered a freshwater invasive rather than marine. Since its introduction into North American waters, it has not been found to colonize other locations besides the Laurentian Great Lakes and no adverse effects of its colonization have been reported.

Case Study: Freshwater Rhodophyta

The Batrachospermales is a good case study because the members primarily inhabit rivers, there is much molecular data for a number of taxa and the biogeography trends may be similar to other algal groups. Systematists studying a specific taxonomic group produce much of the biogeography data available. Systematists tend to be conservative in describing new taxa and tend to look to previous names when studying new specimens (Entwisle 1993; Tyler 1996). This is certainly the case with members of the Batrachospermales. Most of the species epithets come from species first named from Europe (i.e., Bory de Saint Vincent 1797; Sirodot 1884). When researchers encounter specimens from other continents the tendency was to apply European names, as was applied with the Australian flora (Entwisle 1993). Eventually, researchers encounter too much morphological variation to be encompassed by a single species and came to the realization that the European species names do not suit and new species must be described (i.e., Entwisle and Foard 1997). Although this revolution in naming new species took place prior to the widespread use of molecular data, the molecular data have provided evidence for some taxa as either cosmopolitan or endemic in their distribution (Vis and Entwisle 2000; Entwisle et al. 2004; Salomaki et al. 2014).

The Batrachospermales has members that are cosmopolitan, restricted in their distribution or endemic to a continent. Data are incomplete, but several species are well documented. The Australian flora appears to be composed primarily of Australasian endemics, with a few cosmopolitan species (Vis and Entwisle 2000; Entwisle et al. 2004). The two notable cosmopolitan species are *Sirodotia suecica* and *Batrachospermum atrum* (Hudson) Harvey. *Sirodotia suecica* appears to be a true cosmopolitan species having little genetic differentiation among specimens from North America, Europe, Africa, and Australasia (Lam et al. 2012). It should be noted that numerous studies of *Sirodotia* have been conducted in Brazil, South America and there is a *Sirodotia* species commonly collected, yet to date *S. suecica* has not been detected in that flora. *Batrachospermum atrum* was first described from Great Britain and specimens from there and Australasia have shown little genetic variation (Vis and Entwisle 2000).

The South American flora perhaps contains more endemic species than previously thought. There are numerous species of *Kumanoa* that have been described only from South America (i.e., *K. amazonensis, K. cipoensis* (Kumano and Necchi) Entwisle, M.L. Vis, W.B. Chiasson, Necchi and A.R. Sherwood, *K. equisetoidea* (Kumano and Necchi) Entwisle, M.L. Vis, W.B. Chiasson, Necchi and A.R. Sherwood) (Necchi and Vis 2012). Those species (i.e., *K. abilii* (M.P. Reis) Necchi and M.L. Vis, *K. capensis* (Starmach ex Necchi and Kumano) Necchi and MLVis, and *K. curvata* (Z.-X. Shi) M.L. Vis, Necchi, W.B. Chiasson and Entwisle) that occur in South America and on other continents have only been linked by morphology with no molecular data from the other continent such that it may just represent convergent morphology. Only for *K. ambigua* (Montagne) Entwisle, M.L. Vis, W.B. Chiasson, Necchi, and A.R. Sherwood is there DNA sequence data for specimens from North and South America; thus far they appear genetically similar. In addition, recent data have shown specimens of *Batrachospermum macrosporum* Montagne from Asia (Malaysia) to be genetically similar to specimens from Brazil in South America (Johnston et al. 2014). As more sequence data become available for specimens of the same species, it will be interesting to see if the flora of South America is composed of primarily endemics and if it shares species with which continent is it most allied.

The North American flora shares some species with Europe, but that is changing with molecular data. Most notably, *Batrachospermum gelatinosum* and *Sirodotia suecica* have been observed to be genetically similar and abundant on both continents (House et al. 2010; Lam et al. 2013; Keil et al. 2015). However, one recent molecular systematic study has highlighted that the floras are distinct (Salomaki et al. 2014). In the newly-erected genus, *Sheathia*, there were four species recognized from North America with only one (*S. involuta*) having sequence data from Europe and even then it was very restricted in its European distribution to a small portion of Poland. Likewise, there were three species in Europe that were not detected in North America (Salomaki et al. 2014). There are unpublished data on *Batrachospermum* section *Turfosa*, and the genera *Lemanea* and *Paralemanea* that also suggest the floras of these continents are not as similar as would be inferred by morphological traits and European names given to the North American specimens. It will be interesting when the molecular systematics studies are completed to see how many species are shared as the genus-level data suggests these two continents are quite similar (see below).

It is difficult to assess the floras of Africa and Asia since there are few molecular data for specimens from the Batrachospermales. From Africa, there have been two species of *Sirodotia*, *S. suecica*, and *S.* cf. *huillensis* (Lam et al. 2012). One species is cosmopolitan and the other potentially endemic. Likewise, there have been a few more data on specimens from Asia and these show endemics and cosmopolitan species (Ji et al. 2014; Johnston et al. 2014). However, there may be more endemism than cosmopolitanism due to broad species descriptions and much genetic variation such as in *Sheathia arcuata* (Vis et al. 2010).

We are continuing to gain insights on species-level biogeography in the Batrachospermales as more regions are sampled and more molecular data for species are gathered. The biogeographic trends at a higher taxonomic level may be more feasible without molecular data. Since the most recent assessments of the Batrachospermales clearly show *Batrachospermum* as paraphyletic and suggest that each section will eventually be described as a genus, the sections were treated as genera for this analysis (Entwisle et al. 2009). Of the 18 taxa, 13 occur on more than one continent (Table 10.1). Only *Batrachospermum* sections *Acarposporophytum, Gonimopropagulum* and *Balliopsis, Petrohua, Psilosiphon*, and *Tuomeya* with single species are confined to a single continent. *Batrachospermum* section *Setacea, Kumanoa*, and *Sirodotia* have been reported on six continents (Kumano 2002 and references therein). All other taxa are on two to four continents.

The taxa are somewhat evenly distributed among continents ranging from seve to 12 taxa per continent (Fig. 10.1). Both Africa and Australia have only eight and

Table 10.1 Continental distributions for *Batrachospermum* sections and genera in the Batrachospermales

Taxon	Europe	Asia	North America	South America	Australia	Africa
B. section *Acarposporophytum*				X		
B. section *Aristata*		X		X	X	X
B. section *Batrachospermum*	X	X	X			
B. section *Gonimopropagulum*						X
B. section *Macrosporum*		X	X	X		X
B. section *Setacea*	X	X	X	X	X	X
B. section *Turfosa*	X	X	X	X		
B. section *Virescentia*	X	X	X	X		
Balliopsis		X	X	X		
Kumanoa	X	X	X	X	X	X
Lemanea	X	X	X			
Nothocladus					X	X
Paralemanea	X		X			X
Petrohua				X		
Psilosiphon					X	
Sheathia	X	X	X		X	
Sirodotia	X	X	X	X	X	X
Tuomeya			X			

Fig. 10.1 Map of continents showing the number of *Batrachospermum* sections and genera as provided in Table 10.1

Table 10.2 Sorenson's similarity index of continental floras based on *Batrachospermum* sections and genera in the Batrachospermales

Continent	Europe	Asia	North America	South America	Australia	Africa
Europe	–					
Asia	0.80	–				
North America	0.86	0.87	–			
South America	0.53	0.76	0.64	–		
Australia	0.50	0.56	0.42	0.47	–	
Africa	0.47	0.53	0.50	0.56	0.67	–

seven taxa, respectively; for Africa the low number may be due to a lower sampling effort on the continent, whereas the low number may be attributable to the isolation of Australia rather than sampling effort. The distribution of taxa on each continent (38–67 %) may suggest that the Batrachospermales are an evolutionarily old lineage, but with no fossil record or molecular clock data it is difficult to confirm. It might be just as likely that the taxa are good dispersers, although no desiccation-resistant stage is known (Sheath 1984).

The floras of each continent were compared (Table 10.2). The Sorenson's similarity index showed North America and Australia to be the least similar and North America and Asia to be the most similar. North America and Europe were also very similar. A dendrogram produced from the index showed North America, Asia and

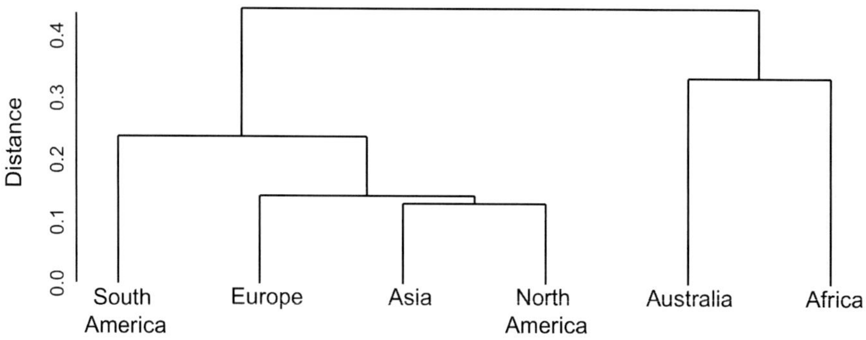

Fig. 10.2 Dendrogram showing the relationship of continental floras of *Batrachospermum* sections and genera based Sorenson's similarity index as provided in Table 10.2

Europe to be similar followed by South America whereas Africa and Australia more closely related to each other than to the other continents (Fig. 10.2).

In the next few years, as systematics studies by international collaborators are finished, a clearer picture of the global biogeography of freshwater red algal species in the Batrachospermales will emerge. It will be informative to compare the species level floras of each continent with the genus level floras. If other taxa show the same trend as *Sheathia* then the species and genus-level floras will show quite different patterns as the genus is shared between Europe and North America, but few of the species (Salomaki et al. 2014).

Conclusion

Although many insights have been gained from the work on the biogeography of river algae, there is much to be done to truly understand these organisms. Our knowledge of their dispersal mechanisms is rudimentary and would greatly benefit from more experimental work on modes of dispersal as well as physiological traits, such as desiccation tolerance, of individual species. There are regions such as Hawaii and Brazil, which have received much attention in the past, but many other geographic regions (e.g., African Continent, China) for which we have little data on their flora and still less on their possible biogeographic patterns.

Unfortunately, there is no easy way to rectify our lack of biogeographic information. To make significant advancements in biogeography of freshwater river algae, the algal community needs to come together and propose large-scale initiatives to make data more available. For example, the Global Biodiversity Information Facility (GBIF—http://www.gbif.org/) makes biodiversity data on any group of organism freely available. Currently the records of freshwater algae are too sporadic and the taxonomy is outdated, but if there were a significant effort put forth to populate this database by the phycological community, it could be an enduring resource. Similar smaller scale initiatives exist such as the taxon-based scaled chrysophytes of Europe database (www.chrysophytes.eu) and the regional based Hawaiian Freshwater Algal Database (algae.manoa.hawaii.edu/hfwadb/) (Skaloud et al. 2013; Sherwood et al. 2012). However, the phycological community needs to find a way to pool such databases into larger scale initiatives.

If the aforementioned data were to be collected, collated, and curated, key biogeographical questions could be addressed. For example, there appears to be little synthetic data on the river algal floras of continents. There is not the ability to compare the floras of taxonomic groups for a continent to examine overall patterns. Do the reds and greens show similar patterns in genus or species richness on the different continents? Do various algal groups employ different modes of dispersal and does that influence different patterns we may see among taxonomic groups. Do river macro- and microalgae show different geographic patterns and if so what are the key drivers of the difference? These are just a few questions that could be addressed with some new studies, but also with just more availability of previous data collected.

Acknowledgements I wish to acknowledge Orlando Necchi for his advice while writing this chapter. I would like to thank Bob Sheath who first introduced me to stream macroalgae and their biogeography. Conversations with Tim Entwisle regarding biogeography, systematics and naming of taxa have very much informed my own thinking. Sam Drerup prepared the dendrogram in R. Thank you to everyone who has sent me specimens of freshwater red algae; these have been invaluable to our understanding of freshwater red algal biogeography.

References

Abarca N, Jahn R, Zimmermann J et al (2014) Does the Cosmopolitan Diatom *Gomphonema parvulum* (Kützing) Kützing Have a Biogeography? PLoS One 9, e86885

Antoniades D, Douglas MSV, Smol JP (2009) Biogeographic distributions and environmental controls of stream diatoms in the Canadian Arctic Archipelago. Botany-Botanique 87:443–454

Atkinson KM (1970) Dispersal of phytoplankton by ducks. Wildfowl 21:110–111

Atkinson KM (1971) Further experiments in dispersal of phytoplankton by birds. Wildfowl 22:98–99

Avise JC (2000) Phylogeography the history and formation of species. Harvard University Press, Cambridge Massachusetts

Boedeker C, Sviridenko BF (2012) *Cladophora koktschetavensis* from Kazakhstan is a synonym of *Aegagropila linnaei* (Cladophorales, Chlorophyta) and fills in the gap in the disjunct distribution of a widespread genotype. Aquat Bot 101:64–68

Boedeker C, Eggert A, Immers A et al (2010) Biogeography of *Aegagropila linnaei* (Cladophorophyceae, Chlorophyta): a widespread freshwater alga with low effective dispersal potential shows a glacial imprint in its distribution. J Biogeogr 37:1491–1503

Boo SM, Kim HS, Shin W et al (2010) Complex phylogeographic patterns in the freshwater alga *Synura* provide new insights into ubiquity vs. endemism in microbial eukaryotes. Mol Ecol 19:4328–4338

Bottin M, Soininen J, Ferrol M et al (2014) Do spatial patterns of benthic diatom assemblages vary across regions and years? Freshw Sci 33:402–416

Bouchard G, Gajewski K, Hamilton PB (2004) Freshwater diatom biogeography in the Canadian Arctic Archipelago. J Biogeogr 31:1955–1973

Branco CCZ, Necchi O Jr (1996) Distribution of stream macroalgae in the eastern Atlantic Rainforest of Sao Paulo State, southeastern Brazil. Hydrobiologia 333:139–150

Brown RM (1971) The distribution of airborne algae and fern spores across the island of Oahu, Hawaii. In: Parker BC, Brown RM (eds) Contributions in Phycology. Allen press, Lawrence Kansas, pp 175–188

Brown RM, Larson DA, Bold HC (1964) Airborne algae: their abundance and heterogeneity. Science 143:583–585

Cantonati M, Rott E, Spitale D et al (2012) Are benthic algae related to spring types? Freshw Sci 31:481–498

Carlile AL, Sherwood AR (2013) Phylogenetic affinities and distribution of the Hawaiian freshwater red algae (Rhodophyta). Phycologia 52:309–319

Casamatta DA, Vis ML, Sheath RG (2003) Cryptic species in cyanobacterial systematics: a case study of *Phormidium retzii* (Oscillatoriales) using 16S rDNA and RAPD analyses. Aquat Bot 77:295–309

Charalambidou I, Santamaria L (2005) Field evidence for the potential of waterbirds as dispersers of aquatic organisms. Wetlands 25:252–258

Chattova B, Lebouvier M, Van de Vijver B (2014) Freshwater diatom communities from Ile Amsterdam (TAAF, southern Indian Ocean). Fottea 14:101–119

Chiasson WB, Johanson KG, Sherwood AR et al (2007) Phylogenetic affinities of the form taxon *Chantransia pygmaea* (Rhodophyta) specimens from the Hawaiian Islands. Phycologia 46:257–262

Coesel PFM, Krienitz L (2008) Diversity and geographic distribution of desmids and other coccoid green algae. Biodivers Conserv 17:381–392

Bory de Saint-Vincent JBGM (1797) Mémoire sur les genres *Conferva* et *Byssus*, du chevalier O. Linné. Bordeaux, Louis Cavazza, 58 p

Després VR, Huffman JA, Burrows SM et al (2012) Primary biological aerosol particles in the atmosphere: a review. Tellus B 64:1–58

Dodds WK, Gudder DA (1992) The ecology of *Cladophora*. J Phycol 28:415–427

Edlund MB, Jahn R (2001) Report of a workshop on Biogeography and Endemism of Diatoms. In: Edonomou-Amilli A (ed) Proceedings of the XVIth International Diatom Symposium. Amvrosieu Press, Athens, pp 575–587

Ehrenberg CG (1849) Passatstaubun Blutregen. Abh Königl Akad Wiss Berlin 1847:26–60

Eloranta P, Kwandrans J, Kusel-Feltzmann E (2011) Rhodophyceae and Phaeophyceae. In: Schagerl M (ed) Süßwasserflora von Mitteleuropa Band 7. Spectrum Akademischer Verlag, Heidelberg, pp 1–155

Entwisle TJ (1989) Macroalgae in the Yarra River basin: flora and distribution. Proc R Soc Vic 101:1–76

Entwisle TJ (1993) The discovery of batrachospermalean taxa (Rhodophyta) in Australia and New Zealand. Muelleria 8:5–16

Entwisle TJ (2007) Biogeography of freshwater macroalgae. In: McCarthy PM, Orchard AE (eds) Algae of Australia: Introduction. CSIRO publishing, Melbourne, pp 566–579

Entwisle TJ, Foard HJ (1997) *Batrachospermum* (Batrachospermales, Rhodophyta) in Australia and New Zealand: new taxa and emended circumscriptions in sections Aristata, *Batrachospermum*, Turfosa and Virescentia. Aust Syst Bot 10:331–380

Entwisle TJ, Vis ML, McPherson H (2004) *Batrachospermum pseudogelatinosum* (Batrachospermales, Rhodophyta), a polyecious, paraspecies from Australia and New Zealand. Aust J Bot 17:17–28

Figuerola J, Green AJ (2002) Dispersal of aquatic organisms by waterbirds: a review of past research and priorities for future studies. Freshw Biol 47:483–494

Filkin NR, Sherwood AR, Vis ML (2003) Macroalgae from 23 stream segments in the Hawaiian Islands. Pac Sci 57:421–432

Finlay BJ (2002) Global dispersal of free-living microbial eukaryote species. Science 296:1061–1063
Frisch D, Green AJ, Figuerola J (2007) High dispersal capacity of a broad spectrum of aquatic invertebrates via waterbirds. Aquat Sci 69:568–574
Fuelling LJ, Adams JA, Badik KJ et al (2012) Occurrence of Freshwater Red Algal Chantransia on Rusty Crayfish. Nov Hedw 94:355–366
Furlotte AE, Ferguson JA, Wee JL (2000) A floristic and biogeographic survey of the Synurophyceae from Southeastern Australia. Nord J Bot 20:247–256
García-Fernández M-E, Vis ML, Aboal M (2015) *Kumanoa mahlacensis* S. Kumano and W.A. Bowden-Kerby (Batrachospermales, Rhodophyta) in a Mediterranean wetland, a new species for the European continental algal flora. An Jard Bot Madrid 72: e0182015. DOI: 10.3989/ajbm.2324
Gardavsky A (1993) *Rhizoclonium fractiflexum* sp. nova, a new member of the Cladophorales (Chlorophyta) described from freshwater aquaria. Arch Protistenk 143:125–136
Geis IW, Schumacher GI, Raynal DI et al (1981) Distribution of *Nitellopsis obtusa* (Charophyceae, Characeae) in the St. Lawrence River: a new record for North America. Phycologia 20:211–214
Geissler U, Gerloff J (1965) Das Vorkommenvon Diatomeeenin menschlichen Organenundinder Luft. Nov Hedw 10:565–577
Genitsaris S, Moustaka-Gouni M, Kormas KA (2011) Airborne microeukaryote colonists in experimental water containers: diversity, succession, life histories and established food webs. Aquat Microb Ecol 62:139–151
Genitsaris S, Kormas KA, Christak U et al (2014) Molecular diversity reveals previously undetected air-dispersed protist colonists in a Mediterranean area. Sci Total Environ 478:70–79
Green AJ, Jenkins KM, Bell D et al (2008) The potential role of waterbirds in dispersing invertebrates and plants in arid Australia. Freshw Biol 53:380–392
Hall SA (1998) Atmospheric transport of freshwater algae *Pediastrum* in the American southwest—biogeographic implications. Grana 37:374–375
Hambrook JA, Sheath RG (1987) Grazing of fresh-water Rhodophyta. J Phycol 23:656–662
Heino J, Bini LM, Karjalainen SM et al (2010) Geographical patterns of micro-organismal community structure: are diatoms ubiquitously distributed across boreal streams? Oikos 119:129–137
Higgins SN, Malkin SY, Howell ET et al (2008) An ecological review of *Cladophora glomerata* (Chlorophyta) in the Laurentian Great Lakes. J Phycol 44:839–854
House DL, Sherwood AR, Vis ML (2008) Comparison of three organelle markers for phylogeographic inference in *Batrachospermum helminthosum* (Batrachospermales, Rhodophyta) from North America. Phycol Res 56:69–75
House DL, VandenBroek AM, Vis ML (2010) Intraspecific genetic variation of Batrachospermum gelatinosum (Batrachospermales, Rhodophyta) in Eastern North America. Phycologia 49:501–507
Ji L, Xie SL, Feng J et al (2014) Molecular systematics of four endemic Batrachospermaceae (Rhodophyta) species in China with multilocus data. J Syst Evol 52:92–100
Johnson JL, Fawley MW, Fawley KP (2007) The diversity of *Scenedesmus* and *Desmodesmus* (Chlorophyceae) in Itasca State Park, Minnesota, USA. Phycologia 46:214–229
Johnston ET, Lim P-E, Buhari N et al (2014) Diversity of freshwater red algae (Rhodophyta) in Malaysia and Indonesia from morphological and molecular data. Phycologia 53:329–341
Juttner I, Chimonides PDJ, Ormerod SJ et al (2010) Ecology and biogeography of Himalayan diatoms: distribution along gradients of altitude, stream habitat and water chemistry. Fund Appl Limnol 177:293–311
Keil EJ, Macy TR, Kwandrans J et al (2015) Phylogeography of *Batrachospermum gelatinosum* (Batrachospermales, Rhodophyta) shows postglacial expansion in Europe. Phycologia 54:176–182

Kilroy C, Unwin M (2011) The arrival and spread of the bloom-forming, freshwater diatom, *Didymosphenia geminata*, in New Zealand. Aquat Invas 6:249–262

Kociolek JP, Spaulding SA (2000) Freshwater diatom biogeography. Nov Hedw 71:223–241

Kristiansen J (1996) Dispersal of freshwater algae—a review. Hydrobiologia 336:151–157

Kristiansen J (2001) Biogeography of silica-scaled chrysophytes. Nov Hedw 122:23–39

Kumano S (2002) Freshwater Red Algae of the World. Biopress Ltd., Bristol

Kumano S, Bowden-Kerby WA (1986) Studies on the freshwater Rhodophyta of Micronesia. I. Six new species of Batrachospermum Roth. Jap J Phycol 34:107–128

Lam DW, Entwisle TJ, Eloranta P et al (2012) Circumscription of species in the genus *Sirodotia* (Batrachospermales, Rhodophyta) based on molecular and morphological data. Eur J Phycol 47:42–50

Lam DW, García-Fernández M, Aboal M et al (2013) *Polysiphonia subtilissima* (Ceramiales, Rhodophyta) from freshwater habitats in North America and Europe is confirmed as conspecific with marine collections. Phycologia 52:156–160

Leone PB, Cerda J, Sala S et al (2014) Mink (*Neovison vison*) as a natural vector in the dispersal of the diatom *Didymosphenia geminata*. Diatom Res 29:259–266

Liu J, Soininen J, Han BP et al (2013) Effects of connectivity, dispersal directionality and functional traits on the metacommunity structure of river benthic diatoms. J Biogeogr 40:2238–2248

Lomolino MV, Riddle BR, Whittaker RJ et al (2010) Biogeography, 4th edn. Sinauer Associates, Sunderland

Maguire B (1963) The passive dispersal of small aquatic organisms and their colonization of isolated bodies of water. Ecol Monogr 33:161–185

Marks JC, Cummings MP (1996) DNA sequence variation in the ribosomal internal transcribed spacer region of freshwater *Cladophora* species (Chlorophyta). J Phycol 32:1035–1042

McClintic AS, Casamatta DA, Vis ML (2003) A survey of the algae from montane cloud forest and alpine streams in Bolivia: Macroalgae and associated microalgae. Nov Hedw 76:363–379

Meiers ST, Proctor VW, Chapman RL (1999) Phylogeny and biogeography of *Chara* (Charophyta) inferred from 18S rDNA sequences. Aust J Bot 47:347–360

Mills EL, Leach JH, Carlton JT et al (1993) Exotic species in the Great Lakes: a history of biotic crises and anthropogenic introductions. J Great Lakes Res 19:1–54

Morales EA, Fernandez E, Kociolek PJ (2009) Epilithic diatoms (Bacillariophyta) from cloud forest and alpine streams in Bolivia, South America 3: diatoms from Sehuencas, Carrasco National Park, Department of Cochabamba. Acta Bot Croat 68:263–283

Moreno JL, Aboal M, Monteagudo L (2012) On the presence of *Nostochopsis lobata* Wood ex Bornet et Flahault in Spain: morphological, ecological and biogeographical aspects. Nov Hedw 95:373–390

Müller KM, Cole KM, Sheath RG (2003) Systematics of *Bangia* (Bangiales, Rhodophyta) in North America II. Biogeographic trends in karyology: chromosome numbers and linkage with gene sequence phylogenetic trees. Phycologia 42:209–219

Naselli-Flores L, Padisák J (2016) Preface: Biogeography and spatial patterns of biodiversity of freshwater phytoplankton. Hydrobiologia 764:1–2

Necchi O Jr (1990) Revision of the genus *Batrachospermum* Roth (Rhodophyta, Batrachospermales) in Brazil. Biblio Phycol 84:1–201

Necchi O Jr, Vis ML (2012) Monograph of the genus *Kumanoa* (Rhodophyta, Batrachospermales). J Cramer, Stuttgart

Necchi O Jr, Pascoaloto D, Branco CCZ et al (1997) Stream macroalgal flora from the northwest region of São Paulo State, southeastern Brazil. Algol Stud 84:91–112

Necchi O Jr, Branco CCZ, Branco LHZ (1999) Distribution of Rhodophyta in streams from Sao Paulo State, southeastern Brazil. Arch Hydrobiol 147:73–89

Necchi O Jr, Vis ML, Oliveira MC (2010) Phylogenetic relationships in *Kumanoa* (Batrachospermales, Rhodophyta) species in Brazil with the proposal of *Kumanoa amazonensis* sp nov. Phycologia 49:97–103

Necchi O Jr, Silva A, Fo G, Salomaki ED et al (2013) Global sampling reveals low genetic diversity within *Compsopogon* (Compsopogonales, Rhodophyta). Eur J Phycol 48:152–162
Novis PM (2004) New records of *Spirogyra* and *Zygnema* (Charophyceae, Chlorophyta) in New Zealand. New Zealand J Bot 42:139–152
Oberholster PJ, Botha A-M, Muller K et al (2005) Assessment of the genetic diversity of geographyically unrelated *Microcystis aeruginosa* strains using amplified fragment length polymorphisms (AFLPs). Afr J Biotech 4:389–399
Patino R, Dawson D, VanLandeghem MM (2014) Retrospective analysis of associations between water quality and toxic blooms of golden alga (*Prymnesium parvum*) in Texas reservoirs: Implications for understanding dispersal mechanisms and impacts of climate change. Harmful Algae 33:1–11
Potapova MG, Charles DF (2002) Benthic diatoms in USA rivers: distributions along spatial and environmental gradients. J Biogeogr 29:167–187
Potapova M, Charles DF (2003) Distribution of benthic diatoms in US rivers in relation to conductivity and ionic composition. Freshw Biol 48:1311–1328
Proctor VW (1963) Viability of *Chara* oospores taken from migratory water birds. Ecology 43:528–529
Rimet F (2012) Recent views on river pollution and diatoms. Hydrobiologia 683:1–24
Roscher JP (1967) Alga dispersal by muskrat intestinal contents. Trans Am Microsc Soc 86:497–498
Rott E, Holzinger A, Gesierich D et al (2010) Cell morphology, ultrastructure, and calcification pattern of *Oocardium stratum*, a peculiar lotic desmid. Protoplasma 243:39–50
Rott E, Hotzy R, Cantonati M et al (2012) Calcification types of *Oocardium stratum* Nageli and microhabitat conditions in springs of the Alps. Freshw Sci 31:610–624
Sabater S, Roca JR (1992) Ecological and biogeographical aspects of diatom distribution in Pyrenean springs. Br Phycol J 27:203–213
Sahu N, Tangutur AD (2015) Airborne algae: overview of the current status and its implications on the environment. Aeobiologia 31:89–97
Salomaki ED, Kwandrans J, Eloranta P, Vis ML (2014) Molecular and morphological evidence for Sheathia gen. nov. and three new species. J Phycol 50:526–542
Saunders GW, Necchi O Jr (2002) Nuclear rDNA sequences from *Ballia prieurii* support recognition of *Balliopsis* gen. nov. in the Batrachospermales (Florideophyceae, Rhodophyta). Phycologia 41:61–67
Schlichting H (1960) The role of waterfowl in the dispersal of algae. Trans Am Microsc Soc 79:160–166
Schlichting H (1961) Viable species of algae and protozoa in the atmosphere. Lloydia 24:81–88
Schloesser DW, Hudson PL, Nichols SJ (1986) Distribution and habitat of *Nitellopsis obtusa* (Characeae) in the Laurentian Great Lakes. Hydrobiologia 133:91–96
Sharma NK, Rai AK, Singh S et al (2007) Airborne algae: their present status and relevance. J Phycol 43:615–627
Shea TB, Sheath RG, Chhun A et al (2014) Distribution, seasonality and putative origin of the non-native red alga *Bangia atropurpurea* (Bangiales, Rhodophyta) in the Laurentian Great Lakes. J Great Lakes Res 40:27–34
Sheath RG (1984) The biology of freshwater red algae. Prog Phycol Res 3:89–157
Sheath RG (1987) Invasions into the Laurentian Great Lakes by marine algae. Arch Hydrobiol 25:165–187
Sheath RG, Cole KM (1992) Biogeography of stream macroalgae in North America. J Phycol 28:448–460
Sheath RG, Muller KM (1997) Distribution of stream macroalgae in four high arctic drainage basins. Arctic 50:355–364
Sheath RG, Vis ML (2013) Biogeography of Freshwater Algae. In: eLS. John Wiley and Sons, Ltd DOI: 10.1002/9780470015902.a0003279.pub3
Sheath RG, Vis ML, Cole KM (1993) Distribution and systematics of the freshwater red algal family Thoreaceae in North America. Eur J Phycol 28:231–242

Sheath RG, Vis ML, Hambrook JA et al (1996) Tundra stream macroalgae of North America: Composition, distribution and physiological adaptations. Hydrobiologia 336:67–82
Sherwood AR (2006) Stream macroalgae of the Hawaiian Islands: A floristic survey. Pac Sci 60:191–205
Sherwood AR, Wang N, Carlile AL et al (2012) The Hawaiian freshwater algal database (HfwADB): a laboratory LIMS and online biodiversity resource. BMC Ecol 12:1–7
Sherwood AR, Carlile AL, Newmann JM et al (2014a) The Hawaiian freshwater algae biodiversity survey (2009–2014): systematic and biogeographic trends with an emphasis on the macroalgae. BCM Ecol 14:28
Sherwood AR, Conklin KY, Liddy ZJ (2014b) What's in the air? Preliminary analyses of Hawaiian airborne algae and land plant spores reveal a diverse and abundant flora. Phycologia 53:579–582
Sherwood AR, Jones CA, Conklin KY (2014c) A new species of *Kumanoa* (Batrachospermales, Rhodophyta) from Koke'e State Park, Kaua'i, Hawaii. Pac Sci 68:577–585
Sirodot S (1884) Les Batrachospermes: Organisation, Fonctions, Développement, Classification, Paris, Librairie de l'Académie de Médecine, G. Masson, 299 p., 50 pl.
Siver PA, Lott AM (2012) Biogeographic patterns in scaled chrysophytes from the east coast of North America. Freshw Biol 57:451–466
Skaloud P, Skaloudova M, Pichrtova M et al (2013) www.chrysophytes.eu—a database on distribution and ecology of silica-scaled chrysophytes in Europe. Nov Hedw 142:141–146
Smucker NJ, Edlund MB, Vis ML (2008) The distribution, morphology, and ecology of a non-native species, *Thalassiosira lacustris* (Bacillariophyceae), from benthic stream habitats in North America. Nov Hedw 87:201–220
Soininen J (2007) Environmental and spatial control of freshwater diatoms—a review. Diatom Res 22:473–490
Stancheva R, Sheath RG, Hall JD (2012) Systematics of the genus *Zygnema* (Zygnematophyceae, Charophyta). J Phycol 48:409–422
Stancheva R, Hall JD, McCourt RM et al (2013) Identity and phylogenetic placement of *Spirogyra* species (Zygnematophyceae, Charophyta) from California streams and elsewhere. J Phycol 49:588–607
Stomp M, Huisman J, Mittelbach GG et al (2011) Large-scale biodiversity patterns in freshwater phytoplankton. Ecology 92:2096–2107
Strunecky O, Komarek J, Elster J (2012) Biogeography of *Phormidium autumnale* (Oscillatoriales, Cyanobacteria) in western and central Spitsbergen. Polish Polar Res 33:369–382
Tyler PA (1996) Endemism in freshwater algae. Hydrobiologia 336:127–135
Van Eaton AR, Harper MA, Wilson CJN (2013) High-flying diatoms: widespread dispersal of microorganisms in an explosive volcanic eruption. Geology 41:1187–1190
Van Gremberghe I, Van der Gucht K, Vanormelingen P et al (2011) Genetic diversity of *Microcystis* blooms (Cyanobacteria) in recently constructed reservoirs in Tigray (Northern Ethiopia) assessed by rDNA ITS. Aquat Ecol 45:289–306
van Overeem MA (1937) On green organisms occurring in the lower troposphere. Travaux Bot Neerlandais 34:388–442
Velasquez GT (1940) On the viability of algae obtained from the digestive tract of the Gizzard Shad, *Dorosoma cepedianum*. Am Midl Nat 22:376–412
Verb RG, Bixby RJ, Casamatta DA et al (2006) Checklist of algal taxa from Ohio's unglaciated Western Allegheny Plateau exclusive of the Ohio River', Ohio Biological Survey Miscellaneous Contribution No. 11.
Vis ML, Entwisle TJ (2000) Insights into Batrachospermales (Rhodophyta) phylogeny from *rbc*L sequence data of Australian taxa. J Phycol 36:1175–1182
Vis ML, Sheath RG (1997) Biogeography of *Batrachospermum gelatinosum* (Batrachospermales, Rhodophyta) in North America based on molecular and morphological data. J Phycol 33:520–526
Vis ML, Sheath RG, Hambrook JA et al (1994) Stream macroalgae of the Hawaiian Islands: a preliminary study. Pac Sci 48:175–187

Vis ML, Sheath RG, Chiasson WB (2004) A survey of the Rhodophyta and associated macroalgae from coastal streams in French Guiana. Cryptogam Algol 25:161–174

Vis ML, Entwisle TJ, West JA et al (2006) *Ptilothamnion richardsii* (Rhodophyta) is a chantransia stage of *Batrachospermum*. Eur J Phycol 41:125–130

Vis ML, Feng J, Chiasson WB et al (2010) Investigation of the molecular and morphological variability in *Batrachospermum arcuatum* (Batrachospermales, Rhodophyta) from geographically distant locations. Phycologia 49:545–553

Wang W-L, Chou J-Y (2004) Biogeography of *Lychnothamnus barbatus* (Charophyta): molecular and morphological comparisons with emphasis on a newly discovered population from Taiwan. Crypto Algol 27:461–471

Wehr JD (2015) Brown Algae. In: Wehr JD, Sheath RG, Kociolek JP (eds) Freshwater algae of North America, ecology and classification. Academic, San Diego, pp 851–871

Wehr JD, Perrone AA (2003) A new record of *Heribaudiella fluviatilis*, a freshwater brown alga (Phaeophyceae), from Oregon. Western North Am Nat 63:517–523

Wehr JD, Stancheva R, Truhn C et al (2013) Discovery of the rare freshwater brown alga *Pleurocladia lacustris* (Ectocarpales, Phaeophyceae) in California streams. Western North Am Nat 73:148–157

West Virginia Department of Environmental Protection (http://www.dep.wv.gov/WWE/watershed/wqmonitoring/Pages/DunkardCreekFishKillInformation.aspx) Accessed 23 August 2014

Zamor RM, Franssen NR, Porter C et al (2014) Rapid recovery of a fish assemblage following an ecosystem disruptive algal bloom. Freshw Sci 33:390–401

Zulkifly SB, Graham JM, Young EB et al (2013) The genus *Cladophora* Kützing (Ulvophyceae) as a globally distributed ecological engineer. J Phycol 49:1–17

Chapter 11
Diatoms as Bioindicators in Rivers

Eduardo A. Lobo, Carla Giselda Heinrich, Marilia Schuch, Carlos Eduardo Wetzel, and Luc Ector

Abstract Diatoms have been widely used to detect changes in streams and rivers water quality due to their specific sensibility to a variety of ecological conditions. Their tolerances and preferences for pH, conductivity, salinity, humidity, organic matter, saprobity, trophic state, oxygen requirements, nutrients, and current velocity in freshwater streams, rivers, lakes, wetlands, and estuaries have been defined, and diatoms have also been used in paleolimnological studies. Biotic indices using diatoms based on the relative abundance of the species weighted by their autoecological values have been developed worldwide, though indices of biotic integrity based on periphyton, diatoms, non-diatom "soft" algae, including cyanobacteria, macroalgae, and macrophytes assemblages have been also developed for biological monitoring. A new approach for water quality evaluation utilizing diatoms has been increasing significantly in recent years, by applying molecular techniques using DNA sequences. Molecular identification has the potential to provide revolutionary discoveries in taxonomy that may have great benefits for bioassessment. This chapter provides an overview of the state of the art of studies related to river quality evaluation using epilithic diatom communities worldwide. Most studies highlight the use of biotic indices to summarize floristic data to assess pollution effects on aquatic communities.

Keywords Biotic indices • Epilithic diatom • Molecular analysis • River ecosystem • Water quality

E.A. Lobo (✉) • C.G. Heinrich • M. Schuch
Laboratory of Limnology, University of Santa Cruz do Sul,
Av. Independência, 2293, Universitário, Santa Cruz do Sul, RS 96815-900, Brazil
e-mail: lobo@unisc.br

C.E. Wetzel • L. Ector
Department Environmental Research and Innovation (ERIN), Luxembourg Institute of Science and Technology (LIST), 41 rue du Brill, 4422 Belvaux, Luxembourg

O. Necchi Jr. (ed.), *River Algae*, DOI 10.1007/978-3-319-31984-1_11

Table 11.1 Relation between saprobic index (SI) and water quality (Pantle and Buck 1955)

SI	Pollution levels
1.0–1.5	Oligosaprobic (negligible pollution)
1.5–2.5	β-Mesosaprobic (weak organic pollution)
2.5–3.5	α-Mesosaprobic (strong organic pollution)
3.5–4.0	Polysaprobic (very strong organic pollution)

Biological Evaluation of Water Quality

Among the first attempts at a biological evaluation of water quality is the remarkable work of Kolkwitz and Marsson (1908), which created the first saprobic system through the recognition of a large number of indicator organisms for specific areas that were organically polluted. According to Sládeček (1973), a saprobic system can be defined as a system of aquatic organisms (bacteria, plants and animals) that indicate different water quality levels by their presence. The concept of biological indicators was originated from the saprobic system proposed by Kolkwitz and Marsson (1908), who developed the idea of saprobity in rivers as a measure of the degree of contamination by organic matter (primarily sewage) and the resulting decrease in dissolved oxygen. The authors recognized five pollution zones using characteristic species, which were termed katharobic (clean unpolluted waters), oligosaprobic (a zone of high oxygen content and diverse flora and fauna), α- and β-mesosaprobic (zones with an intermediate degree of pollution), and polysaprobic (a zone grossly polluted by organic matter).

Another important contribution to the system of saprobic organisms was made by Fjerdingstad (1964), who increased the number of characteristic zones of pollution stages from the five originally recognized to a total of nine. In this system, the author completely omitted the animal component of the communities, characterizing the flowing waters according to the phytobenthos (algae and partly bacteria). Numerous variations have been introduced since Kolkwitz and Marsson (1908) first proposed their system (e.g., Butcher 1947; Kolkwitz 1950; Liebmann 1951; Sládeček 1965), but many are terminological variants of the basic classification originally proposed. Since the sixth decade of the past century, attempts to evolve numerical systems for evaluating water quality on the basis of saprobic organism systems have been developed. Among the several proposals, one of the most important contributions is the saprobic index "SI" introduced by Pantle and Buck (1955). The equation for the saprobic index (SI) is given by

$$\mathrm{SI} = \frac{\sum (sh)}{\sum h}$$

where "*s*" is the saprobic value of individual species, and "*h*" is the frequency of occurrence of each species (rare, $h=1$; frequent, $h=3$; and abundant, $h=5$). The SI is given as a weighted average of "*s*" of all taxa with "*h*" as the weight. The value of SI ranges from 1 to 4 for aquatic environments (Table 11.1). Another important contribution is that by Zelinka and Marvan (1961), who introduced the term "saprobic valency" of the species. This concept is based on the notion that no single indicator species will be representative of only one saprobic zone; instead, its distribution will follow a normal curve over a range of zones reflecting its tolerance. The shape and area of this distribution curve defines the "saprobic valency" of the species. Among the list of saprobic values that have been published, one of the most notable is that of Sládeček (1973), which contains information for approximately 2000 species.

Different biological communities have been used to assess and monitor the quality of freshwater. Among them, epilithic diatoms have been widely used as efficient indicators for evaluating water quality, considering that they respond quickly to environmental changes, especially organic pollution and eutrophication, with a broad spectrum of tolerance, from oligotrophic to eutrophic conditions (Rimet 2012; Álvarez-Blanco et al. 2013; Lobo 2013; Lobo et al. 2014, 2015). Additionally, diatoms are one of the key groups of organisms recommended by the Water Framework Directive introduced in the European Union in 2000 (European Union 2000) for the identification of ecological quality gradients in rivers.

Difficulties of using epilithic diatoms in biomonitoring should also be addressed. These difficulties include the problem of the live/dead cell ratio in samples and the necessity of a good taxonomic knowledge for their identification. Concerning the first problem, experiments were conducted by Gillett et al. (2009), which showed that counts performed with live diatoms (with chloroplasts) were not significantly different from counts performed by conventional methods (cleaning of samples with sulfuric and hydrochloric acid and mounting them on permanent slides). As for the second problem, little attention has been focused on the uncertainties inherent to the use of diatoms to evaluate streams and rivers water quality, such as species identification and counting (Besse-Lototskaya et al. 2006). In fact, Morales et al. (2001) noted that identification of diatom frustules is a vital step for ecological analyses, because the remarkable difference between the use of light microscopy (LM) and scanning electron microscopy (SEM) reveals inconsistencies in diatom identification, especially at low magnification using LM. However, SEM research has had a marked impact on diatom taxonomy, rendering traditional identification methods insufficient for the recognition of newly created taxa. Many diagnostic characteristics of small representatives of these newly created taxa cannot be recognized even when using high LM resolution. In this manner, the loss of taxonomic resolution at the LM level could lead to overestimation of geographical distributions and ranges of tolerance of environmental parameters.

In terms of counting, when identification problems arise using LM, adjustments using SEM counting can be made, as described by Lobo et al. (1990). Working at the Chikuma River, Nagano prefecture, Japan, the authors had difficulty in identifying two *Nitzschia* species, which are often found together in the same sample using LM. They have similar shape, size, striations, and fibulae density; however, they belong to different ecological groups for water quality evaluation. This is the case of *Nitzschia hantzschiana* Rabenhorst and *Nitzschia romana* Grunow. According to Kobayasi and Mayama (1989), the former belongs to Group B (less pollution-tolerant taxa), while the latter belongs to Group C (pollution-sensitive taxa); therefore, an accurate distinction between them is extremely important for biomonitoring. This distinction, however, is possible only by examining the SEM structure. In this case, by using SEM, a clear distinction between these two species can be observed in the striae on the raphe canal. *N. romana* has bifurcated striae with furcate branches composed of two areolae, while *N. hantzschiana* has bifurcated striae with furcate branches composed of a single areola (Kobayasi 1985). From the relative abundance (percentage) of each taxon obtained in SEM counting, a new number of valves is estimated from the total sum of valves recorded in LM counting. Thus, the total number of individuals remains the same, and only the distribution of the relative abundances of each taxon is adjusted.

The use of SEM as a necessary complement to LM for diatom identification and counting in ecological research is highly recommendable as noted by Lobo et al. (1990). Additionally, the use of diatom collections is also of great interest to diatom taxonomists in the resolution of taxonomic issues (e.g., Wetzel and Ector 2014a, b; Wetzel et al. 2015). Through resolving the taxonomic aspects of a given species, its autecology can also be clarified. A clear taxonomy and autecology have favorable repercussions on applied fields such as biomonitoring of freshwaters (KAHLERT et al. 2009; Dreßler et al. 2015).

Studies Concerning Water Quality Evaluation Using Diatoms

The approaches related to water quality evaluation in streams and rivers using diatoms were divided into four categories by Lobo et al. (1995): biotic indices, multivariate analyses, diversity indices, and species-abundance relationship analyses (Fig. 11.1). Diatom communities can be described by their species composition and diversity. Species composition may be analyzed using biotic indices (e.g., Pantle and Buck's Index) and also by multivariate methods, such as two-way indicator species analysis (TWINSPAN) for classification or principal component analysis (PCA) for ordination. Then, the assemblage can be described by numeric indices and/or species compositional patterns. Similarly, the species diversity can be analyzed with diversity indices, such as Shannon's Index, and also by species-abundance relationship models, such as the log-normal distribution. Then, the assemblage can be described by species diversity measures and/or species-abundance patterns. Note that in this scheme, the species composition not only consists of a list of species that constitutes the assemblage but also involves their relative abundance, thus incorporating aspects of the community structure. A structural component of the assemblage is also incorporated in species diversity measures, particularly the species abundance approach.

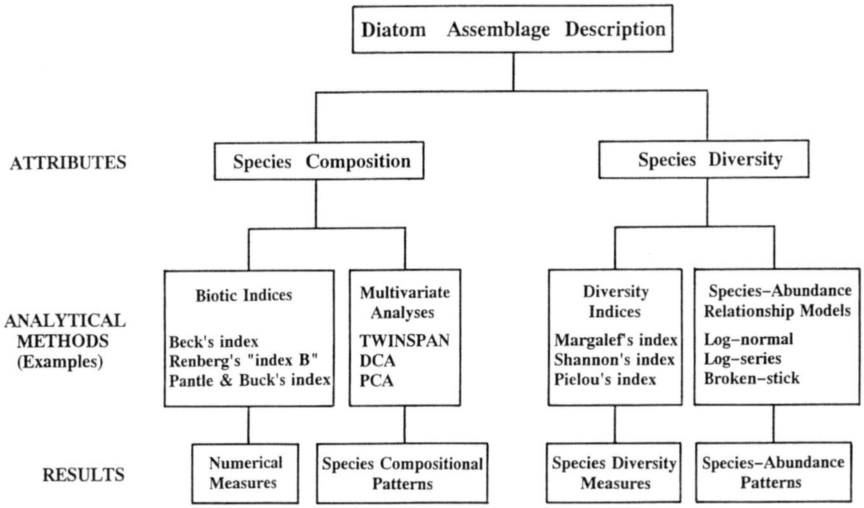

Fig. 11.1 Scheme for describing diatom assemblages (LOBO et al. 1995)

The Biotic Indices Approach

According to Stevenson et al. (2010), at least two basic approaches have been developed to assess environmental conditions in rivers and streams using diatoms. The first approach utilizes the weighted average autecological indices which are defined as indices using an ordinal scale for the species traits with six or less ranks for a specific environmental stressor (e.g., pH, nutrient requirements, organic pollution). Diatom-based autecological indices can be particularly valuable in stream and river assessments because one-time assay of species composition of diatom assemblages in streams could provide better characterizations of physical and chemical conditions than one-time measurement of those conditions. According to the same authors, the second approach establishes that managing stream and river ecosystems calls for an assessment of integrity of the ecosystem and a diagnosis of causes of degradation. Indices of biotic integrity (IBI), often called multimetric indices of biological condition (IBC), have been recently developed for diatoms (e.g., Hill et al. 2000, 2003; Fore and Grafe 2002; Kelly et al. 2009).

Thus, in this chapter we basically provide an overview of the state-of-the-art of studies related to river quality evaluation based on biotic indices using epilithic diatom communities worldwide. In fact, the use of biotic indices to summarize floristic and faunistic data in a numerical form to assess pollution effects on aquatic communities has drawn the attention of scientists since the beginning of the twentieth century. The indices consider the sensitivity or tolerance of individual species or groups to pollution. A value is assigned to each species or group based on its tolerance, and the sum of the values within a sample is used to produce a mathematical expression (index) of pollution for the investigated site.

Table 11.2 Diatoms groups according to their distribution along the organic pollution gradient in Japan (Kobayasi and Mayama 1989)

Groups	Tolerance levels
A	(Most pollution-tolerant group)
B	(Less pollution-tolerant group)
C	(Pollution-sensitive group)

The diatom indices are based on quantitative studies (relative abundances or density) of individual sites ranging from clean to poor quality classes. From such data, tables have been produced to show the tolerance of diatom taxa to environmental pollution factors (e.g., Lange-Bertalot 1979; Leclercq and Maquet 1987; Descy and Coste 1990). These types of classifications are essential to calculate numerical indices of water pollution, such as Pantle and Buck's index (Pantle and Buck 1955).

In Japan, the biotic indices approach has been one of the most widely used methods, and some numerical indices for monitoring flowing waters have been devised considering the tolerance of diatom taxa to organic pollution. For example, Watanabe et al. (1985) developed a numerical analysis of the tolerance of diatoms to organic water pollution, classifying them into three groups: saprophilous (pollution-sensitive species), eurysaprobic (indifferent species) and saproxenous (pollution-tolerant species). Based on this classification, the Diatom Assemblage Index to organic water pollution (DAIpo) was proposed. This index has been widely used as a numerical index for monitoring organic pollution in Japanese freshwaters (e.g., Watanabe et al. 1985, 1988; Watanabe and Asai 1992).

Towards the end of the 1980s, a practical method for evaluating water quality was proposed by Kobayasi and Mayama (1989). With this method, diatoms are classified into three differentiated groups according to their distribution along the organic pollution gradient (Table 11.2). Based on this ecological grouping, Pantle and Buck's (1955) saprobic index is employed to express the degree of pollution. This classification is considered to be equivalent to categories 1–3 of the diatom grouping developed in Germany by Lange-Bertalot (1979); however, noticeable differences were detected in the species composition within the groups, likely because Japanese rivers, where the method was developed, are shallower and steeper than European rivers, and consequently the epilithic diatom communities show different ecological adaptabilities to the morphometric and hydrological characteristics of Japanese rivers. Katoh (1991), after comparing the ability of several biotic indices to indicate the levels of water pollution, concluded that Pantle and Buck's index is the best water pollution index calculated from diatom community data.

Mayama et al. (2011) developed the SimRiver application software to evaluate water quality in Japan using diatoms. The program allows the investigator to create river environments in which the land use, population density, presence of sewage treatment plants, and season of the year are variables that can be manipulated (Fig. 11.2). According to Lobo et al. (2014), the investigator evaluates the environmental impact using the diatom community generated by the computer, which is affected by the river water quality in the electronic system. Therefore, investigators

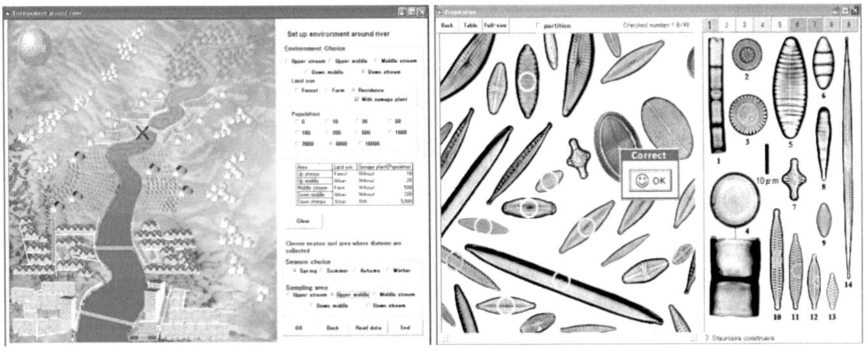

Fig. 11.2 Screen view of the software SimRiver, version 3 (English version). A view showing the set up window for environment along a river (*left*). A view of a computer-synthesized diatom slide equipped with a built-in identification support system (*right*).

that utilize the SimRiver program can determine the relationship between human activities, river water quality, and the diatom communities.

The SimRiver program was originally produced in Japanese, under the coordination of Professor Dr. Shigeki Mayama from the Tokyo Gakugei University of Japan. For international use, however, it was necessary to prepare new versions in other languages. Despite the generalized use of English worldwide, we believe that the use of native languages is important to promote the nation's conscience regarding their aquatic environments. For that reason, we initiated the production of multilingual editions of SimRiver and their support tools based in the original version in Japanese. Currently, researchers of the following countries are participating in this international cooperation network: Brazil, Canada, China, Denmark, Germany, Greece, India, Indonesia, Iran, Japan, Korea, Luxembourg, the Netherlands, the Philippines, Poland, Russia, South Africa, Thailand, Tunisia, Turkey, Ukraine, and the USA. The final products are available at http://www.u-gakugei.ac.jp/~diatom/. In addition, an educational video called "Diatoms: Sampling and Observation, Part 1 to 3" (Fig. 11.3), and a database of photographs showing polluted rivers were produced in Afrikaans, Arabic, Bahasa Indonesia, Chinese, Danish, Dutch, Filipino, French, German, Greek, Hindu, Iranian, Japanese, Kannada, Korean, Marathi, Polish, Portuguese, Russian, Spanish, Tamil, Thai, Turkish, and Ukrainian, which can be accessed at the same site mentioned above.

Diatom Ecological Preferences

Diatoms have been widely used to detect changes in stream and river water quality due to the specific sensibility of this group to a variety of ecological conditions, such as organic matter, pH, salinity, and nutrients (Lange-Bertalot 1979; Ter Braak and Van Dam 1989; Van Dam et al. 1994; Kelly and Whitton 1995). Among these

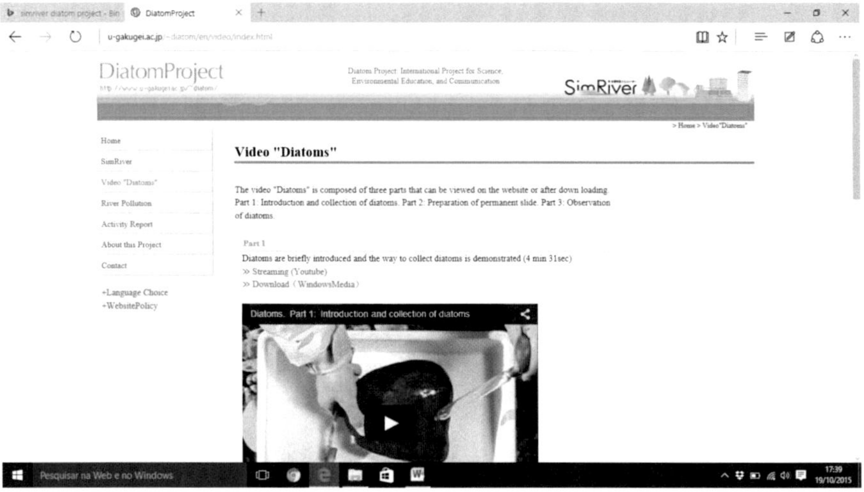

Fig. 11.3 Educational video called "Diatoms: sampling and observation, Part 1 to 3. Part 1: Introduction and diatom sampling". Part 2: Preparation of permanent slides. Part 3: Diatom observation

ecological conditions, the analysis of the relationship between the diatom community and pH has been one of the most studied approaches. This relationship was first studied in lakes by Lowe (1974), Arzet et al. (1986), Dixit et al. (1990), Eloranta (1990), Smith (1990), Battarbee et al. (1997), Van Dam (1997) and Coring (1999). Indices and pH reconstruction models in paleolimnological studies using diatoms have been conducted by Renberg and Hellberg (1982), Ter Braak and Van Dam (1989), Birks et al. (1990), Håkansson (1993) and Vyverman et al. (1995). Similarly, Ziemann (1971, 1991) developed the halobiont index, which utilizes the ecological preferences related to salinity to assess the salt concentration in rivers. Other studies have used classifications related to salinity in lakes and estuaries based on the diatom composition (Juggins 1992; Cumming and Smol 1993; Snoeijs 1994; Wilson et al. 1994, 1997; Gell 1997; Roberts and Mcminn 1998; Underwood et al. 1998; Campeau et al. 1999; Potapova 2011).

Denys (1991a, b) defined the autoecology of 980 fossil diatoms based on 800 samples collected from cores and outcroppings in Holocene deposits along the coastal plain of Belgium. The study defined the tolerances and preferences for saprobity, trophic state, salinity, pH, nitrogen consumption, oxygen requirements, and current velocity. Van Dam et al. (1994) determined the environmental preferences for saprobity, trophic state, salinity, pH, nitrogen, oxygen, and humidity for 948 diatom species in freshwater and lightly brackish lakes in Holland. In a similar manner, Hofmann (1994), working with nine alkaline lakes in the Alps Region of Bavaria (Germany), described preferences for the trophic state, saprobity, conductivity and pH of 200 species among the 487 identified taxa.

Rott et al. (2003) created an extended database of samples collected in 450 sampling points in rivers, including approximately 1000 species from nine different classes of algae. Levels of saprobity (Rott et al. 1997) and trophy (Rott et al. 1999) were defined for 650 diatom species. In South Africa, the main studies on diatom taxonomy and ecology were conducted by Cholnoky (e.g., 1959; 1963). Through his intensive and extensive studies, he created a diatom collection that is housed in the National Institute for Water Research (CSIR) in Pretoria and is used as a reference for research of diatoms in this country (Taylor et al. 2005). Starting in the last decade, research related to diatoms in South Africa focused on the ecological aspect of the community to assess water quality, and testing was conducted by applying the European indices under South African conditions (Bate et al. 2004).

Approaches Developed in USA

Rimet (2012) made extensive revisions to describe the state of the art on studies of diatoms and river pollution between 1999 and 2009. Using a lexical analysis, the author aimed to identify similar groups (related themes) between papers published that used key words in the abstract such as "diatoms," "periphyton," "water quality evaluation," and "bioindicators," among others. In total, 226 indexed papers were selected. The results of this review indicated that the papers were published in 75 different journals, highlighting "*Hydrobiologia*" as the journal that published the highest number of papers (16.7 % of the total). Seven groups of working themes were identified: Group 1, predictive models (16 papers); Group 2, spatial structure of the diatom community (12 papers); Group 3, approaches using diatoms (18 papers); Group 4, biotic indices of diatoms (101 papers); Group 5, biotic indices of diatoms in South Africa (10 papers); Group 6, ecotoxicology (39 papers); and Group 7, effect of land use on the hydrographical basin (34 papers).

Concerning Group 4 (biotic indices of diatoms with 101 papers), as the principal focus of this chapter, the author cited the work of Bahls et al. (2008), who gave a historical overview of diatom bioassessment in the state of Montana (USA). This study highlighted that the metrics most commonly used in the 1970s were the Shannon diversity; a pollution index based on the classification by Lange-Bertalot (1979); a sedimentation index based on the percentage of diatoms in the motile genera *Navicula* Bory, *Nitzschia* Hassall, and *Surirella* Turpin; and a community similarity index (Whittaker 1952). Additional metrics were used in the 1990s, including species richness, a disturbance index using the percentage of occurrence of the species *Achnanthidium minutissimum* (Kützing) Czarnecki, and the percent of abnormal valves. The authors concluded that none of the existing biocriteria can be used to reliably distinguish between negative environmental conditions (disturbed) and positive conditions (undisturbed) due to a high level of metric sensitivity (even the best sites show some degree of disturbance) and a high degree of overlap in the distribution of streams within each of these groups. These metrics, however, can be used to diagnose the cause of disturbances, but require interpretation by a qualified diatom ecologist. Beginning in 2000, authors incorporate the concept of an "ecoregion," defined as areas with similar geology, climate, vegetation and other factors that affect water quality and aquatic biological communities. Ecoregions serve as a spatial framework for monitoring, assessing, and managing water quality, including the development of biological criteria. It is important to note that some ecoregional classifications work better with diatom communities than others; for example, Leland and Porter (2000) demonstrated that ionic concentrations and substratum types were important factors determining diatom community variation in the Illinois river. Geology is often cited as a main factor in explaining diatom community variability on a large geographical scale, as demonstrated by Urrea and Sabater (2009) in Spain, but it is also important when considering small areas, as noted by Robinson and Kawecka (2005) in Swiss alpine streams.

Potapova and Charles (2007) and Potapova et al. (2004) worked with the US Geological Survey National Water-Quality Assessment program data to create diatom metrics for monitoring eutrophication. They showed that these metrics provide better assessments in USA rivers than similar metrics developed for European inland waters. According to Kelly et al. (1998), this is due not only to the floristic differences among regions, but also to the environmental differences that modify species responses to water-quality characteristics. Their results showed that metrics created specifically for USA rivers are better suited for water-quality assessment than those developed for other geographic areas. This is consistent with findings of several authors that diatom indices/metrics developed in certain parts of Europe are not effective when used in other areas of the same continent (Kelly et al. 1998; Pipp 2002; Rott et al. 2003).

According to Smucker and Vis (2013), benthic diatoms are frequently used for biological monitoring of rivers and streams throughout the USA (e.g., Stevenson et al. 2008; Danielson et al. 2012). Indices of biotic integrity based on periphyton, diatoms, non-diatom "soft" algae and cyanobacteria assemblages have been also developed in USA (e.g., Hill et al. 2000, 2003; Griffith et al. 2005; Zalack et al. 2010; Fetscher et al. 2014). Smucker et al. (2013) used a variety of statistic tests to characterize ecological responses and to develop concentration-based nutrient criteria for rivers in Connecticut, where urbanization is the primary cause of watershed alteration. Their results indicated that management practices and decisions at the watershed scale will likely be important for improving degraded streams and conserving high quality streams.

Surface-sediment samples of lakes have also been evaluated, as demonstrated by Stevenson et al. (2013) by developing a multimetric index (MMIs) of Lake Diatom condition (LDCI), based on lake samples from the continental USA. According to the authors, the LDCI is one of the first diatom-based MMIs of biological condition of lakes, and the first based on surface-sediment diatoms. Stevenson (2014) reviewed the use of algae in bioassessment highlighting that we need to better understand the effect of human activities and resulting global change on the biodiversity of algae and the significance of loss in biodiversity for ecosystem goods and services. He states that large-scale assessment programs are providing data with sufficient detail, sample size, and scale to fuel an explosion in ecological knowledge. In combination with experimental and modeling approaches, these survey data should greatly advance our understanding of algal ecology and leading to new concepts in algal assessment and aquatic resource management.

Approaches Developed in Europe

According to Ector and Rimet (2005) and Rimet (2012), many biotic indices were developed before 1999, such us the Trophic Diatom Index (TDI, Kelly and Whitton 1995) in Great Britain; the Generic Diatom Index (GDI, Rumeau and Coste 1988), the Specific Pollution-sensitive Index (SPI, Cemagref 1982), and the Biological Diatom Index (BDI, Lenoir and Coste 1996) in France; the Eutrophication Pollution Diatom Index (EPI-D, Dell'Uomo 1996) in Italy; the Rott Trophic Index (TI, Rott et al. 1999) in Austria; the Schiefele and Kohmann Trophic Index (Schiefele and Kohmann 1993) in Germany; and the CCE Index (Descy and Coste 1991) in France and Belgium. Even considering that these diatom indices were developed in different regions (e.g., USA, Europe, Japan), the results of pollution biomonitoring were significant and demonstrated the robustness of the diatom indices in other regions they were tested: Africa (Bellinger et al. 2006), Malaysia (Wan Maznah and Mansor 2002), Turkey (Kalyoncu et al. 2009), and Vietnam (Duong et al. 2007).

An extensive survey of sampling sites ranging from undisturbed locations to heavily disturbed sites, covering all river types was made by Tornés et al. (2007) in NE Spain to determine indicator taxa for different ecological statuses of streams and to identify type-specific taxa for high ecological status. The results indicated that the main gradient shows a clear separation of sites in relation to the degree of human influence: polluted rivers (mainly located in the lowlands) differ from rivers in mountainous areas and in the Pyrenees. The authors pointed out that the use of diatom indices as an ecological tool needs to take into account the different autoecological characteristics of the diatom taxa in different regions, and be adapted if it is to provide a reliable diagnosis of specific river systems.

In Europe the Water Framework Directive (WFD) (European Union 2000) establishes a community framework for acting in the field of water policy and has as primary objective of achieving a "good ecological status" of aquatic ecosystems (surface waters, estuarine, coastal and groundwater) by the year of 2015. According to Kelly et al. (2008), good ecological status occurs when the values of biological quality elements for the surface water body type show low levels of distortion resulting from human activity, but deviate only slightly from those normally associated with the surface water body type under undisturbed conditions, which requires the establishment of a network of "reference sites" to be used as a comparison pattern relative to the measured deviations. Five ecological status classes were defined: high, good, moderate, poor, and bad, by comparison to the biota expected in water bodies subject to no environmental impact or minimal anthropogenic alterations. WFD states that for the purpose of classification of ecological status, the results shall be expressed as an Ecological Quality Ratio (EQR) Mancini (2003). The ratio shall be expressed as a numerical value between zero and one, with high ecological

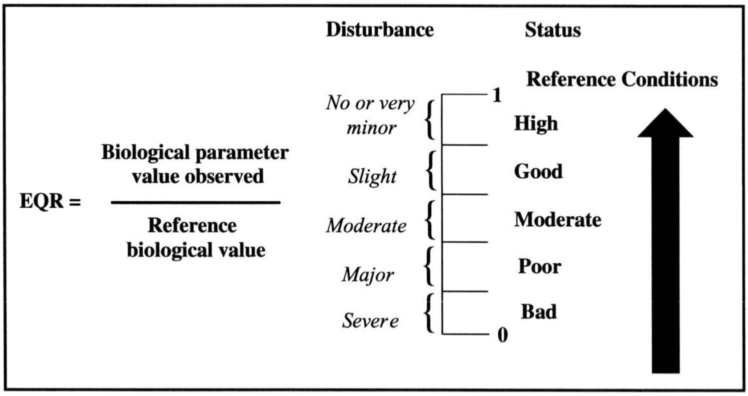

Fig. 11.4 Ecological quality class setting by the Ecological Quality Ratio (EQR). Modified from Mancini (2006)

status represented by values close to one and bad ecological status by values close to zero (Fig. 11.4). Different elements of the aquatic ecosystems can be used as indicators of freshwater biological quality including benthic invertebrates, macrophytes, benthic algae (diatoms), fish and phytoplankton. In addition to these strictly biological elements, relevance is also given to the hydromorphological, physical, and chemical elements that support the biological elements Mancini (2006).

The search for reference sites in European streams and rivers is not straightforward due to the long history of human settlement across much of the continent (Feio et al. 2014). The underlying assumption is that a site free from anthropogenic influences will have a "natural biota," and, in theory, references sites could be selected from a pool of these sites with no human influences (Stoddard et al. 2006). WFD establish that protected areas are the starting point for monitoring plans and must achieve higher levels of biological quality, considering that the main role of protected areas is the maintenance and improvement of the ecological condition of these biological features. In fact, protected areas favor the preservation of biodiversity by preventing the loss of species and play an important ecological role for ecosystem characterization due to their function as a "biological corridor" and source of faunistic recolonization (Pressey et al. 1993). In general, a decrease of the biological quality and function of lotic ecosystems can be observed; therefore, the interest in better organization and planning of protected areas along with the wildlife conservation is increasing (Lowe 2002). According to Osborne and Kovacic (1993), the restoration and sustainable management of protected areas to recover the natural functions of the rivers and reduce pollution sources are essential to improve the biological quality of running waters.

To reach the "good ecological status" of the rivers established in WFD, the definition of a network of "reference sites" is needed. Thus, Beltrami et al. (2012) described diatom assemblages of reference conditions for different stream types of Northern Italy. To characterize the reference assemblages, indicator species were obtained by using the Indicator Species Analysis (IndVal), developed by Dufrêne and Legendre (1997). This analysis establishes indicator values of different species combining information on relative abundances of species in a particular group of samples and the relative frequency of the species present in this group. The results indicated that altitude and geology resulted to be the most important factors influencing diatom assemblage composition, and were used to describe four new stream types. According to the authors, the definition of reference assemblages for each group could be a first step toward the formulation of a new metric that could take in consideration the presence/absence of reference species, following the example of other European countries. Indicator Species Analysis (IndVal) has been used frequently to define indicator species that characterizing different environmental groups worldwide (e.g., Cantonati et al. 2012; Della Bella et al. 2012 in Italy; Bere and Tundisi 2011 in Brazil; Hwang et al. 2011; Cho et al. 2014 in Korea; Lane and Brown 2007 in USA; Szczepocka et al. 2015 in Poland).

In a comparative survey carried out by Hering et al. (2006), periphytic diatoms, macrophytes, benthic macroinvertebrates and fish assemblages were sampled with standard methods in 185 rivers and streams in 9 European countries to determine their response to degradation. Their results indicated that diatom metrics were most strongly correlated to eutrophication gradients in mountain and lowland streams, followed by invertebrate metrics. Responses of the four groups of organisms to other gradients were less strong; all groups responded to varying degrees to land use changes, hydromorphological degradation on the microhabitat scale and general degradation gradients. Fish and macrophyte metrics generally showed a poor response to degradation gradients in mountain rivers and a strong response in lowland rivers.

Approaches Developed in Asia and Oceania

According to Lee et al. (2011), the use of freshwater diatoms in evaluating stream and river ecosystems is quite recent in Korea, although the first report of Korean freshwater diatoms dates from the early 1900s. In the process of developing diatom criteria and assessment tools, the trophic diatom index (TDI) developed by Kelly and Whitton (1995) was selected, highlighting that the values of sensitivity and indicators given in the original TDI were tested with the diatoms occurring in a variety of Korean streams and rivers, and the values for major taxa were amended.

Bae et al. (2011) investigated the relationships of three major aquatic assemblages (diatom, macroinvertebrate, and fish) and environmental variables, including hydrology, land cover, and water quality variables on multiple scales. Samples were collected at 720 sampling sites on the Korean nationwide scale. Their results support the concept of multi-scale habitat filters and functional organization in rivers and streams, and are consistent with the recommended use of multiple biological indices with more than one assemblage for the assessment of the biotic integrity of aquatic ecosystems. Lee et al. (2011) provided an overview of the development and application of the National Aquatic Ecological Monitoring Program (NAEMP) in

Korea, which uses biological and habitat-riparian criteria for stream/river and watershed management. NAEMP is promising in that the results have already had great impacts on policy making and scientific research relevant to lotic water environment and watershed management in Korea.

In China, Tan et al. (2015) developed a benthic diatom index of biotic integrity for ecosystem health assessment in the upper Han River. The assessment further revealed that the main reason for river's degradation was nutrient enrichment through agricultural land use. The authors concluded that diatoms are particularly suitable river health indicators in this study area, corroborating the results of previous studies (e.g. Li et al. 2009; Tan et al. 2013, 2014).

Oeding and Taffs (2015) pointed out that the use of diatoms as bioindicators of water quality is common in temperate regions worldwide. However, less attention has been given to subtropical regions, particularly in Australia. For that reason, the authors studied the relationship between the diatom assemblages and water quality of a subtropical east Australian riverine system to identify a correlation between community composition and water chemistry and to assess the applicability of temperate region based diatom indices to a subtropical river system. They concluded that diatoms are effective indicators of water quality; however, further research is required to develop a diatom biological index applicable to subtropical east Australian river systems to improve the effectiveness of environmental monitoring and sustainable river management.

Approaches Developed in South/Central America

According to Lobo et al. (2004a) diatom communities have been utilized for monitoring and evaluating water quality of streams and rivers in South and Central American countries. The first studies were published only in the second half of the twentieth century and were restricted to research groups located, basically, in Argentina and Brazil. In Argentina, the Pampean plain contains 21 million inhabitants and concentrates the majority of industrial and farming activities, and it is exposed to the most intense use of fertilizers in the country. Furthermore, the increase of urbanization that progressively occupied cultivatable land and caused the displacement of livestock from their traditional areas to marginal lands situated in floodplains has increased the incidence of erosion and the input of particulate material into waterways (INDEC 2010). The streambed of the Pampean lotic ecosystems is composed of fine sediments (clay and silt) that are colonized by an epipelic biofilm in which diatoms are the dominant group, constituting the basal trophic levels for extensive food webs (López van Oosterom et al. 2013). Thus, the design of a local regional index for assessing eutrophication and organic pollution specifically for the Pampean area became necessary, and this index has been designated IDP (Pampean Diatom Index; Gómez and Licursi 2001). For the proposal of this index, 164 epipelic samples were analyzed from 50 sampling sites with different uses of the surrounding land, and their relationship to physicochemical variables was established. This regional index based on the use of epipelon community is an important tool for the study of streams and rivers of the Pampean Plain, where the use of indices developed for other latitudes not always gives adequate results in terms of water quality evaluation. In summary, plain epipelic diatom biofilms have been used as indicators of natural and anthropogenic disturbances in Argentina Pampean (Gómez and Licursi 2001, 2003; Licursi and Gómez 2002, 2004; Gómez et al. 2003, 2008; Sierra and Gómez 2007). In these studies, ecological, physiological, biochemical, and morphological methods were employed to assess the biotic integrity of the biofilms under study (Gómez et al. 2009).

According to Cochero et al. (2015), although laboratory experiments have shown that each diatom species have different levels of tolerance to different stressors, few studies have been conducted in laboratory settings that analyze the responses of the diatom assemblage to the effects of multiple simultaneous variables. Thus, the authors evaluated some structural responses (such as species composition and diversity) of the diatom assemblage on a short time scale to the effects of the simultaneous increase in four variables that are directly linked to the environmental changes affecting the Pampean rivers and streams: turbidity, nutrients (phosphorous and nitrogen), water velocity, and temperature. The results indicated that in the very short term of the bioassay conducted, the diatom assemblage can modify its structure to respond in a sensitive manner to the abrupt changes in multiple physical-chemical variables.

In Brazil, most investigations on the use of diatoms for water quality assessment have focused on the use of biotic indices, especially in the southern region (e.g. Lobo et al. 2002, 2004a, b, 2010, 2014, 2015; Hermany et al. 2006; Salomoni et al. 2006, 2011; Düpont et al. 2007; Bes et al. 2012; Schuch et al. 2012, 2015; Böhm et al. 2013; Lobo 2013; Heinrich et al. 2014). A detailed review of studies on the use of epilithic diatoms as indicator organisms in lotic systems of southern Brazil can be found in Lobo (2013) and Lobo et al. (2014). The first attempt to classify diatom species according to their tolerance to organic pollution was made in rivers of the southern region (Lobo et al. 1996). However, the first saprobic system was published later by Lobo et al. (2002), classifying species according to their tolerance to organic pollution in streams and rivers of the Guaíba Hydrographical Basin, RS in terms of biochemical oxygen demand (BOD_5).

The Biological Water Quality Index (BWQI) was proposed later, which integrated the effects of organic enrichment based on the classification published in 2002 (Lobo et al. 2002), and eutrophication, from indicative values measured in terms of the phosphate variable (PO_4) (Lobo et al. 2004b). However, the calibration of these indexes to assess water quality became extremely important, because of spatial and temporal adaptations of the species to the environment, reflecting their tolerance to water contamination (Seegert 2000; Niemi and McDonald 2004). In this context, Lobo et al. (2014, 2015) published the Trophic Water Quality Index (TWQI) for subtropical and temperate Brazilian lotic systems based on the tolerances to eutrophication of epilithic diatom species for a number of environmental variables (physical, chemical, and microbiological) from regional studies conducted between 2005 and 2013. A total of 140 biological samples and 211 abiotic samples

Table 11.3 Relationship between the Trophic Water Quality Index (TWQI) and the water quality (Lobo et al. 2015)

TWQI	Pollution levels
1.0–1.5	Oligotrophic (negligible pollution)
1.5–2.5	β-Mesotrophic (moderate pollution)
2.5–3.5	α-Mesotrophic (strong pollution)
3.5–4.0	Eutrophic (excessive pollution)

were collected and analyzed by means of multivariate analysis (cluster analysis and canonical correspondence analysis). The index incorporated a total of 70 taxa, of which 16 species showed a low tolerance to eutrophication, 28 species showed an intermediate tolerance and 26 species showed a high tolerance. The TWQI values range from 1 to 4 in aquatic environments (Table 11.3) and represent a new technological tool for environmental monitoring studies of water quality assessment in temperate and subtropical Brazilian lotic ecosystems. Light microscopy images are presented (Fig. 11.5) for some diatom species tolerant to eutrophication worldwide and currently used to calculate the TWQI in Brazil, following Lange-Bertalot (1979), Kobayasi and Mayama (1989), Van Dam et al. (1994), and Lobo et al. (2015).

In Central America, Michels-Estrada (2003) worked with the species composition and ecological requirements of benthic diatom assemblages from various streams and rivers in Costa Rica. The study pointed out that baseline information on the ecology of aquatic systems is urgently needed in the tropics in order to develop new and effective biological methods of monitoring water quality. Limnological and ecological concepts developed for temperate zones have to be evaluated as to their applicability in tropical conditions.

Genetic Tools for Biomonitoring: New Approach

Visco et al. (2015) pointed out that diatoms are widely used as bioindicators for the assessment of water quality in streams and rivers, highlighting that biotic indices are based on the relative abundance of morphologically identified species weighted by their autoecological values. However, obtaining such indices is time-consuming, costly, and requires excellent taxonomic expertise, which is not always available. In this context, the authors tested the possibility to overcome these limitations using a next-generation sequencing (NGS) approach to identify and quantify diatoms found in environmental DNA and RNA samples from river sites in the Geneva area (Switzerland). Their results indicated that the Swiss Diatom Index shows a significant correlation between morphological and molecular data, demonstrating the potential of the NGS approach for identification and quantification of diatoms in environmental samples. In contrast, Manoylov et al. (2009) worked in a stream benthic community in Cedar River, Michigan (USA), with both bacterial and

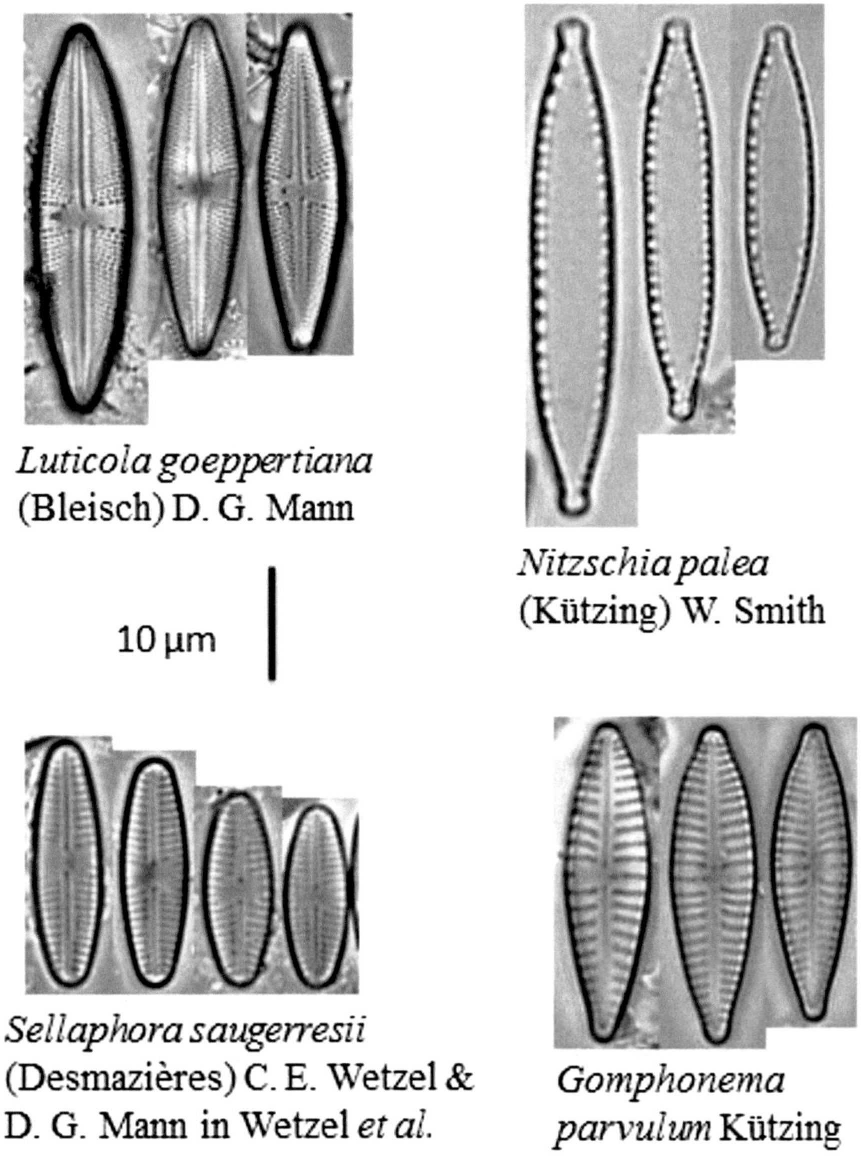

Fig. 11.5 Diatom taxa tolerant to eutrophication worldwide

eukaryotic gene sequences matched with sequences from the database and then were compared with taxa identified with morphological characters. The results indicated that there was no correlation between the dominant taxa identified morphologically and molecularly. The authors concluded that while species-level matching with the two techniques was not as high as expected, the data suggested that with more comprehensive databases there may be a great potential for applications in bioassessment.

Manoylov (2014) argues that currently we cannot adopt molecular bioassessment of algal community composition, considering the low number of samples analysed with molecular tools (e.g., Manoylov et al. 2009; Kermarrec et al. 2013) and the identity of most taxa in the reference libraries of genetic sequences has not been rigorously evaluated. Following the view of Sluys (2013), the author pointed out that the complexity of algal shapes, adaptations, and survival strategies cannot be simply translated into a DNA code because the expression of proteins depends on the unique set of environmental conditions in which algae live. Morphological analyses of algal communities will still remain important as standalone choices for bioassessment, documentation of biodiversity, or an initial stage of proof for the molecular taxonomic identification by expert taxonomists (Lücking 2008). As Bicudo (2004) resolutely quotes "what really matters is that without taxonomy it is impossible to know which species lived yesterday, are living today, and will have the chance to be alive tomorrow in a given area (the balance within a community, rare and dominant species, the cost of the biodiversity, etc.). In conclusion, none of the above will become reality if there is neither taxonomy nor expert taxonomists."

Zimmermann et al. (2015) argue that diatom identification to species level is difficult, time consuming and needs taxonomic expertise. For these reasons, the authors investigated how identification methods based on DNA (metabarcoding using NGS platforms) perform in comparison to morphological diatom identification and proposed a workflow to optimize diatom freshwater quality assessments. The results provide evidence that metabarcoding of diatoms via NGS sequencing has a great potential for water quality assessments and could complement and maybe even improve the identification via light microscopy. According to the authors "another challenge lies in the bioinformatics data handling: the scientist to tackle these challenges in future will neither be a biologist with some insight into informatics nor a computer scientist with biological basics but a true interdisciplinary scientist, who will at least support the communication between the two fields (Yoccoz 2012)."

In short, this alternative approach for water quality evaluation utilizing diatoms by applying molecular techniques using DNA sequences has been increasing significantly in recent years (e.g., Mann et al. 2010; Kermarrec et al. 2011, 2013; Zimmermann et al. 2011), opening new avenues towards the routine application of genetic tools for bioassessment and biomonitoring of aquatic ecosystems.

Final Remarks

This chapter provides an overview of the state of the art of studies related to river quality evaluation using epilithic diatom communities worldwide. The studies highlight the use of biotic indices to summarize floristic data to assess pollution effects on aquatic communities, which has drawn the attention of scientists since the beginning of the twentieth century. Diatoms have been widely used to detect changes in streams and rivers water quality due to the specific sensibility of this group to a variety of ecological conditions. Their tolerances and preferences for pH, conductivity, salinity, humidity, organic matter, saprobity, trophic state, oxygen requirements, nutrients, and current velocity in freshwater streams, rivers, lakes, wetlands, and estuaries have been defined, and diatoms have also been used in paleolimnological studies. Biotic indices using diatoms based on the relative abundance of the species weighted by their autoecological values have been developed worldwide, though indices of biotic integrity based on periphyton, diatoms, non-diatom "soft" algae, including cyanobacteria, macrophytes, macroinvertebrates, and fish assemblages have been also developed for biological monitoring.

A new approach for water quality evaluation utilizing diatoms has been increasing significantly in recent years, by applying molecular techniques using DNA sequences, opening new perspectives towards the routine application of genetic tools for bioassessment and biomonitoring of aquatic ecosystems. On the other hand, some researchers argue that currently we cannot adopt molecular bioassessment of algal community composition, considering that much work is still required to complete the DNA reference libraries. It is also necessary to increase the taxonomic coverage for more taxa and to evaluate variability in sequences within and among taxa to improve both the molecular and morphological identification of diatoms. But molecular identification has the potential to provide revolutionary discoveries in taxonomy that may have great benefits for bioassessment.

Acknowledgements The authors want to thank the Brazilian Research Council (CNPq) for financial support of the research conducted to develop the Trophic Water Quality Index (TWQI) for subtropical and temperate Brazilian lotic systems (MCT/CNPq/Universal—n° 14/2011).

References

Álvarez-Blanco I, Blanco S, Cejudo-Figueiras C et al (2013) The Duero Diatom Index (DDI) for river water quality assessment in NW Spain: design and validation. Environ Monit Assess 185:969–981

Arzet K, Krause-Dellin D, Steinberg C (1986) Acidification of four lakes in the Federal Republic of Germany as reflected by diatoms assemblages, cladocerans remains and sediment chemistry. In: Smol JP, Batarbee RW, Davis RB, Merilainen J (eds) Diatoms and lake acidity. Dr. Junk Publisher, Dordrecht, pp 227–250

Bae MJ, Kwon Y, Hwang SJ et al (2011) Relationships between three major stream assemblages and their environmental factors in multiple spatial scales. Ann Limnol Int J Lim 47:91–105

Bahls LM, Teply RS, Suplee MW (2008) Diatom biocriteria development and water quality assessment in Montana: a brief history and status report. Diatom Res 23:533–540

Bate GC, Smailes PA, Adams JB (2004) Benthic diatoms in the rivers and estuaries of South Africa. WRC Report N° TT 234/04, Water Research Commission, Pretoria

Battarbee RW, Flower RJ, Juggins S et al (1997) The relationship between diatoms and surface water quality in the Hoylandet area of Nord-Trondelag, Norway. Hydrobiologia 348:69–80

Bellinger BJ, Cocquyt C, O'Reilly CM (2006) Benthic diatoms as indicators of eutrophication in tropical streams. Hydrobiologia 573:75–87

Beltrami ME, Ciutti F, Cappelletti C et al (2012) Diatoms from Alto Adige/Südtirol (Northern Italy): characterization of assemblages and their application for biological quality assessment in the context of the Water Framework Directive. Hydrobiologia 695:153–170
Bere T, Tundisi JG (2011) The effects of substrate type on diatom-based multivariate water quality assessment in a tropical river (Monjolinho), São Carlos, SP, Brazil. Water Air Soil Poll 216:391–409
Bes D, Ector L, Torgan LC, Lobo EA (2012) Composition of the epilithic diatom flora from a subtropical river, Southern Brazil. Iheringia Ser Bot 67:93–125
Besse-Lototskaya A, Verdonschot PFM, Sinkeldam JA (2006) Uncertainty in diatom assessment: Sampling, identification and counting variation. Hydrobiologia 566:247–260
Bicudo CEM (2004) Taxonomy. Biota Neotropica 4:1–2
Birks HJB, Juggins S, Line JM (1990) Lake surface-water chemistry reconstructions from paleolimnological data. In: Mason BJ (ed) The surface waters acidification programme. Cambridge University Press, Cambridge, pp 301–313
Böhm JS, Schuch M, Düpont A, Lobo EA (2013) Response of epilithic diatom communities to downstream nutrient increases in Castelhano Stream, Venâncio Aires City, RS, Brazil. J Environ Prot 4:20–26
Butcher RW (1947) Studies in the ecology of rivers: VII. The algae of organically enriched waters. J Ecol 35:186–191
Campeau S, Pienitz R, Héquette A (1999) Diatoms from the Beaufort Sea coast, southern Arctic Ocean (Canada): modern analogues for reconstructing Late Quaternary environments and relative sea levels. Bibliotheca Diatomologica. Schweizerbart Science Publishers, Stuttgart
Cantonati M, Angeli N, Bertuzzi E et al (2012) Diatoms in springs of the Alps: spring types, environmental determinants, and substratum. Freshw Sci 31:499–524
Cemagref (1982) Etude des méthodes biologiques d'appréciation quantitative de la qualité des eaux. Rapport Q.E. Lyon A.F., Bassin Rhône-Méditerranée-Corse, Lyon
Cho IH, Hwang SJ, Kim BH et al (2014) Distribution of epilithic diatom communities in relation to land-use and water quality in the Geum river system, South Korea. J Korean Soc Water Qual 30:283–291
Cholnoky BJ (1959) Neue und seltene Diatomeen aus Afrika IV, Diatomeen aus der Kaap-Provinz. Österr Bot Z 106:1–69
Cholnoky BJ (1963) Beitrage zur Kenntnis der Ökologie der Diatomeen des Swakop-Flusses in Südwest-Afrika. Rev Bras Biol 3:233–260
Cochero J, Licursi M, Gómez N (2015) Changes in the epipelic diatom assemblage in nutrient rich streams due to the variations of simultaneous stressors. Limnologica 51:15–23
Coring E (1999) Situation and developments of algal (diatom)-based techniques for monitoring rivers in Germany. In: Prygiel J, Whitton BA, Bukowska J (eds) Use of Algae for Monitoring Rivers III. Agence de l'Eau Artois-Picardie, Douai, pp 122–127
Cumming BF, Smol JP (1993) Development of diatom-based salinity models for paleoclimatic research from lakes in British Columbia (Canada). Hydrobiologia 269(270):179–196
Danielson TJ, Loftin CS, Tsomides L et al (2012) An algal model for predicting attainment of tiered biological criteria of Maine's streams and rivers. Freshw Sci 31:318–340
Dell'Uomo A (1996) Assessment of water quality of an Apennine river as a pilot study for diatom-based monitoring of Italian watercourses. In: Whitton BA, Rott E (eds) Use of algae for monitoring rivers (II). Universtät Innsbruck, Innsbruck, Institut fur Botanik, pp 65–72
Della Bella V, Pace G, Barile M et al (2012) Benthic diatom assemblages and their response to human stress in small-sized volcanic-siliceous streams of central Italy (Mediterranean eco-region). Hydrobiologia 695:207–222
Denys L (1991a) A check-list of the diatoms in the Holocene deposits of the Western Belgian Coastal plain in a survey of their apparent ecological requirements I: Introduction, ecological code and complete list. Ministère des Affaires Economiques, Service Géologique de Belgique, Brussels

Denys L (1991b) A check-list of the diatoms in the Holocene deposits of the Western Belgian Coastal plain in a survey of their apparent ecological requirements II: Centrales. Ministère des Affaires Economiques, Service Géologique de Belgique, Brussels
Descy JP, Coste M (1990) Utilisation des diatomées benthiques pour l'évaluation de la qualité des eaux courantes. Rapport Final Contract CEE B 71-23, CEMAGREF, Bordeaux
Descy JP, Coste M (1991) A test of methods for assessing water quality based on diatoms. Verh Internat Verein Theor Angew Limnol 24:2112–2116
Dixit SS, Dixit AS, Smol JP (1990) Paleolimnological investigation of three manipulated lakes from Sundbury, Canada. Hydrobiologia 214:245–252
Dreßler M, Verweij G, Kistenich S et al (2015) Applied use of taxonomy: lessons learned from the first German intercalibration exercise for benthic diatoms. Acta Bot Croat 74(2):211–232
Dufrêne M, Legendre P (1997) Species assemblages and indicator species: the need for a flexible asymmetrical approach. Ecol Monogr 61:53–73
Duong TT, Feurtet-Mazel A, Coste M et al (2007) Dynamics of diatom colonization process in some rivers influenced by urban pollution (Hanoi, Vietnam). Ecol Indic 7:839–851
Düpont A, Lobo EA, Costa AB et al (2007) Avaliação da qualidade da água do Arroio do Couto, Santa Cruz do Sul, RS, Brasil. Cad Pesqu Ser Biol 19:56–74
Ector L, Rimet F (2005) Using bioindicators to assess rivers in Europe: an overview. In: Lek S, Scardi M, Verdonschot PFM, Descy JP, Park YS (eds) Modelling community structure in freshwater ecosystems. Springer Verlag, Heidelberg, pp 7–19
Eloranta P (1990) Periphytic diatoms in the Acidification Project Lakes. In: Kauppi P, Anttila P, Kenttämies K (eds) Acidification in Finland. Springer Verlag, Heidelberg, pp 985–994
Feio MJ, Aguiar FC, Almeida SFP et al (2014) Least disturbed condition for European Mediterranean rivers. Sci Total Environ 476–477:745–756
Fetscher AE, Stancheva R, Kociolek JP et al (2014) Development and comparison of stream indices of biotic integrity using diatoms vs. non-diatom algae vs. a combination. J Appl Phycol 26:433–450
Fjerdingstad E (1964) Pollution of stream estimated by benthal phytomicro-organisms. I. A saprobic system based on communities of organisms and ecological factors. Int Rev Gesamten Hydrobiol 49:63–131
Fore L, Grafe C (2002) Using diatoms to assess the biological condition of large rivers in Idaho (U.S.A.). Freshw Biol 47:2015–2037
Gell PA (1997) The development of a diatom database for inferring lake salinity, western Victoria, Australia: towards a quantitative approach for reconstructing past climates. Aust J Bot 45:389–423
Gillett N, Pan YD, Parker C (2009) Should only live diatoms be used in the bioassessment of small mountain streams? Hydrobiologia 620:135–147
Gómez N, Licursi M (2001) The Pampean Diatom Index (IDP) for assessment of rivers and streams in Argentina. Aquat Ecol 35:173–181
Gómez N, Licursi M (2003) Abnormal forms in Pinnularia gibba (Bacillariophyceae) in a polluted lowland stream from Argentina. Nova Hedwigia 77:389–398
Gómez N, Licursi M, Bauer DE et al (2003) Reseña sobre las modalidades de estudio mediante la utilización de microalgas en la evaluación y monitoreo de algunos sistemas lóticos pampeanos bonaerenses. Bol Soc Argent Bot 38:93–103
Gómez N, Sierra MV, Cortelezzi A et al (2008) Effects of discharges from the textile industry on the biotic integrity of benthic assemblages. Ecotoxicol Environ Saf 69:472–479
Gómez N, Sierra MV, Cochero J et al (2009) Epipelic biofilms as indicators of environmental changes in lowland fluvial systems. In: Bailey WC (ed) Biofilms: formation, development and properties. Nova, La plata, pp 259–290
Griffith MB, Hill BH, McCormick FH et al (2005) Comparative application of indices of biotic integrity based on periphyton, macroinvertebrates, and fish to southern Rocky Mountain streams. Ecol Indic 5:117–136

Håkansson S (1993) Numerical methods for the inference of pH variations in mesotrophic and eutrophic lakes in southern Sweden—a progress report. Diatom Res 8:349–370
Heinrich CG, Leal VL, Schuch M et al (2014) Epilithic diatoms in headwater areas of the hydrographical sub-basin of the Andreas Stream, RS, Brazil, and their relation with eutrophication processes. Acta Limnol Bras 26:347–355
Hering D, Johnson RK, Kramm S et al (2006) Assessment of European streams with diatoms, macrophytes, macroinvertebrates and fish: a comparative metric-based analysis of organism response to stress. Freshw Biol 51:1757–1785
Hermany G, Schwarzbold A, Lobo EA et al (2006) Ecology of the epilithic diatom community in a low-order stream system of the Guaíba hydrographical region: subsidies to the environmental monitoring of southern Brazilian aquatic systems. Acta Limnol Bras 18:9–27
Hill BH, Herlihy AT, Kaufmann PR et al (2000) Use of periphyton assemblage data as an index of biotic integrity. J N Am Benthol Soc 19:50–67
Hill BH, Herlihy AT, Kaufmann PR et al (2003) Assessment of streams of the eastern United States using a periphyton index of biotic integrity. Ecol Indic 2:325–338
Hofmann G (1994) Aufwuchs-Diatomeen in Seen und ihre Eignung als Indikatoren der Trophie. Bibliotheca Diatomologica. Schweizerbart Science Publishers, Stuttgart
Hwang SJ, Kim NY, Yoon SA et al (2011) Distribution of benthic diatoms in Korean rivers and streams in relation to environmental variables. Ann Limnol Int J Lim 47:15–33
INDEC (2010) Censo Nacional de Población, Hogares y Viviendas 2010—Instituto Nacional de Estadística y Censos. http://www.indec.gov.ar. Accessed 18 August 2015
Juggins S (1992) Diatoms in the Thames Estuary. Ecology, Palaeoecology, and Salinity Transfer Function. Bibliotheca Diatomologica. Schweizerbart Science Publishers, Stuttgart, England
Kahlert M, Albert RL, Anttila EL et al (2009) Harmonization is more important than experience — results of the first Nordic–Baltic diatom intercalibration exercise 2007 (stream monitoring). J Appl Phycol 21:471–482
Kalyoncu H, Cicek NL, Akkoz C et al (2009) Comparative performance of diatom indices in aquatic pollution assessment. Afr J Agric Res 4:1032–1040
Katoh K (1991) A comparative study on some pollution indices using diatoms. Diatom 6:11–17
Kelly MG, Whitton BA (1995) The trophic diatom index: a new index for monitoring eutrophication in rivers. J Appl Phycol 7:433–444
Kelly MG, Cazaubon A, Coring E et al (1998) Recommendations for the routine sampling of diatoms for water quality assessments in Europe. J Appl Phycol 10:215–224
Kelly M, Juggins S, Guthrie R et al (2008) Assessment of ecological status in U.K. rivers using diatoms. Freshw Biol 53:403–422
Kelly M, Bennett C, Coste M et al (2009) A comparison of national approaches to setting ecological status boundaries in phytobenthos assessment for the European Water Framework Directive: results of an intercalibration exercise. Hydrobiologia 621:169–182
Kermarrec L, Ector L, Bouchez A et al (2011) A preliminary phylogenetic analysis of the Cymbellales based on 18S rDNA gene sequencing. Diatom Res 26:305–315
Kermarrec L, Franc A, Rimet F et al (2013) Next generation sequencing to inventory taxonomic diversity in eukaryotic communities: a test for freshwater diatoms. Mol Ecol Resour 13:607–619
Kobayasi H (1985) Ultrastructural differences in certain taxonomically difficult species of Nitzschia section Lanceolatae in Japan. In: Hara H (ed) Origin and evolution of diversity in plants and plant community. Academia, Tokyo, pp 304–313
Kobayasi H, Mayama S (1989) Evaluation of river water quality by diatoms. Korean J Phycol 4:121–133
Kolkwitz R (1950) Ökologie der Saprobien. Schriftenreihe des Vereins für Wasser. Boden und Lufthygiene, Stuttgart
Kolkwitz R, Marsson M (1908) Ökologie der tierischen Saprobien. Beiträge zur Lehre von des biologischen Gewasserbeurteilung. Int Rev Gesamten Hydrobiol 2:126–152

Lane CR, Brown MT (2007) Diatoms as indicators of isolated herbaceous wetland condition in Florida, USA. Ecol Indic 7:521–540

Lange-Bertalot H (1979) Pollution tolerance of diatoms as a criterion for water quality estimation. Nova Hedwigia 64:285–304

Leclercq L, Maquet B (1987) Deux nouveaux indices chimique et diatomique de qualité d'eau courante: application au Samson et à ses affluents (Bassin de la Meuse Belge); comparaison avec d'autres indices chimiques, biocénotiques et diatomiques (document de travail 28). Institut Royal des Sciences Naturelles de Belgique, Bruxelles

Lee SW, Hwang SJ, Lee JK et al (2011) Overview and application of the National Aquatic Ecological Monitoring Program (NAEMP) in Korea. Ann Limnol Int J Lim 47:3–14

Leland HV, Porter SD (2000) Distribution of benthic algae in the upper Illinois River basin in relation to geology and land use. Freshw Biol 44:279–301

Lenoir A, Coste M (1996) Development of a practical diatom index of overall water quality applicable to the French National Water Board Network. In: Whitton BA, Rott E (eds) Use of Algae for Monitoring River II. Universität Innsbruck, Innsbruck, Institut für Botanik, pp 29–43

Li S, Gu S, Tan X et al (2009) Water quality in the upper Han River basin, China: the impacts of land use/land cover in riparian buffer zone. J Hazard Mater 65:317–324

Licursi M, Gómez N (2002) Benthic diatom and some environmental condition in three lowland streams of Pampean Plain. Ann Limnol 38:109–118

Licursi M, Gómez N (2004) Aplicación de índices bióticos en la evaluación de la calidad del agua en sistemas lóticos de la llanura pampeana a partir del empleo de diatomeas. Biol Acuat 21:31–49

Liebmann H (1951) Handbuch der Frischwasser- und Abswasserbiologie. Verlag Oldenbourg, München

Lobo EA (2013) O perifíton como indicador da qualidade da água. In: Schwarzbold A, Burliga AL, Torgan LC (eds) Ecologia do Perifíton. RiMa Editora, São Carlos, pp 205–233

Lobo EA, Kitazawa S, Kobayasi H (1990) The use of scanning electron microscopy as a necessary complement of light microscopy diatom examination for ecological studies. Diatom 5:33–43

Lobo EA, Katoh K, Aruga Y (1995) Response of epilithic diatom assemblages to water pollution in rivers in the Tokyo Metropolitan area. Freshw Biol 34:191–204

Lobo EA, Callegaro VLM, Oliveira MA et al (1996) Pollution tolerant diatoms from lotic systems in the Jacuí Basin, Rio Grande do Sul, Brasil. Iheringia Ser Bot 47:45–72

Lobo EA, Callegaro VLM, Bender EP (2002) Utilização de Algas Diatomáceas Epilíticas como Indicadores da Qualidade da Água em Rios e Arroios da Região Hidrográfica do Guaíba, RS, Brasil. EDUNISC, Santa Cruz do Sul

Lobo EA, Callegaro VLM, Hermany G et al (2004a) Review of the use of microalgae in South America for monitoring rivers, with special reference to diatoms. Vie Milieu 54:105–114

Lobo EA, Callegaro VLM, Hermany G et al (2004b) Use of epilithic diatoms as bioindicator from lotic systems in southern Brazil, with special emphasis on eutrophication. Acta Limnol Bras 16:25–40

Lobo EA, Wetzel CE, Ector L et al (2010) Response of epilithic diatom communities to environmental gradients in subtropical temperate Brazilian rivers. Limnetica 29:323–340

Lobo EA, Wetzel CE, Schuch M et al (2014) Diatomáceas epilíticas como indicadores da qualidade da água em sistemas lóticos subtropicais e temperados brasileiros. EDUNISC, Santa Cruz do Sul

Lobo EA, Schuch M, Heinrich CG et al (2015) Development of the Trophic Water Quality Index (TWQI) for subtropical temperate Brazilian lotic systems. Environ Monit Assess 187:1–13

López van Oosterom MV, Ocón CS, Brancolini F et al (2013) Trophic relationships between macroinvertebrates and fish in a pampean lowland stream (Argentina). Iheringia Ser Zool 103:57–65

Lowe RL (1974) Environmental Requirements and Pollution Tolerance of Freshwater Diatoms. National Environmental Research Center, Cincinnati

Lowe WH (2002) Landscape-scale spatial population dynamics in human-impacted stream systems. Environ Manage 30:225–233
Lücking R (2008) Taxonomy: a discipline on the brink of extinction. Are DNA barcode scanners the future of biodiversity research? Arch Sci 61:75–88
Mancini L (2003) Bioindicatori: necessità di nuovi sviluppi a seguito della attuazione del decreto legislative 152/99 e del recepimento della direttiva 2000/60/CE Water Framework Directive. Atti dela 7th Conferenza Nazionale dele Agenzie Ambientali, Milano
Mancini L (2006) Organization of biological monitoring in the European Union. In: Ziglio G, Siligardi M, Flaim F (eds) Biological Monitoring of Rivers: Applications and Perspectives. Water Quality Measurements Series. John Wiley & Sons, Chichester, pp 171–202
Mann DG, Sato S, Trobajo R et al (2010) DNA barcoding for species identification and discovery in diatoms. Cryptogamie Algol 31:557–577
Manoylov KM (2014) Taxonomic identification of algae (morphological and molecular): species concepts, methodologies, and their implications for ecological bioassessment. J Phycol 50:409–424
Manoylov KM, Marsh T, Stevenson RJ (2009) Testing molecular tools for assessment of taxonomic composition of a benthic algal community. Nova Hedwigia 135:121–136
Mayama S, Katoh K, Omori H et al (2011) Progress towards construction of an international web-based educational system featuring improved SimRiver for understanding of the river environment. Asian J Biol Educ 5:2–14
Michels-Estrada A (2003) Ökologie und Verbreitung von Kieselalgen in Flieβgewässern Costa Rica als Grundlage für eine biologische Gewässergütebeurteilung in den Tropen. Diss Bot 377:244–257
Morales EA, Siver PA, Trainor FR (2001) Identification of diatoms (Bacillariophyceae) during ecological assessments: comparison between light microscopy and scanning electron microscopy techniques. P Acad Nat Sci Phila 151:95–103
Niemi GJ, McDonald ME (2004) Application of ecological indicators. Annu Rev Ecol Evol Syst 35:89–111
Oeding S, Taffs KH (2015) Are diatoms a reliable and valuable bio-indicator to assess sub-tropical river ecosystem health? Hydrobiologia 758:151–169
Osborne LL, Kovacic DA (1993) Riparian vegetated buffer strips in water quality restoration and stream managements. Freshw Biol 29:243–258
Pantle R, Buck H (1955) Die biologische Überwachung der Gewässer und die Darstellung der Ergebnisse. Gas Wasserfach Wasser Abwasser 96:609–620
Pipp E (2002) A regional diatom-based trophic state indication system for running water sites in Upper Austria and its over-regional applicability. Verh Internat Verein Limnol 27:3376–3380
Potapova MG (2011) Patterns of diatom distribution in relation to salinity. In: Seckbach J, Kociolek JP (eds) The Diatom World, Cellular Origin. Life in Extreme Habitats and Astrobiology. Springer, Dordrecht, pp 313–332
Potapova M, Charles DF (2007) Diatom metrics for monitoring eutrophication in rivers of the United States. Ecol Indic 7:48–70
Potapova MG, Charles DF, Ponader KC et al (2004) Quantifying species indicator values for trophic diatom indices: a comparison of approaches. Hydrobiologia 517:25–41
Pressey RL, Humphries CJ, Margules CR et al (1993) Beyond opportunism: key principles for systematic reserve selection. Trends Ecol Evol 8:124–128
Renberg I, Hellberg T (1982) The pH history of lakes in southwestern Sweden, as calculated from the subfossil diatom flora of the sediments. Ambio 11:341–348
Rimet F (2012) Recent views on river pollution and diatoms. Hydrobiologia 683:1–24
Roberts D, McMinn A (1998) A weighted-averaging regression and calibration model for inferring lake water salinity from fossil diatom assemblages in saline lakes of the Vestfold Hills: a new tool for interpreting Holocene lake histories in Antarctica. J Paleolimnol 19:57–78
Robinson CT, Kawecka B (2005) Benthic diatoms of an Alpine stream/lake network in Switzerland. Aquat Sci 67:492–506

Rott E, Hofmann G, Pall K et al (1997) Indikationslisten für Aufwuschalgen in Österreischen Fliessgewässern. Teil 1: Saprobielle Indikation. Bundesministerium für Land und Forstwirtschaft, Wasserwirtschaftkataster.
Rott E, Pfister P, Van Dam H, Pipp E, Pall K, Binder N, Ortler K (1999) Indikationslisten für Aufwuchalgen in Österreischen Fliessgewässern. Teil 2: Trophieindikation und autökologische Anmerkungen, Bundesministerium für Land und Forstwirtschaft, Wasserwirtschaftkataster.
Rott E, Pipp E, Pfister P (2003) Diatom methods developed for river quality assessment in Austria and a cross-check against numerical trophic indication methods used in Europe. Algol Stud 110:91–115
Rumeau A, Coste M (1988) Initiation à la systématique des diatomées d'eau douce pour l'utilisation pratique d'un indice diatomique générique. Bulletin Français de la Pêche et de la Pisciculture. Conseil supérieur de la pêche, Paris
Salomoni SE, Rocha O, Callegaro VL, Lobo EA (2006) Epilithic diatoms as indicators of water quality in the Gravataí river, Rio Grande do Sul, Brazil. Hydrobiologia 555:233–246
Salomoni SE, Rocha O, Hermany G et al (2011) Application of water quality biological indices using diatoms as bioindicators in Gravataí River, RS, Brazil. Braz J Biol 71:949–959
Schiefele S, Kohmann F (1993) Bioindikation der Trophie in Fliessgewässern. In: Umweltforschungsplan des Bundesministers für Umwelt, Naturschutz und Reaktorsicherheit. Forschungsbericht, Nr. 102 01 504. Bayerisches Landesamt für Wasserwirtschaft, München
Schuch M, Abreu E Jr, Lobo EA (2012) Water quality evaluation of urban streams in Santa Cruz do Sul city, RS, Brazil. Bioikos 26:3–12
Schuch M, Oliveira MA, Lobo EA (2015) Spatial response of epilithic diatom communities to downstream nutrient increases. Water Environ Res 87:547–558
Seegert G (2000) The development, use, and misuse of biocriteria with an emphasis on the index of biotic integrity. Environ Sci Policy 3:51–58
Sierra MV, Gómez N (2007) Structural characteristics and oxygen consumption of the epipelic biofilms in three lowland streams exposed to different land uses. Water Air Soil Pollut 186:115–127
Sládeček V (1965) The future of the saprobity system. Hydrobiologia 25:518–537
Sládeček V (1973) System of water quality from the biological point of view. Archiv für Hydrobiologie Ergebnisse Limnol 7:1–218
Sluys R (2013) The unappreciated, fundamentally analytical nature of taxonomy and the implications for the inventory of biodiversity. Biodivers Conserv 22:1095–1105
Smith MA (1990) The ecophysiology of epilithic diatom communities of acid lakes in Galoway, southwest Scotland. Phil Trans R Soc Lond 327:251–256
Smucker NJ, Vis ML (2013) Can pollution severity affect diatom succession in streams and could it matter for stream assessments? J Freshw Ecol 28:329–338
Smucker NJ, Becker M, Detenbeck NE et al (2013) Using algal metrics and biomass to evaluate multiple ways of defining concentration-based nutrient criteria in streams and their ecological relevance. Ecol Indic 32:51–61
Snoeijs P (1994) Distribution of epiphytic diatom species composition, diversity and biomass on different macroalgal hosts along seasonal and salinity gradients in the Baltic Sea. Diatom Res 9:189–211
Stevenson J (2014) Ecological assessments with algae: a review and synthesis. J Phycol 50:437–461
Stevenson RJ, Pan Y, Manoylov KM et al (2008) Development of diatom indicators of ecological conditions for streams of the western US. J North Am Benthol Soc 27:1000–1016
Stevenson RJ, Pan Y, Van Dam H (2010) Assessing environmental conditions in rivers and streams with diatoms. In: Smol JP, Stoermer EF (eds) The Diatoms: applications for the environmental and earth sciences, 2nd edn. Cambridge University Press, Cambridge, pp 57–85
Stevenson RJ, Zalack JT, Wolin J (2013) A multimetric index of lake diatom condition based on surface-sediment assemblages. Freshw Sci 32:1005–1025

Stoddard JL, Larsen DP, Hawkins CP et al (2006) Setting expectations for the ecological condition of streams: the concept of reference condition. Ecol Appl 16:1267–1276
Szczepocka E, Kruk A, Rakowska B (2015) Can tolerant diatom taxa be used for effective assessments of human pressure? River Res Appl 31:368–378
Tan X, Sheldon F, Bunn ES et al (2013) Using diatom indices for water quality assessment in a subtropical river, China. Environ Sci Pollut Res 20:4164–4175
Tan X, Ma P, Xia X, Zhang Q (2014) Spatial pattern of benthic diatoms and water quality assessment using diatom indices in a subtropical river, China. Clean - Soil, Air, Water 42:20–28
Tan X, Ma P, Bunn SE, Zhang Q (2015) Development of a benthic diatom index of biotic integrity (BD-IBI) for ecosystem health assessment of human dominant subtropical rivers. China J Environ Manage 151:286–294
Taylor JC, de la Rey PA, van Rensburg L (2005) Recommendations for the collection, preparation and enumeration of diatoms from riverine habitats for water quality monitoring in South Africa. Afr J Aquat Sci 30:65–75
Ter Braak CFJ, Van Dam H (1989) Inferring pH from diatoms: a comparison of old and new calibration methods. Hydrobiologia 178:209–223
Tornés E, Cambra J, Gomà J et al (2007) Indicator taxa of benthic diatom communities: a case study in Mediterranean streams. Ann Limnol-Int J Lim 43:1–11
Underwood GJ, Philips JS, Saunders K (1998) Distribution of estuarine benthic diatom species along salinity and nutrient gradients. Eur J Phycol 33:173–183
Union E (2000) Directive 2000/60/EC of the European Parliament and of the Council of 23 October 2000 establishing a framework for Community action in the field of water policy. Off J Eur Commun 327:1–73
Urrea G, Sabater S (2009) Epilithic diatom assemblages and their relationship to environmental characteristics in an agricultural watershed (Guadiana River, SW Spain). Ecol Indic 9:693–703
Van Dam H (1997) Partial recovery of moorland pools from acidification: indications by chemistry and diatoms. Neth J of Aquatic Ecol 30:203–218
Van Dam H, Mertens A, Sinkeldam J (1994) A coded checklist and ecological indicator values of freshwater diatoms from The Netherlands. Neth J Aquatic Ecol 28:117–133
Visco JA, Apothéloz-Perret-Gentil L, Cordonier A et al (2015) Environmental monitoring: inferring the diatom index from next-generation sequencing data. Environ Sci Technol 49:7606–7613
Vyverman W, Vyverman R, Hodgson D et al (1995) Diatoms from Tasmanian mountain lakes: a reference data-set (TASDIAT) for environmental reconstruction and a systematic and autecological study. Bibliotheca Diatomologica. Schweizerbart Science Publishers, Stuttgart
Wan Maznah WO, Mansor M (2002) Aquatic pollution assessment based on attached diatom communities in the Pinang River Basin, Malaysia. Hydrobiologia 487:229–241
Watanabe T, Asai K (1992) Simulation of organic water pollution using highly prevailing diatom taxa (1). Diatom assemblage in which the leading taxon belongs to *Achnanthes, Anomoeoneis, Aulacoseira* or *Melosira*. Diatom 7:13–19
Watanabe T, Asai K, Houki A (1985) Epilithic diatom assemblage index to organic pollution (DAIpo) and its ecological significance. Annual Report of Graduate Division of Human Culture, Doctoral Degree Program, Nara Women's University, Nara
Watanabe T, Asai K, Houki A (1988) Numerical water quality monitoring of organic pollution using diatom assemblages. In: Round FE (ed) Proceedings of the Ninth International Diatom Symposium 1986. Koeltz Scientific Books, Koenigstein
Wetzel CE, Ector L (2014a) Taxonomy, distribution and autecology of *Planothidium bagualensis* sp. nov. (Bacillariophyta) a common monoraphid species from southern Brazilian rivers. Phytotaxa 156:201–210
Wetzel CE, Ector L (2014b) *Planothidium lagerheimii* comb. nov. (Bacillariophyta, Achnanthales) a forgotten diatom from South America. Phytotaxa 188:261–267

Wetzel CE, Ector L, Van de Vijver B et al (2015) Morphology, typification and critical analysis of some ecologically important small naviculoid species (Bacillariophyta). Fottea 15:203–234
Whittaker RH (1952) A study of summer foliage insect communities in the Great Smoky Mountains. Ecol Monogr 22:1–144
Wilson SE, Cumming BF, Smol JP (1994) Diatom-salinity relationships in 111 lakes from the Interior Plateau of British Columbia, Canada: the development of diatom-based models for paleosalinity reconstructions. J Paleolimnol 12:197–221
Wilson SE, Smol JP, Sauchyn DJ (1997) A Holocene paleosalinity diatom record from southwestern Saskatchewan, Canada: Harris lake revisited. J Paleolimnol 17:23–31
Yoccoz NG (2012) The future of environmental DNA in ecology. Mol Ecol 21:2031–2038
Zalack JT, Smucker NJ, Vis ML (2010) Development of a diatom index of biotic integrity for acid mine drainage impacted streams. Ecol Indic 10:287–295
Zelinka M, Marvan P (1961) Zur Präzisierung der biologischen Klassifikation der Reinheit fliessender Gewässer. Arch Hydrobiol 57:389–407
Ziemann H (1971) Die Wirkung des Salzgehaltes auf die Diatomenflora als Grundlage für eine biologische Analyse und Klassifikation der Binnengewässer. Limnologica 8:505–525
Ziemann H (1991) Veränderungen der Diatomeenflora der Werra unter dem Einfluss des Salzgehaltes. Acta Hydrochim Hydrobiol 19:159–174
Zimmermann J, Jahn R, Gemeinholzer B (2011) Barcoding diatoms: evaluation of the V4 subregion on the 18S rRNA gene, including new primers and protocols. Org Divers Evol 11:173–192
Zimmermann J, Glöockner G, Jahn R et al (2015) Metabarcoding vs. morphological identification to assess diatom diversity in environmental studies. Mol Ecol Resour 15:526–542

Index

O. Necchi Jr. (ed.), *River Algae*, DOI 10.1007/978-3-319-31984-1